Novel Delivery Systems for Transdermal and Intradermal Drug Delivery

ADVANCES IN PHARMACEUTICAL TECHNOLOGY

A Wiley Book Series

Series Editors:
Dennis Douroumis, University of Greenwich, UK
Alfred Fahr, Friedrich–Schiller University of Jena, Germany
Jürgen Siepmann, University of Lille, France
Martin Snowden, University of Greenwich, UK
Vladimir Torchilin, Northeastern University, USA

Titles in the Series

Hot-Melt Extrusion: Pharmaceutical Applications
Edited by Dionysios Douroumis

Drug Delivery Strategies for Poorly Water-Soluble Drugs
Edited by Dionysios Douroumis and Alfred Fahr

Forthcoming titles:

In Vitro Drug Release Testing of Special Dosage Forms
Edited by Nikoletta Fotaki and Sandra Klein

Novel Delivery Systems for Transdermal and Intradermal Drug Delivery

RYAN F. DONNELLY and **THAKUR RAGHU RAJ SINGH**

School of Pharmacy, Queen's University Belfast, UK

WILEY

Registered Office
John Wiley & Sons, Ltd, The Atrium, Southern Gate, Chichester, West Sussex, PO19 8SQ, United Kingdom

For details of our global editorial offices, for customer services and for information about how to apply for permission to reuse the copyright material in this book please see our website at www.wiley.com.

Library of Congress Cataloging-in-Publication Data

Novel delivery systems for transdermal and intradermal drug delivery / Ryan F. Donnelly, Thakur Raghu Raj Singh.
    pages   cm
  Includes bibliographical references and index.
  ISBN 978-1-118-73451-3 (cloth)
1. Drug delivery systems.   2. Transdermal medication.   3. Injections, Intradermal.   I. Donnelly, Ryan F., editor.
II. Singh, Thakur Raghu Raj, editor.
  RS199.5.N68 2015
  615′.6–dc23
                                         2015015480

A catalogue record for this book is available from the British Library.

Set in 10/12pt Times by SPi Global, Pondicherry, India
Printed and bound in Singapore by Markono Print Media Pte Ltd

1   2015

Unfortunately during the preparation of this book, one of the authors, Dr Sian Lim died in a tragic cycling accident and left behind a wife Evelyn, a daughter Caelyn and a son Elijah. Sian was a brilliant scientist and an expert formulator who was involved in the development of over 20 topical and transdermal medicines that are now on the market. We would like to dedicate this book to his memory.

Contents

About the Editors

Ryan F. Donnelly

Ryan Donnelly graduated with a BSc (First Class) in Pharmacy from Queen's University Belfast in 1999 and was awarded the Pharmaceutical Society of Northern Ireland's Gold Medal. Following a year of pre-registration training spent in Community Pharmacy, he returned to the School of Pharmacy to undertake a PhD in Pharmaceutics. He graduated in 2003 and, after a short period of post-doctoral research, was appointed to a Lectureship in Pharmaceutics in January 2004. He was promoted to Senior Lecturer in 2009, Reader in 2011 and, in 2013, to a Chair in Pharmaceutical Technology.

Professor Donnelly's research is centered on design and physicochemical characterisation of advanced polymeric drug delivery systems for transdermal and topical drug delivery, with a strong emphasis on improving therapeutic outcomes for patients. His bioadhesive patch design was used in successful photodynamic therapy of over 100 patients and this technology has now been licensed to Swedish Pharma AB, for whom Professor Donnelly acts as a Technical Director. Currently, Professor Donnelly's group is focused on novel polymeric microneedle arrays for transdermal administration of 'difficult-to-deliver' drugs and intradermal delivery of vaccines and photosensitisers. His work has attracted funding of approximately £4.5 million, from a wide range of sources, including BBSRC, EPSRC, MRC, the Wellcome Trust, Action Medical Research, the Royal Society and the pharmaceutical and medical devices industries.

Still at a relatively early stage of his career, he has authored over 350 peer-reviewed publications, including 4 patent applications, 3 textbooks and approximately 120 full papers. He has been an invited speaker at numerous national and international conferences. Professor Donnelly is the Associate Editor of *Recent Patents on Drug Delivery & Formulation* and a member of the Editorial Advisory Boards of *The American Journal of Pharmacology and Toxicology*, *Pharmaceutical Technology Europe*, *Expert Review of Medical Devices* and *Journal of Pharmacy & Bioallied Sciences* and is Visiting Scientist at the Norwegian Institute for Cancer Research, where he is Associate Member of the Radiation Biology Group.

His work has attracted numerous awards, including the BBSRC Innovator of the Year Award and the American Association of Pharmaceutical Scientists *Pharmaceutical Research* Meritorious Manuscript Award in 2013, the GSK Emerging Scientist Award in 2012; he is a previous winner of the Royal Pharmaceutical Society's Science Award (2011), the Queen's Improvement to Society Award (2011), an Innovation Leader Award from the NHS Research & Development Office (2009) and a Research Scholarship from the Research Council of Norway (2004). In 2013, he was listed in the 40 most influential business leaders in Northern Ireland under the age of 40 by Belfast Media Group. Professor Donnelly's microneedles work has featured on the front cover of *Journal of Controlled Release* and *BBSRC Business* and he has represented BBSRC and the Royal Society of Chemistry at Parliamentary Receptions at Westminster and Stormont, respectively. He has been extensively involved in activities promoting public engagement with science through regular interviews on television and radio and online platforms, such as You Tube, Twitter and Tumblr. His *Pharmacists in Schools* Programme has made over 100 school visits and his work featured at the 2014 Great British Bioscience Festival.

Thakur Raghu Raj Singh

Thakur Raghu Raj Singh is Lecturer in Pharmaceutics at the School of Pharmacy, Queen's University Belfast. Dr Singh's research interests lie in the design and physicochemical characterisation of advanced polymeric drug delivery systems for ocular, transdermal and topical applications. In particular, his current research involves fabrication and design of novel long-acting injectable and implantable drug delivery systems for treating chronic ocular diseases. Dr Singh has authored over 90 scientific publications, including 40 full papers and a textbook on microneedles. He has been an invited speaker at a number of national/international meetings.

Dr Singh is currently an Editorial Board Member of the *International Journal of Pharmacy* and *Chronicles of Pharmacy* and Scientific Advisor to the editors of the *Journal of Pharmaceutical Sciences*. He is a reviewer for at least 18 other international scientific journals. Following his appointment as Lecturer in August 2010, he has secured funding of approximately £560 000 from Invest NI, WHO and industry. Dr Singh's group is currently working on design and development of injectable *in situ* implant-forming systems for ocular drug delivery, funded by Invest Northern Ireland, and on industrial development of novel non-aqueous-based protein eye drops for the treatment of age-related macular degeneration and diabetic retinopathy.

Contributors

Marc. B. Brown, MedPharm Ltd, Guildford, UK, and School of Pharmacy, University of Hertfordshire, UK

Andrzej M. Bugaj, College of Health, Beauty Care and Education, Poznań, Poland

Francesco Caserta, Department of Pharmacy, University of Hertfordshire, UK

Aaron J. Courtenay, School of Pharmacy, Queen's University Belfast, UK

Ryan F. Donnelly, School of Pharmacy, Queen's University Belfast, UK

Charles Evans, MedPharm Ltd, Guildford, UK, and School of Pharmacy, University of Hertfordshire, UK

Chirag Gujral, School of Pharmacy, Queen's University Belfast, UK

Jonathan Hadgraft, Department of Pharmaceutics, UCL School of Pharmacy, UK

Mary-Carmel Kearney, School of Pharmacy, Queen's University Belfast, UK

Adrien Kissenpfennig, The Centre for Infection & Immunity, Queen's University Belfast, UK

Majella E. Lane, Department of Pharmaceutics, UCL School of Pharmacy, UK

Jon Lenn, Stiefel, A GSK Company, USA

Cui Lili, Department of Inorganic Chemistry, School of Pharmacy, Second Military Medical University, Shanghai, China

Sian Lim, MedPharm Ltd, Guildford, UK

Rita Mateus, Department of Pharmaceutics, UCL School of Pharmacy, UK

Abhijeet Maurya, School of Pharmacy, The University of Mississippi, USA

William J. McAuley, Department of Pharmacy, University of Hertfordshire, UK

Gary P.J. Moss, School of Pharmacy, Keele University, UK

S. Narasimha Murthy, School of Pharmacy, The University of Mississippi, USA

Lars Norlén, Department of Cell and Molecular Biology (CMB), Karolinska Institute, Stockholm, Sweden and Dermatology Clinic, Karolinska University Hospital, Sweden

Helen L. Quinn, School of Pharmacy, Queen's University Belfast, UK

Thakur Raghu Raj Singh, School of Pharmacy, Queen's University Belfast, UK

Venkata K. Yellepeddi, College of Pharmacy, Roseman University of Health Sciences, South Jordan, UT, USA and College of Pharmacy, University of Utah, USA

Marija Zaric, The Centre for Infection & Immunity, Queen's University Belfast, UK

Advances in Pharmaceutical Technology: Series Preface

The series *Advances in Pharmaceutical Technology* covers the principles, methods and technologies that the pharmaceutical industry uses to turn a candidate molecule or new chemical entity into a final drug form and hence a new medicine. The series will explore means of optimizing the therapeutic performance of a drug molecule by designing and manufacturing the best and most innovative of new formulations. The processes associated with the testing of new drugs, the key steps involved in the clinical trials process and the most recent approaches utilized in the manufacture of new medicinal products will all be reported. The focus of the series will very much be on new and emerging technologies and the latest methods used in the drug development process.

The topics covered by the series include the following:

Formulation: The manufacture of tablets in all forms (caplets, dispersible, fast-melting) will be described, as will capsules, suppositories, solutions, suspensions and emulsions, aerosols and sprays, injections, powders, ointments and creams, sustained release and the latest transdermal products. The developments in engineering associated with fluid, powder and solids handling, solubility enhancement, colloidal systems including the stability of emulsions and suspensions will also be reported within the series. The influence of formulation design on the bioavailability of a drug will be discussed and the importance of formulation with respect to the development of an optimal final new medicinal product will be clearly illustrated.

Drug Delivery: The use of various excipients and their role in drug delivery will be reviewed. Amongst the topics to be reported and discussed will be a critical appraisal of the current range of modified-release dosage forms currently in use and also those under development. The design and mechanism(s) of controlled release systems including macromolecular drug delivery, microparticulate controlled drug delivery, the delivery of biopharmaceuticals, delivery vehicles created for gastrointestinal tract targeted delivery, transdermal delivery and systems designed specifically for drug delivery to the lung will all be reviewed and critically appraised. Further site-specific systems used for the delivery of drugs across the blood–brain barrier including dendrimers, hydrogels and new innovative biomaterials will be reported.

Manufacturing: The key elements of the manufacturing steps involved in the production of new medicines will be explored in this series. The importance of crystallisation; batch and continuous processing, seeding; and mixing including a description of the key engineering principles relevant to the manufacture of new medicines will all be reviewed and reported. The fundamental processes of quality control including good laboratory practice, good manufacturing practice, Quality by Design, the Deming Cycle, Regulatory requirements and the design of appropriate robust statistical sampling procedures for the control of raw materials will all be an integral part of this book series.

An evaluation of the current analytical methods used to determine drug stability, the quantitative identification of impurities, contaminants and adulterants in pharmaceutical materials will be described as will the production of therapeutic bio-macromolecules, bacteria, viruses, yeasts, moulds, prions and toxins through chemical synthesis and emerging synthetic/molecular biology techniques. The importance of packaging including the compatibility of materials in contact with drug products and their barrier properties will also be explored.

Advances in Pharmaceutical Technology is intended as a comprehensive one-stop shop for those interested in the development and manufacture of new medicines. The series will appeal to those working in the pharmaceutical and related industries, both large and small, and will also be valuable to those who are studying and learning about the drug development process and the translation of those drugs into new life saving and life enriching medicines.

Dennis Douroumis
Alfred Fahr
Jürgen Siepmann
Martin Snowden
Vladimir Torchilin

Preface

Medicines have been delivered across the skin since ancient times. However, the first rigorous scientific studies involving transdermal delivery seeking to determine what caused skin to have barrier properties that prevent molecular permeation were not carried out until the 1920s. Rein proposed that a layer of cells joining the skin's *stratum corneum* (*SC*) to the epidermis posed the major resistance to transdermal transport. Blank modified this hypothesis after removing sequential layers of *SC* from the surface of skin and showing that the rate of water loss from skin increased dramatically once the *SC* was removed. Finally, Scheuplein and colleagues showed that transdermal permeation was limited by the *SC* by a passive process. Despite the significant barrier properties of skin, Michaels and coworkers measured apparent diffusion coefficients of model drugs in the *SC* and showed that some drugs had significant permeability. This led to the active development of transdermal patches in the 1970s, which yielded the first patch approved by the United States Food and Drug Administration in 1979. It was a 3-day patch that delivered scopolamine to treat motion sickness. In 1981, patches for nitroglycerin were approved. Understanding of the barrier properties of skin and how they can be chemically manipulated was greatly enhanced in the 1980s and early 1990s through the work of Maibach, Barry, Guy, Potts and Hadgraft. Today there are a number of transdermal patches marketed for delivery of drugs such as clonidine, fentanyl, lidocaine, nicotine, nitroglycerin, oestradiol, oxybutynin, scopolamine and testosterone. There are also combination patches for contraception, as well as hormone replacement.

Recently, the transdermal route has vied with oral treatment as the most successful innovative research area in drug delivery. In the United States (the most important pharmaceutical market), out of 129 API delivery products under clinical evaluation, 51 are transdermal or dermal systems; 30% of 77 candidate products in preclinical development represent such API delivery. The worldwide transdermal patch market approaches $20 billion, yet is based on only 20 drugs. This rather limited number of drug substances is attributed to the excellent barrier function of the skin, which is accomplished almost entirely by the outermost 10–15 μm (in the dry state) of tissue, the SC. Before being taken up by blood vessels in the upper dermis and prior to entering the systemic circulation, substances permeating the skin must cross the SC and the viable epidermis. There are three possible pathways leading to the capillary network: through hair follicles with associated sebaceous glands, *via* sweat ducts or across continuous SC between these appendages. As the fractional appendageal area available for transport is only about 0.1%, this route usually contributes negligibly to apparent steady state drug flux. The intact SC thus provides the main barrier to exogenous substances, including drugs. The corneocytes of hydrated keratin are analogous to 'bricks', embedded in a 'mortar' composed of highly organised, multiple lipid bilayers of ceramides, fatty acids, cholesterol and its esters. These bilayers form regions of semicrystalline gel and liquid crystal domains. Most molecules penetrate through skin *via* this intercellular microroute. Facilitation of drug penetration through the SC may involve bypass or reversible disruption of its elegant molecular architecture. The ideal properties of a molecule penetrating intact SC well are as follows:

- Molecular mass less than 600 Da
- Adequate solubility in both oil and water so that the membrane concentration gradient, which is the driving force for passive drug diffusion along a concentration gradient, may be high
- Partition coefficient such that the drug can diffuse out of the vehicle, partition into, and move across the SC, without becoming sequestered within it
- Low melting point, correlating with good solubility, as predicted by ideal solubility theory.

Clearly, many drug molecules do not meet these criteria. This is especially true for biopharmaceutical drugs, which are becoming increasingly important in therapeutics and diagnostics of a wide range of illnesses. Drugs that suffer poor oral bioavailability or susceptibility to first-pass metabolism, and are thus often ideal candidates for transdermal delivery, may fail to realise their clinical application because they do not meet one or more of the above conditions. Examples include peptides, proteins and vaccines which, due to their large molecular size and susceptibility to acid destruction in the stomach, cannot be given orally and, hence, must be dosed parenterally. Such agents are currently precluded from successful transdermal administration, not only by their large sizes but also by their extreme hydrophilicities. Several approaches have been used to enhance the transport of drugs through the SC. However, in many cases, only moderate success has been achieved and each approach is associated with significant problems. Chemical penetration enhancers allow only a modest improvement in penetration. Chemical modification to increase lipophilicity is not always possible and, in any case, necessitates additional studies for regulatory approval, due to generation of new chemical entities. Significant enhancement in delivery of a large number of drugs has been reported using iontophoresis. However, specialized devices are required and the agents delivered tend to accumulate in the skin appendages. The method is presently best-suited to acute applications. Electroporation and sonophoresis are known to increase transdermal delivery. However, they both cause pain and local skin reactions and sonophoresis can cause breakdown of the therapeutic entity. Techniques aimed at removing the SC barrier such as tape-stripping and suction/laser/thermal ablation are impractical, while needle-free injections have so far failed to replace conventional needle-based insulin delivery. Clearly, robust alternative strategies are required to enhance drug transport across the *SC* and thus widen the range of drug substances amenable to transdermal delivery.

Recently, nanoparticulate and super-saturated delivery systems have been extensively investigated. Nanoparticles of various designs and compositions have been studied and, while successful transdermal delivery is often claimed, therapeutically useful plasma concentrations are rarely achieved. This is understandable, given the size of solid nanoparticles. So called ultra-deformable particles may act more as penetration enhancers, due to their lipid content, while solid nanoparticles may find use in controlling the rate or extending the duration of topical delivery. Super-saturated delivery systems, such as 'spray-on' patches may prove useful in enhancing delivery efficiency and reducing lag times.

Amongst the more promising transdermal delivery systems to emerge in the past few decades are microneedle (MN) arrays. MN arrays are minimally invasive devices that can be used to bypass the SC barrier and thus achieve transdermal drug delivery. MNs (50–900 µm in height, up to 2000 MN cm^{-2}) in various geometries and materials (silicon, metal, polymer) have been produced using recently-developed microfabrication techniques. Silicon MNs arrays are prepared by modification of the dry- or wet-etching processes employed in microchip manufacture. Metal MNs are produced by electrodeposition in defined polymeric moulds or photochemical etching of needle shapes into a flat metal sheet and then bending these down at right angles to the sheet. Polymeric MNs have been manufactured by micromoulding of molten/dissolved polymers. MNs are applied to the skin surface and pierce the epidermis (devoid of nociceptors), creating microscopic holes through which drugs diffuse to the dermal microcirculation. MNs are long enough to penetrate to the dermis but are short and narrow enough to avoid stimulation of dermal nerves. Solid MNs puncture skin prior to application of a drug-loaded patch or are pre-coated with drug prior to insertion. Hollow bore MNs allow diffusion or pressure-driven flow of drugs through a central lumen, while polymeric drug-containing MNs release their payload as they biodegrade in the viable skin layers. *In vivo* studies using solid MNs have demonstrated delivery of oligonucleotides, desmopressin and human growth hormone, reduction of blood glucose levels from insulin delivery, increase in skin transfection with DNA and enhanced elicitation of immune response from delivery of DNA and protein antigens. Hollow MNs have also been shown to deliver insulin and reduce blood glucose levels. MN arrays do not cause pain on application and no reports of development of skin infection currently exist. Recently, MNs have been considered for a range of other applications, in addition to transdermal and intradermal drug/vaccine delivery. These include minimally-invasive therapeutic drug monitoring, as a stimulus for collagen remodelling in anti-ageing strategies and for delivery of active cosmeceutical ingredients. MN technology is likely to find ever-increasing utility in the healthcare field as further advancements are made. However, some significant barriers will need to be overcome before we see the first MN-based drug delivery or monitoring device on the market. Regulators, for example, will need to be convinced that MN puncture of skin does not lead to skin infections or any long-term skin problems. MN will also need to be capable of economic mass production.

In this book, we review the work that has been carried out recently on innovative transdermal delivery systems in both the academic and industrial sectors. We have looked in detail at both *in vitro* and *in vivo* studies and covered the important area skin characterisation, since thorough understanding of this is vital when designing delivery systems to overcome its barrier function. We also consider safety and public perception aspects of new delivery systems and discuss potentially-novel applications of these exciting technologies moving forwards. Since scientists in the cosmetics field have borrowed techniques and formulation designs from the transdermal field, we also look at the recent innovations in this area. Importantly, the final chapter discusses the process of commercialisation of skin delivery systems. It is our hope that this book will serve as a comprehensive overview of the field and hence that it will be of use to those new to transdermal delivery, as well as people already engaged in work in this area.

We are indebted to the contributors for their hard work, openness to suggestions for directions of their chapters and prompt delivery of the chapters. Editing this text took considerable time and we would like to thank our families for their patience and support throughout the project. We are also grateful to the members of the Microneedles Group at Queen's for their hard work and imagination in the lab: Dr Maeliosa McCrudden, Dr Ester Caffarel-Salvador, Dr Rebecca Lutton, Dr Eneko Larraneta, Dr Aaron Brady, Patricia Gonzalez-Vazquez, Eva Vicente-Perez, Joakim Kennedy, Helen Quinn, Aaron Courtenay, Mary-Carmel Kearney and Steven Fallows. Gratitude is also due to the members of the Ocular Delivery Group: Dr Chirag Gujral, Dr Hannah McMillan, Dr Ismaeil Tekko, Samer Adwan and Katie McAvoy. We would like to acknowledge BBSRC, EPSRC, MRC, the Wellcome Trust, PATH, Action Medical Research and the Royal Society for funding our work in this area. Sarah Tilley Keegan and Rebecca Stubbs from John Wiley & Sons provided considerable help and encouragement as we completed this project and their support and guidance are greatly appreciated.

Ryan Donnelly and Raj Singh
Belfast, 2014

1

Introduction

Gary P.J. Moss

School of Pharmacy, Keele University, Keele, UK

The skin is the most physiologically complex and diverse organ of the human body. It has many roles, including the regulation of temperature, mechanical and protective functions. This latter function includes the regulation of water ingress and egress, as well as the prevention of entry into the body of exogenous chemical and biological entities.

The skin is the largest organ of the body, accounting on average for approximately 10% of body mass. It receives approximately one-third of the blood circulating throughout the body and has a surface area of approximately 2–3 m² [1]. It provides a robust, flexible and self-repairing barrier to the external environment and protects internal body organs and fluids from external influences, harmful molecules and micro-organisms. Its permeability limits excessive water loss and exercises temperature regulation over the body. The skin forms an extensive sensory surface, transmitting sensations such as heat, cold, touch, pressure and pain to the central nervous system. The skin is a multi-layered organ consisting of three main histological layers: the epidermis, the dermis and the subcutis. Mammalian skin is a stratified epithelium, and each layer will be considered individually, below, progressing from the innermost tissues to the outermost.

1.1 The Subcutis (Subcutaneous Fat Layer)

Immediately beneath the epidermis and dermis lies the subcutaneous fatty tissue layer (or subcutis or hypodermis). This layer provides support and cushioning for the overlying skin, as well as attachment to deeper tissues. It acts as a depository for fat and contains

Novel Delivery Systems for Transdermal and Intradermal Drug Delivery, First Edition.
Ryan F. Donnelly and Thakur Raghu Raj Singh.
© 2015 John Wiley & Sons, Ltd. Published 2015 by John Wiley & Sons, Ltd.

blood vessels that supply the skin. It also acts as a heat insulator and a shock absorber. The subcutis is variable in thickness, ranging from a few centimetres thick in some regions, such as the abdominal wall, to areas where there is little or no fat, and the subcutis may be difficult to distinguish, such as the eyelid or scrotum. It is often difficult to distinguish the subcutis from the dermis, as both are irregular connective tissues, but the subcutis is generally looser and contains a higher proportion of adipose cells. The deeper layers of the subcutis are fully continuous, with layers of deep fascia surrounding muscles and periosteum.

1.2   The Dermis

The dermis, or corium, lies immediately below the dermo-epidermal junction. It is 10–20 times thicker than the epidermis and ranges from 0.1 to 0.5 cm in thickness, depending on its location in the body. It is a robust and durable tissue that provides flexibility and tensile strength. It protects the body from injury and infection and provides nutrition for the epidermis and acts as a water storage organ. The main feature of the dermis is a matrix of mechanically strong fibrous proteins, consisting mainly of collagen, but with elastin embedded in a gel-like mix of mucopolysaccharides [2]. Embedded within this matrix are various structures, including nerve tissues, vascular and lymphatic systems and the base of various skin appendages. The upper section of the dermis consists of loose connective tissue and a superficial, finely structured papillary layer which progresses upwards into the epidermis. The lower dermis is a coarse, fibrous layer which is the main supporting structural layer of the skin. The transition between epidermal and dermal structures occurs at the dermo-epidermal junction. Both the epidermis and dermis vary greatly in structure, with the former being mostly cellular in construction, whereas the latter contains few cells, other than mast cells. The dermis is the locus of the blood vessels in the skin, extending to within 0.2 mm of the skin surface and derived from the arterial and venous systems in the subcutaneous tissue. The blood vessels supply the hair follicles, glandular skin appendages and subcutaneous fat, as well as the dermis itself [1].

The vasculature of the skin is responsible for regulating the skin temperature, supplying nutrients and oxygen to the skin, removing toxins and waste products and for assisting in wound repair. Clearly, the vasculature also plays an important role in the removal of locally absorbed chemicals, carrying them into the systemic circulation. The blood supply to the skin can sit relatively close to the skin surface, meaning that exogenous penetrants are removed into the circulation from around the dermo-epidermal junction. Thus, for percutaneous absorption into the systemic circulation, including transdermal drug delivery, the blood supply to the skin facilitates the maintenance of a concentration gradient between the material applied to the external skin surface and the vasculature, across the skin barrier. Such clearance may also be facilitated by the lymphatic system, which is similarly located at a comparable distance from the exterior of the skin to the blood supply [3, 4].

1.3   Skin Appendages

Associated with the skin are several types of appendages, including hair follicles and their associated sebaceous glands (Figure 1.1) and eccrine and apocrine sweat glands.

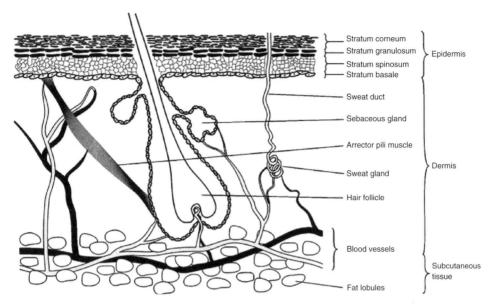

Stratum corneum
Stratum granulosum } Epidermis
Stratum spinosum
Stratum basale

Sweat duct

Sebaceous gland

Arrector pili muscle

Dermis

Sweat gland

Hair follicle

Blood vessels

Subcutaneous tissue

Fat lobules

Figure 1.1   *Schematic diagram of the skin. Reproduced with permission from Ref. [5].*

On average, human skin contains 40–70 hair follicles and 200–250 sweat ducts/cm² of skin. The skin appendages occupy approximately 0.1% of the total human skin surface [4, 6], although this varies from region to region. Hairs are formed from compacted plates of keratinocytes and reside in hair follicles formed as an epidermal invagination. The associated sebaceous glands (Figure 1.1) are formed as outgrowths of the follicle and secrete an oily material – sebum – onto the skin surface. Sebum is a combination of various lipids and acts as a plasticiser for the *stratum corneum*, maintaining an acidic mantle of approximately pH 5 [6]. The eccrine glands are principally concerned with temperature control and are responsible for secretion and evaporation of sweat when stimulated by an increase in external temperature or emotional factors. These glands commonly occupy only 10^{-4} of the total skin area, and extend well into the dermis. Whereas eccrine glands are found throughout the body, apocrine glands are located in specific regions, such as the axillae and anogenital regions. Similar to eccrine glands, they descend into the dermis.

1.4   The Subcutaneous Sensory Mechanism

The extensive size of the skin lends itself to act as a major source of sensory input for the sensory nervous system. It provides information about the environment from both direct contact and from more remote sources, such as the effect of radiation on skin temperature. Cutaneous fibres within the dermis form a plexus lying parallel to the surface of the skin. This plexus is composed of unmyelinated and myelinated fibres, organised in the same manner as the parent nerve trunks. The dermal networks send twisted extensions into the papillae, where they form loops which return to the superficial part of the plexus. From the plexus some individual fibres extend to supply particular locations. The terminal branches of each fibre interconnect with and superimpose themselves on each other [7] in such a way that every area in the skin is supplied by several different fibres. Each of these fibres ends in at least one particular receptor. Most of the cutaneous receptors can be excited by various stimuli, but it is the different thresholds of the stimuli required to provoke responses that yields specifically to these receptors [8].

There are three main categories of cutaneous receptor which are distinguished by their different sensitivities to stimuli: mechanoreceptors, thermoreceptors and nociceptors. Mechanoreceptors have high sensitivities to indentation or pressure on the skin, or to movement of the hairs. This group may be further subdivided into the rapidly adapting (RA) and slowly adapting (SA) receptor types. The RA mechanoreceptors include Pacinian corpuscles, found in both hairy and glabrous skin, and Meissner's corpuscles, located in the glabrous skin of primates and hair follicle afferent units found only in hairy skin. Pacinian corpuscles are small pearl-shaped structures found in the deeper layers of the skin. They are 0.5–2 mm long and are composed of an 'onion-like' lamellar structure which is formed from non-nervous tissue. They contain an elongated nerve ending at its core which is not derived from the dermal plexus. The most important characteristic of the Pacinian corpuscle is its ability to detect mechanical vibrations at high frequencies, which may be relayed at greater than 100 Hz/s. Such frequencies are often sensed in traumatised or unanaesthesised skin [9, 10]. The Meissner's corpuscle is an encapsulated myelinated receptor which resides in the dermis of the human glabrous skin. It is tucked into the dermal papillae that fill the grooves formed by epidermal ridges. The entire corpuscle is surrounded by connective tissue, continuous with the perineurium, which is attached to the basal projections of the epidermal cells by elastin fibrils. The Meissner's corpuscle discriminates highly localised sensations of touch, especially in the palmar regions where they are found in their highest density [11].

Hair follicle receptors are myelinated fibres, circumferentially arranged around the hair root sheath below the sebaceous gland which innervate hair follicles. Large hair follicles can be supplied by up to 28 fibres. The hair is well placed in its follicle to stimulate the nerve collar and is primarily associated with tactile sensations [12]. SA mechanoreceptors respond during skin displacement. They also maintain a discharge of impulses when the skin is held in a new position [8]. These receptors include the Ruffini endings and the C-mechanoreceptors. The Ruffini endings are encapsulated receptors which are found in the dermis of both hairy and glabrous skin and provide a continuous indication of the intensity of the steady pressure or tension within the skin [9]. C-mechanoreceptors have small receptive fields (about 6 mm²) in hairy skin and may emit a slowly adapting discharge when the skin is indented or when hairs are moved. Repetitive stimulation will, however, produce a rapid fall in excitability, and the receptors will fail to respond after 20–30 s because the receptor terminals have become inexcitable [8].

Thermoreceptors are characterised by a continuous discharge of impulses at a given constant skin temperature which increases or decreases when temperature is raised or lowered. The receptive fields of the thermoreceptor are spot-like and cover an area of no more than 1 mm². Thermoreceptors are classed as either 'cold' or 'warm' receptors, with 'cold' receptors lying more superficially in the skin than 'warm' receptors. The depth of 'cold' and 'warm' receptors was estimated at about 0.15 and 0.6 mm, respectively, below the surface. The firing frequency accelerates in 'cold' receptors when the temperature is falling – and vice versa for the warm receptors. Such dynamic sensitivity is high and permits the receptors' response to relatively slow (<1°C in 30 s) and small changes in skin temperature [8].

Damaging or potentially damaging excitation of thermo- and mechanoreceptors is not necessary for such receptors to reach maximum activation, indicating their inability to control pain. They do, however, contribute to the sensory quality of perceived pain. The receptor systems that detect and signal high intensities of stimulation form a distinct class of sense peripheral organs called 'nociceptors'. They have unencapsulated nerve endings and exhibit the smallest identified structures [9–15]. Nociceptors generally reside at the dermo-epidermal junction, and are either *mechanical nociceptors*, which respond to pin-pricks, squeezing and to crushing of the skin, or *thermal* (or mechanothermal) *nociceptors* which respond to severe mechanical stimuli and to a wide range of skin temperatures.

1.5   The Epidermis

The epidermis is the outermost layer of the skin. It is the thinnest part of the skin, with its thickness varying around the body – for example, the thickest skin is commonly found on the weight-bearing planter surfaces (feet and hands, ~0.8 mm) and the thinnest skin is normally found on the eyelids and scrotum (0.06 mm) [5]. Despite the extensive vasculature present in deeper tissues such as the dermis, the epidermis has no blood supply and passage of materials into or out of it is usually by a process of diffusion across the dermo-epidermal layer. It is essentially a stratified epithelium, consisting of four, or often five, distinct layers.

1.6   The *stratum germinativum*

The deepest layer of the epidermis is the *stratum germinativum*, or basal layer. This metabolically active layer contains cells that are similar to those found in other tissues in the body, as they contain organelles such as mitochondria and ribosomes. It is often single celled in thickness and contains cuboid or columnar-to-oval-shaped cells which rest upon the basal lamina. The basal cells are continually undergoing mitosis, as they provide replacement cells for the higher (outer) epidermis. Basal keratinocytes are connected to the dermo-epidermal membrane by hemidesmosomes, which connect the basal cells to the basement membrane. Throughout the basal layer and higher layers of the epidermis, such as the *stratum spinosum*, keratinocyte cells are connected together by desmosomes. The basal layer is also the location of other cells, including melanocytes, Langerhans cells and Merkel cells. The basal cells become flatter and more granular as they move up through the epidermis.

1.7   The *stratum spinosum*

Immediately above the *stratum germinativum* is the *stratum spinosum*, or prickle cell layer. It is often described, in conjunction with the basal layer, as the Malpighian layer. It is several (usually between two and six) layers thick and forged from cells of irregular morphology, varying from columnar to polyhedral in structure as this layer progresses outward. Each cell possesses distinct tonofilamental desmosomes, characterised as prickles or spines, which extend from the surface of the cell in all directions and which help to maintain a distance of approximately 20 nm between cells. The prickles of adjacent cells link via inter-cellular bridges, providing improved structural rigidity and increasing the resistance of the skin to abrasion. Though lacking in mitosis, the prickle cell layer is metabolically active.

1.8   The *stratum granulosum*

The next epidermal tier is the *stratum granulosum*, or granular layer. It usually one to three layers deep and consists of several layers of flattened, granular cells whose cytoplasm contains characteristic granules of keratohyalin, which is responsible for their appearance. It is produced by the actively metabolising cells and is believed to be a precursor of keratin. The *stratum granulosum* is the skin layer where degradation of cell components becomes significant, resulting in a decrease in metabolic activity which eventually ceases towards the top of this layer due to the degeneration of cell nuclei, leaving them unable to carry out important metabolic reactions.

1.9   The *stratum lucidum*

The layer above the *stratum granulosum*, the *stratum lucidum*, is easily observed on thick skin, but may be missing from thinner skin, hence the often differing descriptions of the epidermis as having four or five layers. It is often considered that the *stratum lucidum* is functionally indistinct from the *stratum corneum* and that it may be an artefact of tissue preparation and cell differentiation, rather than a morphologically distinct layer. The cells are elongated, translucent and mostly lack either nuclei or cytoplasmic organelles. The *stratum lucidum* exhibits an increase in keratinisation consistent with the progression of cell flattening from the bottom to the top of the epidermis.

1.10   The *stratum corneum*

The outermost layer of the skin is the *stratum corneum*, often called the horny layer. It is the final result of cell differentiation and compaction prior to desquamation and removal from the body. While it is an epidermal layer it is often considered a separate layer of the skin and is often described as such. It consists of a compacted, dehydrated and keratinised multilayer, which is, on average, 15–20 cells thick; that is, around 10 μm in thickness when dry, although it can swell to many times its thickness when wet. The formation of keratin and the resultant death of these cells are part of the process of keratinisation, or cornification. The *stratum corneum* is, in effect, the outer envelope of the body. In areas where the *stratum lucidum* is apparent, the *stratum corneum* is much thicker, being designed to cope with the effects of weight support and pressure. Its thickness also mirrors that of the viable epidermis around the body. Thus, the epidermis in those regions, such as the palms and soles, can be up to 800 μm in thickness, compared to 75–150 μm in other areas. Cells of the *stratum corneum* are physiologically inactive, continually shedding and replenishing themselves from the upward migration of cells from the underlying epidermal layers [1].

The *stratum corneum* is predominant rate-limiting membrane of the skin, and is responsible for regulation of water loss from the body as well as limiting the ingress of harmful materials from the external environment.

The *stratum corneum* is described as the main rate-limiting barrier of the skin with regard to the viable epidermis and dermis [16]. It consists of two alternating amorphous lipophilic and hydrophilic layers, and is comparatively – with regard to the rest of the epidermal layers – more lipophilic. The hydrophilic cells of the *stratum corneum* consist mainly of corneocytes, natural moisturising agents. The water content of the *stratum corneum* is highly variable, depending on both moisture content of the external environment of the body and the location on the body from where the skin is obtained. It varies with the position of the tissue, with the water content generally decreasing as the external interface is approached. The *stratum corneum* has been shown to possess 40% water by weight in an environment where the relative humidity is between 33 and 50%. It has also been estimated that, by weight, the *stratum corneum* is further composed of 40% protein, mostly keratin, and 15–20% lipid, predominately triglycerides, cholesterol, fatty acids and phospholipids [17]. These lipids occupy the intercellular space in the *stratum corneum* and originate from several sources, including the discharged lamellae of membrane-coated granules, intercellular cement and the keratinocyte cell envelope.

The *stratum corneum* is an exceedingly dense tissue and may swell to many times its own thickness in water. Its elongated cells, approximately 1 μm in thickness, form a close-packed array of interdigitated cells stacked in vertical columns [18]. Interdigitation between adjacent cells allows the formation of cohesive laminae. Each cell is contained by a largely proteinaceous envelope rather than the conventional lipid bilayer cell membrane. An individual horny cell is approximately 1 μm thick and occupies an area of 700–1200 μm^2. There are approximately 10^5 cells/cm^2. The mechanical strength of the *stratum corneum* is mostly due to the nature of the proteinaceous envelope, the disulphide bonds of the intracellular keratin and the bridges linking cells that are embedded in an intercellular lipid matrix [19].

The *stratum corneum* constantly sheds its outermost layer in a process called desquamation. The daily loss of flakes from the horny layer of the skin is typically not more than 1 g. The desquamation process involves the cleavage of the intercellular bridges, suggesting that there is a certain degree of metabolic activity and regulatory control occurring in what is often considered to be a dead layer. In normal human skin the rate of *stratum corneum* shedding is generally equal to the rate of epidermal cell regeneration, thus maintaining an epidermis of approximately constant thickness. The *stratum germinativum* and *stratum spinosum* generate one new cell layer per day. Typically, differentiation from *stratum basale* to *stratum corneum* takes an average of 14 days. Cell regeneration is a more complex process in the epidermis, including dehydration and polymerisation of the intracellular material, that ultimately produced the cells found in the *stratum corneum* [1].

Classically, the *stratum corneum* skin barrier has classically been described using a 'bricks and mortar' model [20, 21], with the bricks representing the tightly packed corneocytes which are embedded in a 'mortar' of lipid bilayers. These flattened, often hexagonal – but more accurately described as polygonal – highly proteinaceous cells are the final point of keratinocyte differentiation and are interconnected by structures termed 'corneodesmosomes' (Figure 1.2). The 'bricks' are enclosed within a continuous and highly ordered lipid phase, which is lamellar in structure and often described as a lipid bilayer. It is generally understood that the ceramides are the most important component of this phase. Ceramides are polar

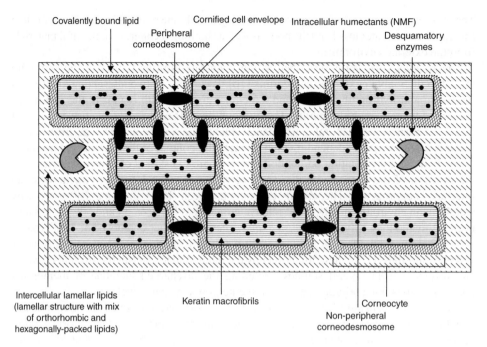

Figure 1.2   *Detailed schematic structure of the stratum corneum. Reproduced with permission from Ref. [23]. © 2010, John Wiley and Sons, Ltd.*

lipids which contain hydroxylated alkyl side chains that, under normal conditions, are packed both hexagonally and orthorhombically. This barrier forms a continuous poly-proteinaceous structure whose thickness and exact composition vary across different body sites. The 'bricks' of the skin barrier may hydrate extensively and cause significant changes in the packing and structure – as well as the permeability – of the *stratum corneum* [22, 23, 25–32]. Thus, it is now understood that the *stratum corneum* does not simply form a homogenous bricks and mortar structure. The corneocytes change in their morphological and biochemical functions as they progress from the lower to higher levels of the *stratum corneum*. Associated with this transition are increases in transglutaminase-mediated protein crosslinking and increased levels of inter-corneocyte ceramides and fatty acids. This results in a progression from fragile to rigid structures (described as the transition from '*stratum compactum*' to '*stratum disjunctum*' [23]) where non-peripheral corneodes-mosomes exhibit a reduction in interdigitation towards the outer layers of the barrier. This is concomitant with an increase in the occurrence of (pro)filaggrin – a protein thought to play a role in the aggregation of keratin filaments within corneocytes [23].

Significant advancements have been made in the characterisation and understanding of the *stratum corneum* structure and barrier function in the past 10 years. For example, new species of ceramides, and the synthetic pathways that generate them, are still being identified and their synthetic pathways are still being characterised [23]. The lamellar arrangement of the *stratum corneum* lipids was characterised by electron microscopy and X-ray diffraction and, more recently, by cryoelectron microscopy [27, 33–39]. This latter technique has proposed the existence of a single-gel phase model for the stratum corneum lipids. The further suggested that cryoelectron microscopy failed to show the expected presence of the trilamellar-conformation long periodicity phase (LPP). Bouwstra and colleagues [40] suggested that the *stratum corneum* lipid phase could be represented by a 'sandwich model'. This model accounts for differences in *stratum corneum* lipid packing – particularly with regard to differing periodicity phases reported in the barrier lipids, highlights the importance of a fluid phase within the *stratum corneum* which may be dictated by the presence of ω–esterified long-chain acylceramides.

However, it is known that the lamellar phase is often missing from the outer layers of the stratum corneum, even in healthy skin. It is now known that other changes also occur [41, 42]. In the most tightly packed lipid barrier – the orthorhombically packed state – the presence of long-chain fatty acids is required to induce the formation of the orthorhombic lattice in ceramide and cholesterol mixtures. Ultimately, the presence of the LPP with orthorhombic packing defines ultimate lipid barrier functionality [43, 44]. The *stratum corneum* is not a homogenous tissue and exhibits characteristic changes as it progresses outwards from the body – often described as the transition from '*stratum compactum*' at the inner base of this layer, to '*stratum disjunctum*' at its outermost layer. Such a transition may be exemplified by a transition in the packing of ceramide sides chains from a transition from a more tightly packed to a less tightly packed hexagonal phase which occurs closer to the skin surface. Further, at the skin surface the lamellar phase is normally missing, becoming amorphous in nature [23, 45, 46].

1.10.1   Routes of Absorption

One of the classic characteristics of the *stratum corneum* barrier function is that the predominant route of absorption is through the lipid layers of this part of the skin. While it is a longer and more tortuous route across the *stratum corneum* compared to the transcellular pathway, it does not require the potential partitioning between the *stratum corneum* lipids and corneocytes, but it relies on partitioning into the stratum corneum lipids from the formulation vehicle and subsequent diffusion across the *stratum corneum*, predominately in a single phase. The other proposed route across the skin, that of permeation via skin appendages such as hair follicles and sweat glands, is limited by the occurrence of such structures as they occupy, on average, approximately 0.1% of the total skin surface and in some cases, such as sweat glands are often morphologically similar to the remainder of the skin surface. This latter point both limits absorption through targeting this route and, in the case of sweat glands in particular, absorption must compete with an opposing outward current. Thus, to understand the absorption process in the context of skin physiology, the *stratum corneum* lipids appear to govern the percutaneous absorption of exogenous chemicals. However, it should be noted that the other potential routes can also play an important part in the overall process of percutaneous absorption [24].

1.10.2   Transdermal Permeation – Mechanisms of Absorption

A chemical must undergo a series of steps if it is to pass across the skin and become systemically available. For example, it will usually be presented to the skin surface in some sort of vehicle, or formulation. This may be a simple aqueous solution, or a complex, multi-phase pharmaceutical or cosmetic formulation, and it may not remain in the same state during the absorption process (e.g. due to volatility of any of its ingredients). The first step in absorption is therefore the partition of the penetrant from the formulation onto the skin surface, where those molecules in contact with the *stratum corneum* will begin to partition. The rate and extent of this process will depend on the physicochemical properties of each penetrant. The tendency for a chemical to penetrate into the skin may be influenced significantly by the nature of the formulation – if the penetrant has a high affinity for the formulation, then it may remain there, whereas if it has a low affinity for the formulation (or a higher affinity for the *stratum corneum*), it may partition into the skin more readily.

Thus, those molecules adjacent to the skin surface will permeate into the *stratum corneum*. Maintenance of permeation is dependent on the random movement of the penetrant from the bulk of the vehicle to the surface of the skin, which again may be influenced by the nature of the formulation. In the case of suspensions, where the penetrant may be presented to the skin surface as a combination of a saturated solution and as an undissolved solid, partitioning into the *stratum corneum* may be further influenced by dissolution within the formulation to maintain a saturated solution.

Once the penetrant has diffused into the *stratum corneum,* it will diffuse through this layer. This may occur via any of the three main routes mentioned earlier (intracellular, intercellular and transappendageal; Figure 1.3), either individually by combined permeation via more than one of these routes. The next significant permeation step will be at the junction of the *stratum corneum* and the viable epidermis, due to the substantial change in the characteristics of the tissue, which is increasingly hydrophobic in nature. This results in a further partitioning step which is followed by further diffusion into the viable epidermis and partitioning between the viable epidermis and the dermis. Finally, partitioning from the dermis to the capillary system will see the penetrant removed into the blood vessels and the systemic circulation.

The transepidermal route, via the intact *stratum corneum*, is the main route through which penetrants may enter, as it provides the major area available to a potential penetrant. The *stratum corneum* has been morphologically and functionally represented by the 'bricks and mortar' model [47]. The 'bricks' of this model, the corneocytes, are fibrous protein networks, whereas the 'mortar' is an intercellular network predominately consisting of

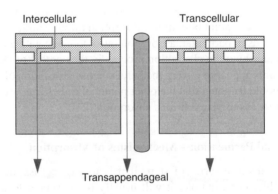

Figure 1.3   *Pathways of drug penetration through skin.*

neutral lipids, as described earlier. Figure 1.2 shows that penetrants may diffuse via a combination of intercellular, transcellular or transappendageal routes.

Successful permeability of the intact *stratum corneum* has been shown to relate predominately to lipophilic materials and depends on the oil/water partitioning property of a particular penetrant. Relationships between permeability and partition coefficients have classically been demonstrated by various investigations which have indicated that the existence of separate lipophilic and hydrophilic pathways is supported by histological and cytochemical studies [48–53].

The other potential route for transdermal penetration is the transappendageal (or 'shunt') route, through hair follicles and sweat ducts. These appendages lack a horny layer and in theory offer low resistance rapid diffusion shunts, allowing penetrants to by-pass the horny layer [54, 55]. By consideration of bulk diffusion relative to shunt diffusion, Scheuplein concluded that transappandageal absorption may be important in the early 'lag' period of the penetration process [51]. Diffusion through glands is generally considered negligible due to the small area they occupy on the surface of the skin, and the current of secretions passing to the outer surface. Valve mechanisms at the openings of glands further lessen their ability as potential routes of access for penetrants. However, whereas Barr demonstrated that areas of the skin rich in eccrine glands do not show significantly greater permeability to chemical penetration [56], more recent studies [55, 57] indicate that glands could contribute to the penetration of hydrocortisone and testosterone. The shunt routes have been shown to potentially play an important route in the delivery of vesicles onto the skin surface [58], although more recent studies, described in the following, suggest that this may be due to deposition of intact particles in skin furrows, or dermatoglyphs. Ultimately, however, as with transepidermal penetration, successful penetration via the shunt route depends upon the physicochemical properties of the penetrant as well as the nature of the *stratum corneum* and may be more successful for some penetrants than for others.

Other factors may additionally influence the penetration process. These include the potential for protein binding, which may occur in the *stratum corneum*, contributing to the reservoir effect associated with that layer. Metabolic activity may see some, or potentially all, of the permeant degraded before it reaches the blood vessels. Further, there is potential for permeants to pass into deeper layers of the skin, including the subcutaneous fatty layer, or even into muscle tissues underlying the skin.

1.11   Theoretical Considerations

Diffusion is 'a process of mass transfer of individual molecules of a substance, brought about by random molecular motion and associated with a concentration gradient' [59]. Passage of a diffusant through a barrier, such as skin, occurs by simple molecular permeation or by movement through pores and channels, as discussed earlier. Diffusion through a non-porous membrane occurs when the penetrant dissolves in the bulk membrane or solvent-filled pores of the membrane. Such diffusion is influenced by the size and physicochemical properties of the penetrant and the nature of the membrane. The three layers of the skin, the epidermis, the dermis and the subcutis, each has their own diffusion coefficient. The diffusion coefficients of all layers other than the *stratum corneum* are generally considered to be negligible so they are normally treated together and represented by a single diffusion coefficient.

The ability of a compound to pass through the skin has been reported in various manners. Total diffusional resistance of the skin is generally attributed to the *stratum corneum* under passive diffusion and, therefore, Fick's first law of diffusion may be applied [60]:

$$J = -D\frac{\partial C}{\partial x}$$

where:

J is the rate of transfer per unit area of the surface (i.e. the flux).

C is the concentration of the diffusing substance.

x is the spatial co-ordinate measured normal to the section.

D is the diffusion coefficient, or diffusivity.

The dermal permeability coefficient, K_p, is defined by the equations:

$$J_{ss} = K_p C_v \tag{1.1}$$

or

$$K_p = \frac{J_{ss}}{C_v} \tag{1.2}$$

Combination of Equations 1.1 and 1.2 gives:

$$K_p = K_m \cdot \frac{D}{h} \tag{1.3}$$

where

K_p is the permeability coefficient (cm/s or cm/h).

C_v represents the concentration of penetrant in the vehicle when sink conditions apply.

J_{ss} is the steady-state flux of the solute.

D is the average diffusion coefficient (cm^2/s or cm^2/h).

K_m represents the partition, or distribution, coefficient between the stratum corneum and the vehicle.

h is the thickness of the skin.

Thickness of the membrane has generally been recognised as being inversely proportional to flux, although it has been suggested that lipid content, and not thickness, was more relevant [21]. Further, the aforementioned model is more appropriate for *in vitro* systems, as it is unlikely to hold for *in vivo* situations due to the low permeability of the *stratum corneum* which ensures steady-state conditions would take a significant time to become established. However, *in vitro* diffusion is still a highly important area of research, particularly in the development of models for percutaneous absorption, providing excellent theoretical and preliminary investigative models of *in vivo* drug permeation.

Standard methods for membrane studies were derived which stated that, in an *in vitro* diffusion experiment, the concentration of the penetrant should be maintained at a constant level in the donor phase and conditions in the receptor phase should be invariable [61–63]. Further, he stated that the composition of the donor phase should be kept steady, avoiding losses in evaporation or diffusion, and that both the donor and receptor phases should be continually stirred throughout the course of an experiment. Variations on such methods have been widely employed since.

Thus, from the viewpoint of the percutaneous absorption of exogenous chemical – into and across – the skin, the *stratum corneum* is essentially a lipidic layer, which interfaces with a predominately aqueous medium sitting beneath it. The transport of lipophilic chemicals predominately occurs via the *stratum corneum*, and as these compounds must transfer directly from this comparatively lipid-rich environment into an aqueous medium, compounds that are highly lipophilic will remain largely in the *stratum corneum*. There have, therefore, been considered to be a number of pathways by which compounds with different physicochemical natures may permeate the skin – the so-called polar and lipophilic pathways [6].

1.12 Physicochemical Properties of the Penetrant

Whether through empirical observation or quantitative modelling, the physicochemical properties of a penetrant are known to significantly influence its ability to penetrate into and across the skin. Such properties should be considered in the context of skin permeability and, initially, partitioning.

1.12.1 Partition Coefficient

The Meyer–Overton theory of absorption states that lipid-soluble molecules will pass through the cell membrane due to its lipid content, whereas water-soluble substances pass after the hydration of the protein portion of the cell wall, leaving it permeable to water-soluble substances. The partition coefficient is the ability of a substance to partition between two immiscible phases, usually octanol/water or heptane/buffer. Somewhat simplistically, a higher partition coefficient represents a more lipophilic molecule, and is usually associated experimentally with an increase in permeation via the lipid domains of the *stratum corneum*. For a chemical to cross the *stratum corneum* it must first partition into this membrane, and this may be the rate-limiting step in the permeation process. Barry [64] determined that the partition coefficient, usually described as log P or log K_{ow}, of a penetrant will influence the path it takes in traversing the skin. In practice, the ideal transdermal penetrant should possess both lipophilic and hydrophilic properties [1, 64–66].

It was determined by Bronaugh and Congdon that, for a series of hair dyes, increasing the lipophilicity of a molecule increased the rate of penetration [67]. Le and Lippold indicated that the maximum flux may be estimated from the penetrant's physicochemical properties, including the partition coefficient [68]. Further, Higo *et al.* demonstrated that skin penetration was dependant on the partition coefficient for a series of salicylic acid derivatives [69].

Generally, the lipid bilayers of the *stratum corneum* provide a rate-limiting barrier to the permeation of predominately lipophilic permeants. However, as predominately hydrophilic permeants will have a comparatively higher tendency to permeate via hydrophilic pathways, such as hydrated keratin-filled keratinoctyes, the effect of partition coefficient for such penetrants is not as clear. For example, the lipid bilayer contains hydrophilic elements (e.g. polar head groups), suggesting that hydrophilic permeants may traverse the skin barrier by a number of different routes. Williams suggested that those permeants with intermediate properties – defined as having a log P of between 1 and 3 – will traverse the skin barrier via both lipid and aqueous pathways but the intercellular route will probably dominate [5]. For lipophilic molecules – those with a log P of greater than 3 – the intercellular pathway will be the predominant route for permeation. Finally, a major consideration for the skin permeation of highly lipophilic molecules is their ability to partition from the lipid domains of the *stratum corneum* and into the predominately hydrophilic tissues of the underlying viable epidermis.

1.12.2 Molecular Size and Shape

Consideration of the size and shape of a molecule are important factors in determining its suitability as a percutaneous penetrant. While molecular volume is the most appropriate term to consider, molecular weight is more frequently used due to convenience and practicality, and assumes that molecules are essentially spherical [5, 70]. An inverse relationship exists between the diffusivity of a molecule and its molecular weight, and as such small molecules may diffuse comparatively faster within a particular medium [71, 72] with a cut-off limit to absorption being generally associated with a molecular weight of 500 Da.

Chemical modifications made to a penetrant molecule can result in substantial changes in its ability to penetrate the skin barrier. For example, Scheuplein and Blank compared the rates of penetration of a series of related compounds, all consisting of four carbon atoms, and varying in the position of either one or two added oxygen atoms, which were present as various functional groups [52].

Table 1.1 illustrates that permeability varies greatly when the functional groups are changed, and the permeability coefficients – determined experimentally – indicate that

Table 1.1 Effect of molecular structure and functional group on in vitro *permeability*[a]

Solute	Molecular Structure	Permeability constant, k_p (cm/h)
Ethyl ether	C—C—O—C—C	15–17
2-Butanone	C—C—C—C (with =O on second C)	4–5
1-Butanol	C—C—C—C—OH	2–4
2-Ethoxyethanol	C—C—O—C—C—OH	0.2–0.3
2,3-Butanediol	C—C—C—C (with OH on second and third C)	0.05

[a] From Ref. [16].

the least permeable molecules are those which are the most polar. It has also been demonstrated that the permeability of steroids decreases when they are modified to incorporate more polar functionalities, such as hydroxyl groups [52].

1.12.3   Applied Concentration/Dose

It is generally recognised that increasing the drug loading of a vehicle increases the amount of drug absorbed across the skin [5, 54, 73, 74]. Further, increasing the surface area available for permeation increases the potential for a topically applied molecule to be absorbed across the skin [72, 75, 76]. Frequency of application will also affect the delivered dose. Although one large application usually results in a higher dose absorbed, a single application may also have a greater toxicological potential compared to frequent, smaller doses [75, 77, 78]. Occlusion and duration of contact can also increase the dose absorbed percutaneously [79, 80].

1.12.4   Solubility and Melting Point

In general, references to the solubility in the context of skin permeability refer to *aqueous solubility*.

The percutaneous penetration of a molecule is greatly influenced by its aqueous solubility and partition coefficient [71, 81, 82]. Generally, lipophilic molecules will penetrate into the *stratum corneum* more rapidly than hydrophilic molecules. However, this needs to be balanced with preferential solubility in deeper layers of the viable epidermis and dermis, as well as the effects of the depletion of the concentration gradient in the vehicle. For example, if a penetrant is relatively lipophilic and is delivered from an aqueous vehicle at either saturated or sub-saturated concentrations, then it may be present in a low concentration in the vehicle, resulting in a diminished concentration gradient as diffusion progresses, and a reduction in the rate of permeation due to donor phase depletion. The partition of the penetrant between the *stratum corneum* and the vehicle is of great importance in transdermal drug delivery. If the drug is more soluble in the *stratum corneum* than the vehicle, then the concentration of that drug in the *stratum corneum* may be greater than in the vehicle at equilibrium. Complete solubilisation in the vehicle has been shown to increase the flux of a penetrant [52, 62, 63, 83–85], although other formulations, including suspensions of largely insoluble drugs, have also been shown to exhibit enhanced permeation [1]. Where drugs are fully solubilised in the formulation the rate of penetration is generally increased by complete diffusion in the vehicle, and may be due to improved diffusion through the vehicle, which replenishes the vehicle/skin interface. Further, melting point is well correlated to aqueous solubility, to the extent that predictive models often employ melting point to determine solubility.

1.12.5   Ionisation

The predominately lipophilic nature of the *stratum corneum* and the largely lipophilic pathway therein infer that the ionised form of a molecule is less likely to permeate the skin than the unionised form. Thus, the degree of penetrant ionisation is essential in determining the successful delivery of drugs by both passive and assisted delivery mechanisms. According to the pH partition theory, if a molecule is unionised, then it may readily penetrate the *stratum corneum* via the intercellular pathway. The basis of this theory is that lipophilic regions of the skin act as barriers to ionised species, and that ionised species may pass through pores [5, 86, 87]. Parry and co-workers [88] showed theoretically, by employing a mathematical model, and experimentally that only unionised species enter and traverse the skin. Roy and Flynn [89] demonstrated that the unionised, free base forms of fentanyl and sufentanil are, respectively, 218 and 100 times more permeable than their ionised counterparts. They concluded that the contribution to the process of passive diffusion by ionised species is negligible.

Nevertheless, such comments should be taken in the wider context of a penetrant's physicochemical properties relative to the available diffusive pathways across the skin. Despite the partition theory, several studies [87, 90–93] have shown that both ionised and unionised molecules can penetrate a lipophilic membrane, although the rates and routes of transport are radically different for both species. Classically, it was suggested that ions, ion pairs and electrolytes, such as sodium and potassium salts, can readily traverse the skin [94]. Larger, ionised compounds may penetrate by mechanisms of either ion-pairing [90, 91, 95–97] or ion-exchange [91, 98, 99]. Thus, the ionisation state of a potential penetrant, in the context of its pK_a and the vehicle pH, will significantly affect the permeability of a molecule into and across the skin [100, 101]. Thus, it follows that the different aqueous solubilities of ionised and unionised species will influence permeability as drug flux is the product of the permeability coefficient, K_p, and the effective drug concentration in its vehicle [5]. Adjustment of the pH will therefore alter the amounts of penetrant available in the free base (unionised) or charged (ionised) forms, consequently affecting concentration, solubility and ultimately the rate of penetration across the skin [5, 93, 100].

1.12.6   Physiological Factors Affecting Percutaneous Absorption

The skin is a diverse tissue whose physiology varies considerably around the body. This inherent variation and a range of physiological factors influence the rate of drug delivery into and across healthy, intact skin.

1.13   Physiological Properties of the Skin

1.13.1   Skin Condition

In general, reference to the skin barrier and skin permeability relates to the ingress of chemicals into and across intact, healthy skin. Such skin, particularly the *stratum corneum*, provides a formidable barrier to the passage of substances applied through the skin. However, various disease states may interrupt the continuity of the *stratum corneum* barrier and may result in a number of physiological changes, such as occasionally increasing vasodilation which may result in an increased permeability. Even if the skin is not broken, irritation and mild trauma may reduce the barrier to absorption. Mechanical damage, such as cuts and abrasions, or chemical burns, from acids, alkalis and aqueous phenols may decrease the barrier properties of the skin and increase the rate of absorption. For example, Barry [102] demonstrated that soaking excised *stratum corneum* in chloroform/methanol mixtures dramatically increased skin permeation due to delipidation and the creation of gaps, or artificial shunts, in the barrier layer. Where the skin is disrupted, it has been shown that absorption of hydrophilic solutes increases significantly more than hydrophobic molecules [103].

1.13.2   Skin Hydration and Occlusion

It has been demonstrated that an increase in the hydration of the skin increases the rate of penetration of most molecules. The exact nature and magnitude of such changes have been attributed to the physicochemical nature of the penetrant and the manner in which excess hydration is induced. Imokawa, for example, concluded that the *stratum corneum* lipids were important in holding water in the skin through the formation of lamellar structures within the *stratum corneum* [104]. Wiedmann suggested that the effective diffusion coefficient across the *stratum corneum* increases with an increase in water content, proposing that the water content of the *stratum corneum* heightens the dynamic motion of epidermal tissue, thereby increasing the effective diffusion coefficient [105]. Diffusion coefficients of skin are also altered by a change in the mobility of skin constituents. The barrier presented by the skin has been shown to decrease rapidly over a short space of time – as little as 10 min – with an increase in hydration [106]. Increased hydration of the skin modifies its rheological properties, altering skin elasticity and increasing its suppleness [106, 107]. Hydration of the skin may also be influenced by the relative humidity of its external environment. Changes in relative humidity have been shown to increase hydration and elevate the rate of diffusion. Fritsch and Stoughton reported increases in acetylsalicylic acid penetration when the *stratum corneum* was fully hydrated, compared to conditions of much lower humidity at the same temperature [108].

Occlusion involves entrapment of water which would normally be lost to the surrounding environment, resulting in a rise in temperature and increased hydration of the skin site [109]. It is normally achieved by placement of a water-impervious dressing on the skin. Increased permeability is also associated with occlusion, where the use of a dressing or formulation which is intrinsically occlusive (i.e. an ointment) is commonly observed to increase permeation as occlusion increases the hydration of the *stratum corneum* [110, 111]. Certain topical formulations may induce occlusion, and increase permeation, by virtue of their high viscosities. The occlusive effect of certain formulations has resulted in an increase of therapeutic activity for a range of drugs, including hydrocortisone [112–114], steroids [115, 116] and citropten [117]. However, Treffel demonstrated that while the rate of penetration of lipophilic citrophen increased under occlusion, amphiphilic caffeine exhibited no such increase when occluded [117]. It has also been shown that volatility of the vehicle in which the penetrant is applied and the physical nature of the penetrant can influence permeation and may not be associated with an increase in permeation under occlusive conditions [118, 119].

1.13.3   Skin Age

The structure and appearance of skin changes significantly with age, but it is often unclear if such changes are as a result of inherent ageing or influenced by environmental factors. With regard specifically to the skin barrier, skin permeability to the ingress of exogenous chemicals is often considered to be lower for infants, and it is often perceived to lessen with age.

The skin of an infant has, compared to adult skin, a higher water content and the *stratum corneum* is not fully developed, leaving it more permeable than fully developed adult skin [120]. The surface-area-to-volume ratio and metabolic activity of an infant or child's skin are much greater than that of an adult, and as such a larger absorption of drug per kilogram of body weight may occur, influencing dosing [121]. For example, Christophers and Kligman [122] and Idson [71] have shown that absorption of topical steroids is greater in children than in adults. This was explained by comparing the decreasing moisture content and increased transepidermal water loss experienced by adult skin at an average age of 40. Further, it was shown that, due to alterations in keratinisation and epidermal cell production which lead to an increase in corneocyte surface area and a decrease in the size of intercellular spaces, the moisture content of human skin decreased with age [123–125]. However, the implications of such a reduction are not clearly decoupled from other effects associated with ageing. It should also be noted that Roy and Flynn concluded that age was not a significant factor in the penetration of fentanyl and sufentanil, suggesting that any age-related permeability effects may not uniformly apply to all penetrants and that, once the *stratum corneum* is fully formed it maintains its barrier function [89]. Nevertheless, other factors may influence the change in permeability. For example, changes to the underlying vasculature may reduce blood flow and thus dermal clearance of topically applied drugs, reducing transdermal flux. However, the main factor in considering skin permeability is the barrier function of the *stratum corneum*.

1.13.4   Regional Variation (Body Site)

Wide variations in absorption rates have been found across different skin sites in the same individual. Additionally, the inherent interpersonal variation of skin means that the most permeable skin sites on one person's body may have the same absorbance as the least permeable site on another person. Although conflicting results have been reported, the permeability rates of molecules can be related to the thickness of the skin at particular points on the body. Wester and Maibach reported that this regional variation in absorption was not necessarily due to *stratum corneum* thickness, and that areas with the same thickness of *stratum corneum* demonstrated different permeability, and areas with different thicknesses of *stratum corneum* demonstrated similar permeability [126]. Thus, while the inherent biological variation of skin ensures that the overall process of skin permeability is multifactorial; trends in the wider literature are apparent and suggest that skin permeability may be ranked by decreasing rates of absorption as follows [21, 50, 89, 114, 127]:

> posterior aricular skin > scrotum > head and neck > abdomen > forearm
>
> > thigh > instep > heel > planter

The regional variability of skin permeability may influence absorption and, thus, site of application. Poorly penetrating molecules, such as scopolamine, have been applied postauricularly where permeability is comparatively high. Thus, site-to-site variation of skin permeability should always be considered in the wider context of the other factors discussed herein, including the variation in permeability within a particular body site and the same body site on different individuals [128, 129].

1.13.5   Race

The few studies that have examined how race may affect skin absorption have shown that there are no substantial differences between the permeability across African, Asian or European skin types [130]. It has been suggested that greater skin pigmentation has been shown to present a more resilient barrier and one that recovers after perturbation more rapidly than more lightly pigmented skin [130, 131]. Significant differences have been observed in *stratum corneum* water content between different races [132]. However, while the latter might expect to manifest itself through different drug absorption profiles the limited amount of research carried out in this field, coupled with the inherent variation in skin permeability described earlier, make it difficult to draw definite conclusions on this subject.

1.13.6   Skin Temperature

The effect of temperature on the physiological structure and activity of the skin is complex, affecting both blood flow and metabolism. It is generally accepted that an increase in temperature will increase the rate of absorption and that a decrease in temperature may lower the rate of absorption by up to one order of magnitude [108, 133–135]. Percutaneous penetration usually occurs within a narrow temperature range, although this may be raised from 32 to 37°C by occlusion [5]. However, it has been suggested that there is no change in the rate of absorption when temperature is increased up to 60°C, beyond which point it has been demonstrated that irreversible structural changes take place in the *stratum corneum*, where lipids may solubilise to an extent, resulting in a decrease in skin impedance and resistance [136]. Further, it has been indicated that changes in temperature will mostly assist molecules that normally penetrate the skin readily [16]. Increased vasodilation is a common phenomenon associated with absorption at higher temperatures.

Passive transport through the skin is temperature dependant, as it is initially a diffusion process. The diffusion constant of a penetrant may be related to the temperature of the environment by the Stokes–Einstein equation:

$$D = \frac{kT}{(6\pi r\eta)} \tag{1.4}$$

where D represents the diffusional constant, k represents the Boltzmann constant, T is the absolute temperature, r represents the hydrodynamic radius of the diffusing drug molecule and η represents viscosity.

1.14   Vehicle Effects

Ultimately, rates of percutaneous penetration rely on the effects that the skin, penetrant and vehicle collectively exert on the diffusion process. The vehicle allows optimisation and control of release at a rate adequate to provide a sufficient therapeutic dose of drug. The physical and chemical nature of the vehicle will influence the extent of drug migration to the skin, and may exacerbate this by exerting changes on the skin physiology in general, and barrier function in particular. The driving force for a drug to diffuse from the vehicle, into and through the skin surface is its thermodynamic activity in the vehicle. As discussed previously, the physicochemical properties of the drug will influence its rate of diffusion. The vehicle must therefore present the drug in a manner which will facilitate its rapid and controlled exit from the vehicle to the skin. The pH of a vehicle will affect the activity coefficient of weakly acidic and basic molecules. For example, it was found that the activity coefficients of weakly acidic compounds were reduced in alkali vehicles [133]. Further, vehicles may affect the skin by hydration and occlusion. Waxes and oil-based vehicles, commonly found, for example, in ointments, increase hydration through occlusion. Aqueous vehicles will occlude the skin less than non-aqueous systems, being generally less occlusive than non-aqueous vehicles, but their aqueous nature may increases hydration at the site of application. Choice of solvent will also affect the drug release from a vehicle; Bronaugh and Franz demonstrated that the release through human skin of caffeine, benzoic acid and testosterone formulated in three vehicles (petroleum, ethylene glycol gel and an aqueous gel) was significantly different [137]. Ethanol has been widely used as a solvent or co-solvent to increase the flux of molecules through the skin [138, 139].

Optimum transdermal delivery is very generally associated with a high concentration of drug in the vehicle, in order to provide a high concentration gradient across the skin. Nevertheless, Idson [140] determined that the ideal vehicle should contain the *lowest possible* concentration of drug, and that all of the drug should be released from the vehicle. Significant success has been observed in the formulation of supersaturated formulations [141–145]. However, no universal vehicle exists for transdermal drug delivery, and the drug carrier must be formulated to consider the physicochemical properties of a particular drug and to maximise its release into and across the skin.

1.15   Modulation and Enhancement of Topical and Transdermal Drug Delivery

Compared to other routes of administration – most notably the oral and parenteral pathways – the transdermal route is limited; this is reflected in the comparative number of drugs available for administration across each route. This is due to the highly efficient manner in which the *stratum corneum* provides a barrier between the body and its external environment. It may also reflect, particularly in transdermal drug delivery, the nature and origin of molecules which were principally designed with other routes of administration in mind.

Much research has therefore been undertaken in improving the transdermal delivery and subsequent bioavailability of numerous drugs for which the transdermal pathway offers many advantages, described above. This work is reviewed excellently elsewhere and is summarised further. The main strategies for enhancing transdermal delivery can be generally divided into two types: chemical and physical enhancement. They are described in the following text.

1.15.1   Chemical Modulation of Permeation

Chemicals incorporated into topical formulations with the express aim of enhancing drug release have been variously labelled penetration enhancers, accelerants or sorption promoters. Katz and Poulsen [83] proposed that the ideal penetration enhancer should be:

- pharmacologically inert,
- non-toxic, irritating or allergenic,
- able to provide immediate onset of penetration enhancement following application,
- able to allow the barrier function of the skin to recover immediately and fully upon removal,
- compatible with a wide range of drugs and excipients,
- able to solubilise drugs,
- clinically compliant and well tolerated by patients,
- inexpensive,
- organoleptically acceptable.

Clearly this is a somewhat idealised list as very few chemicals could fulfil all the aforementioned criteria. Nevertheless, certain molecules do possess several of them and have been incorporated into formulations in order to enhance drug delivery. Such materials usually alter the physiological nature of the *stratum corneum*, either reversibly or irreversibly. Penetration enhancers may promote drug diffusion by either interacting with the skin or promoting release of the drug from the vehicle, or by a combination of these methods [54].

1.15.1.1   Water

The use of water is one of the most widespread and safest methods to enhance skin permeability. This may be achieved by increasing the water content of the *stratum corneum*, either directly or indirectly – for example, by occlusion and reduction in transepidermal water loss. Water exists in a range of states in the skin and may be freely available, bound to skin structural elements or present in skin secretions, such as sebum and the natural moisturising factor (NMF) of the skin [5]. Therefore, the manner in which water increases skin permeability may vary both in mechanism and magnitude depending on the state in which water is present on or in the skin surface. The use of occlusive formulations or dressings, as described above, may reduce transepidermal water loss and increase hydration of the skin, thereby reducing diffusional resistance and increasing permeability.

1.15.1.2   Chemical Penetration Enhancers

Chemical penetration enhancers were classified by Hori and co-workers [146] into three distinct groups depending on their physicochemical characteristics (Figure 1.3). Classical penetration enhancers were placed in group one and were usually aprotic solvents such as dimethylsulfoxide and propylene glycol. Group two consisted of oleic acid and newer and more effective enhancers, such as Azone® (1-dodecylazacycloheptan-2-one). Group three enhancers were mostly organic molecules and were mainly designed for enhancement of specific drug molecules [147].

It has been demonstrated that incorporation of propylene glycol or ethanol, simply as solvents or co-solvents, may enhance the permeability of drugs into and across the skin. Such an increase in flux occurs when presenting saturated solutions of the drug, maximising thermodynamic activity [93, 148]. However, their method of action is unlike that of aprotic solvents, and they do not induce the same changes on the physiology of the skin.

The aprotic solvents employed as penetration enhancers (Figure 1.3) include dimethylsulfoxide (DMSO), dimethylacetamide (DMAC) and dimethylformamide (DMF). They were found to enhance permeation more effectively than other vehicles such as propylene glycol, polyethylene glycol and ethanol [66]. DMSO has been one of the most widely used chemical enhancers in transdermal drug delivery, providing substantially greater enhancement of permeation than either DMAC or DMF. It was first used to increase the rate of transdermal penetration in the 1960s for a wide range of compounds with substantially different physicochemical properties, including water, antibiotics and local anaesthetics [149–152]. Proposed mechanisms of its action include its ability to associate and solubilise skin lipids and proteins, altering the conformation and barrier function of the skin [137, 153]. The high osmotic potential of aprotic solvents allows association and, ultimately, replacement of water in the *stratum corneum*. This distorts the lamellar structure of the sin barrier, resulting in swelling which opens channels in the skin, increasing permeability [154, 155]. Aprotic solvents are, however, irritant to the skin in high concentrations. Topical application of DMSO has been shown to cause halitosis and to leave a bad taste in the mouth of the patient due to the metabolic formation of dimethylsulphide [1].

2-Pyrrolidone and a number of its derivatives have been shown to promote penetration and assist in the establishment of a drug reservoir in skin as they partition into the lipids of the *stratum corneum*. Stoughton demonstrated increased retention of griseofulvin in the *stratum corneum* when presented in a vehicle containing 2-pyrrolidone [156]. 2-Pyrrolidone has also been used to promote the penetration of theophylline, aspirin, ibuprofen, flurbiprofen and caffeine [157–159]. Pyrrolidones, however, produce skin irritation at the high concentrations needed to promote penetration.

Oleic acid (Figure 1.3) is one of a number of long-chain fatty acids that has been used in a range of pharmaceutical applications, including as a skin penetration enhancer. It is classified by Hori *et al.* as a group 2 enhancer, facilitating the permeation of a number of drugs, particularly hydrophilic potential penetrants [5, 146]. It acts in a manner similar to DMSO by mechanisms of lipid fluidisation and, more importantly, phase separation and the formation of novel lipid domains within the skin barrier [160, 161].

Urea, and in some cases its synthetic analogues, acts as both a hydrotrope and a keratolyte [162]. This ability to both hydrate the skin and structurally modify *stratum corneum* lipids has seen urea used as a skin penetration enhancer. For example, formulations containing at least 10% urea have been shown to promote drug penetration by lowering the phase transition temperature of the *stratum corneum*, resulting in fluidisation of the skin lipids at ambient temperatures. Urea has also been shown to increase the hydration of the *stratum corneum* and increase the onset of erythema [163, 164].

Surfactants are found widely in topical many formulations, including pharmaceutical and cosmetic preparations. Anionic surfactants appear to have the greatest activity in terms of increasing skin permeability, followed by cationic and then nonionic surfactants; such comparative activities may also be related to the conformation of surfactants. The anionic laurate moiety has been found to exhibit the greatest promotion effect [165]. Anionic surfactants have high protein binding affinity, leading to gross swelling of the *stratum corneum*. This results in the uncurling of the protein filaments as they bind to the surfactant [166]. While surfactants are widely considered to cause significant localised irritation (erythema and oedema) they exhibit low chronic toxicity. Non-ionic surfactants have little influence on transdermal penetration and are generally well tolerated by the skin. Skin permeation is not promoted with surfactants containing several long chains instead of a single chain or several short chains, suggesting the enhancement ability of a particular surfactant depends on the ability of the molecules to penetrate the lipid membranes of the *stratum corneum* [167]. The tendency of ionic surfactants to cause irritation to the skin has recently seen their use questioned in pharmaceutical applications, particularly in patients with atopic eczema [168–170].

Azone, shown in Figure 1.4, is a novel, non-polar penetration enhancer widely shown to effectively promote the absorption of certain drugs. Azone has exhibited low toxicity and low irritancy to the skin, and has been shown to be effective in low concentrations [171–175]. However, Baker and Hadgraft showed that Azone had no enhancement effect on the delivery of the antiviral drug arildone [176]. Azone's mechanism of action is thought to be due to fluidisation of the intercellular lipid bilayers of the *stratum corneum*. This reduces diffusional resistance of the skin barrier to the permeation of exogenous chemicals with a wide range of physicochemical characteristics [164, 172].

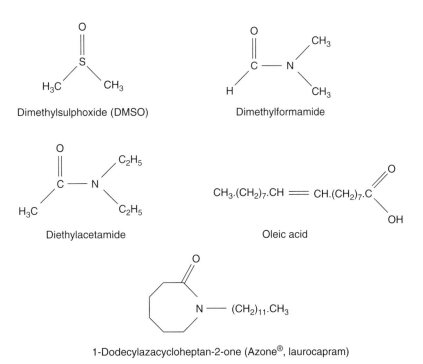

Figure 1.4 *Chemical structures of some penetration enhancers employed for transdermal drug delivery.*

More recently, other chemical classes have been explored for their potential to act as skin penetration enhancers. Terpenes are derived from a number of essential oils. They are based on the isoprene unit (C_5H_8) but are structurally diverse, containing a range of aromatic and aliphatic structures. Williams and Barry examined 17 terpenes for their permeability enhancing effects and found that oxide terpenes and terpenoids had the greatest enhancement effects [177, 178]. Other researchers (i.e. Monti *et al.* [175]) have explored the use of terpenes as enhancers but, so far, they have found few applications in medicinal products [175]. Phospholipids have also found use as enhancers, being formulated as vesicles (liposomes) in order to carry drugs into and across skin. In non-liposomal preparations phospholipids have shown little ability to enhance permeation, although, again, their application in medicinal products has been limited.

A range of other chemicals have found use as enhancers. For example, various phospholipids – often as liposomal preparations [178], HPE-101 (1-[2-(decylthio)ethyl] azacyclopentane-2-one) [179], SEPA™ (2-*N*-nonyl-1,3-dioxolone) [174, 180, 181], *N*-pentyl-*N*-acetylprolinate [182] – have all been investigated for their potential as enhancers. Such compounds have not, as yet, found widespread use with a variety of drugs and vehicles.

1.15.1.3   Prodrugs

Enhancement of permeation has also been achieved by the use of formulation strategies which aim to maximise drug delivery capability of a dosage form. This, in the case of prodrug strategies, may also include the chemical modification of the drug into a form which is more physicochemically amenable to percutaneous absorption. This allows its partitioning, diffusivity and solubility profile to align more fully with the needs of an effective transdermal permeant.

The general strategy for prodrug design is to increase the lipophilicity of the drug by the addition of lipophilic moieties. This is seen, for example, in the use of the lipophilic valerate form of betamethasone-17-valerate which results in improved skin permeation. This increases the permeability of the drug into the *stratum corneum* whereafter it may permeate further into the skin. Further permeation may be accompanied by enzymatic degradation – returning the prodrug back to its parent therapeutic agent – either in the viable tissues of the skin or in the systemic circulation. Despite considerable promise and substantial research [183–185], including the use of co-enhancement strategies (where, e.g., prodrugs are formulated with permeation-enhancing solvents), the clinical use of transdermal prodrugs is limited as such entities are considered to be new chemical entities, thus limiting development opportunities.

1.15.1.4   Ion Pairing

Ionised species seldom penetrate into the relatively lipid *stratum corneum*. The majority of successful transdermal candidates are therefore formulated and delivered as the free acid or base form. However, the use of ion pairing forms a complex in which the charges of each moiety are neutralised. This has been achieved successfully by a number of researchers. For example, the use of simple salts of has been shown to enhance permeation of different drugs [186, 187]. Further, Takahashi and Rytting employed a counter ion with known permeation-enhancing properties, oleic acid [188]. Stott *et al.* used coacervates to enhance absorption, resulting in small increases in permeation [189].

1.15.1.5   Eutectic Mixtures

Modification of physical properties by mixing materials in such a way as to modify their melting point has been a comparatively successful formulation strategy. This has been achieved by the formulation of eutectic mixtures, where the mixing of, in the case of binary systems, two chemicals may inhibit each other's crystal growth, resulting in a reduction of melting point. This has seen significant practical application in EMLA® cream, where a eutectic mixture in a 1:1 ratio of lidocaine and prilocaine reduce their melting points, facilitating an increase in skin permeation. After approximately 60 min this results in clinically relevant, transient anaesthesia. Such strategies have also been more broadly employed in other local anaesthetic systems, such as AmetopTM Gel, where water decreases the melting point of tetracaine to 29°C, enhancing permeation and providing a rapid onset of long-lasting (4–8 h) clinically relevant anaesthesia [1, 74, 100, 135, 190–194].

1.15.1.6   Supersaturation

While the optimum transdermal drug delivery is commonly understood to occur from a saturated solution of the permeant, increasing the concentration of the drug above its saturated solution in a particular solvent further increases the thermodynamic activity of the drug in its formulation. However, the formation of supersaturated states is difficult, as most exhibit poor physical stability and usually result in precipitation and crystallisation of the drug from its vehicle. Addition of a range of excipients – such as those used to stabilise suspensions – may improve stability. These include antinucleating agents and materials to reduce or eliminate aggregation of particles. Despite issues of instability supersaturated formulations have shown great promise in significantly enhancing the skin permeability of a range of drugs [141, 195–197].

1.15.1.7   Vesicles

Encapsulation of active ingredients, in both pharmaceutical and cosmetic applications, has been a popular strategy to enhance deposition and delivery onto and into the skin. Various materials, including drugs, humectants and enzymes have all been delivered in this manner. A wide range of vesicles have been used for topical and transdermal drug delivery systems. These include liposomes [198, 199], non-ionic surfactants, or noisomes, elastic or deformable liposomes [200] and ethosomes. This subject is reviewed in detail elsewhere [5].

   Despite a wide range of vesicle types and some clinical success – as observed in the literature – the widespread use of such technology to enhancing skin drug delivery has not yet been achieved.

   More recently, nanotechnological approaches have been employed to improve percutaneous absorption. For example, the penetration of zinc oxide nanoparticles has been investigated. Multi-photon imaging using time-correlated single photon counting showed that nanoparticles aggregated in the furrows of the skin and that they penetrated laterally from the furrows into the *stratum corneum*, but remained outside of the viable epidermis in non-lesional and lesional tissue [201–203]. Nanoparticulate accumulation in hair follicle following massage has also been observed [204–206].

   Luengo *et al.* used atomic force microscopy to describe topical delivery with 328 nm poly-lactic-*co*-glycolic acid (PLGA) drug-loaded particles [207]. Drug release experiments showed that 100% release was observed after 6 h using human skin *in vitro*; after 24 h the particulate group showed significantly greater delivery in the deep skin layers than the non-particulate control – a 1.8-fold increase.

Other recent studies have shown, for example, the use of capsaicin-loaded nanoparticles in the treatment of pain associated with diabetic neuropathy. Advancements in particle engineering, formulation science and an improved understanding of nanoparticle–skin interactions will undoubtedly lead to important clinically relevant improvements in topical drug delivery [208–210]. In addition, biomimetic nanoparticle engineering has applied principles associated with permeation of nanoscale viruses across compromised skin, such as the human papilloma virus, to designing drug delivery systems applies these 'natural' principles in improving nanoparticle design [211–215].

1.15.2   Physical Methods of Enhancement

As well as formulation-focused chemical methods of enhancement physical methods have also been employed to enhance skin permeation. These methods usually employ a novel device to facilitate delivery and include the use of an electrical current, ultrasound, high-velocity particulate delivery or arrays of microneedles. In general, they may be defined as 'needleless' methods of drug delivery.

1.15.2.1   Ablation

One of the simplest methods of overcoming the challenge to successful drug delivery presented by the *stratum corneum* is to simply remove it. This has been achieved by a number of methods, most notably by Wolf's method of skin tripping with adhesive cellophane [216], so-called 'chemical peels' which may remove the superficial layers of the epidermis, dermabrasion – where the skin is physically abraded to remove the *stratum corneum* – and laser ablation. This last method employs an excimer laser to remove the stratum corneum, resulting in a similar increase in permeability to that observed following tape stripping [217]. However, while subsequent methods have shown similar outcomes, given the overall nature of the methods described they have found little significant use in clinically relevant scenarios.

1.15.2.2   Iontophoresis

The migration of ions or charged drug molecules under an electrical potential gradient allows the transdermal delivery of the salt forms of a drug. Drug delivery assisted in this manner is termed iontophoresis [218]. Iontophoresis is achieved by passing an electric current of the appropriate amplitude through an electrolyte solution containing the charged drug species. Positively charged species, usually salts of weak organic bases, are driven into the skin at the positive electrode, or anode, whereas negatively charged drugs, such as the salts of organic acids, are delivered from the negatively charged cathode. Figure 1.5 shows a typical anodic iontophoresis setup.

Permeation of an ionised species under an applied electric field is due to the electro-chemical potential gradient applied across the skin, increased skin permeability under the applied electric field and a current-induced water transport effect, related to electro-osmotic, convective or iontohydrokinesis effects [219, 220].

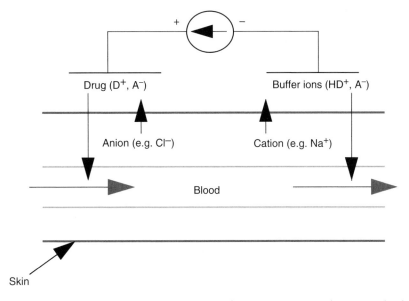

Figure 1.5 *A typical anodic iontophoretic setup, where D⁺ represents the positively charged drug, and A⁻ its counter ion. H⁺ and A⁻ are charged species, usually Na⁺ and Cl⁻, associated with the extracellular fluid beneath the skin.*

Flux under an applied current may be represented by

$$J_{app} = J_p + J_e + J_c \tag{1.5}$$

where J_p is the flux due to passive delivery, and is given by Equation 1.5. J_e, the flux due to electric current facilitation, is given by

$$J_e = \frac{Z_i D_i F}{RT} \cdot C_i \cdot \frac{dE}{h_s} \tag{1.6}$$

where

Z_i is the electric valence of the ionic species i.
D_i is the diffusivity of the ionic species i in the skin.
F is the Faraday constant.
R is the gas constant.
T is the absolute temperature.
C_i is the donor concentration of the ionic species i.
h_s is the concentration gradient across the skin.
dE/h_s is the electric potential gradient across the skin.

J_c, the flux due to convective transport, may be represented by

$$J_c = kC_s I_d \tag{1.7}$$

where

K is the proportionality constant.
C_s is the concentration in the skin tissue.
I_d is the current density.

The *stratum corneum* has a relatively high electrical resistance, but the underlying layers of the skin are relatively conductive. The applied voltage across the skin thus provides the driving force under which a charged drug may penetrate the skin. The main route of penetration is not through the poorly conductive *stratum corneum* lipids, but through the sweat glands, and, to a lesser extent, through the hair follicles and sebaceous glands covering the surface of the skin. Hydration of the *stratum corneum* may allow diffusion via the intercellular route [221–223]. The isoelectric point of skin is between pH 3 and 4. Thus, above pH 4 the skin has a net negative charge due to the carboxylic acid functionalities of skin proteins. This is also the case at physiological pH. Positively charged drugs, therefore, penetrate the skin across the shunt route more rapidly than negatively charged ions, which have to overcome electrostatic repulsion from the charged groups on the skin surface [222].

A number of factors influence successful iontophoretic transport. They include ionisation, pH related to the pK of the penetrant and associated competitive ion effects, electrolysis associated with the passage of an electric current through the body, the electro-osmotic effect, where polarisation and 'solvent drag' can facilitate the delivery of uncharged species into and across the skin [224, 225], the nature of the applied current (being either a direct or alternating current), and permselectivity, where the valence of a penetrant will influence its permeation. Finally, unlike passive diffusion-facilitated percutaneous penetration molecular size appears to not limit iontophoretic delivery.

The use of iontophoresis has fluctuated over the years, its failure to establish itself as a major method for percutaneous delivery of drugs being mainly due to the potential hazards from burns and mild shocks associated with the applied current, which could be reduced by the use of an alternating current. However, resurgence in the 1990s saw iontophoresis applied to the delivery of anti-inflammatory, local anaesthetic and ophthalmic drugs and for the diagnosis of cystic fibrosis [220] and blood sugar levels – the reverse iontophoresis of the GlucoWatchTM device, associated with the electro-osmotic effect. To date, however, relatively few products of significant clinical relevance have found their way onto the marketplace, despite the publication of a substantial body of research in this field.

1.15.2.3   *Phonophoresis (Sonophoresis)*

Phonophoresis, or sonophoresis, entails the use of ultrasound to promote enhanced percutaneous penetration by the direct transfer of ultrasonic energy through the drug-carrying vehicle. Phonophoretic delivery may be accompanied by rubbing the skin site (innuction). Such an action by itself may promote drug diffusion in the absence of phonophoresis, and suitable controls are thus needed to ensure enhancement of penetration is due solely to the applied ultrasound. McElnay *et al.* investigated the use of phonophoresis for enhanced absorption of local anaesthetics, and found that phonophoresis did not significantly increase the onset of anaesthesia [226]. Recent studies, however, have indicated that sonophoretic treatment significantly increased drug permeation, compared to control [227, 228].

1.15.2.4   Particulate Delivery

The PowderJectTM system – and related technologies – employed a gas 'gun' to fire solid particle into the skin, thus avoiding the need for painful injections. The advantages of such delivery systems include the ability to deliver fine particles in a pain-free manner (thus having concomitant improvements in compliance and a reduction in patient discomfort or distress due to 'needle phobia'), to avoid needle-stick injuries and the ability to provide solid particles to specific regions of the skin, improving formulation stability and drug targeting.

However, the clinical development of such technologies has been blighted by a number of issues, including consistency of dosing, which may be less accurate than from a needle and which might be tissue-dependant (i.e. dosing into muscle tissue), and which may also be influenced by physiological issues, such as regional changes in *stratum corneum* thickness, technique-dependant issues including 'bounce-off' if the device is incorrectly positioned during use, and environmental factors, including skin hydration.

Despite substantial investment and promising results in early stage clinical studies as recently as 2007 for a range of drugs, no products of this type have as yet been released onto the market.

1.15.2.5   Microneedles

Probably the most significant advancement in dermal and transdermal drug delivery in recent years has been the development of microneedle technology. Microneedles were developed from silicone microfabrication technologies [229] and offer the potential of pain-free delivery of a wide range of drugs into and across the skin. The aim of microneedle devices is to pierce the superficial tissues of the skin, including the *stratum corneum*, in order to facilitate the ingress of therapeutic agents from the delivery device. The depth of delivery is important as, in piercing only the superficial layers of the skin the needles should avoid contact with the pain receptors, which are located at or just below the dermo-epidermal junction [1, 230]. A subsequent study indicated that microneedle delivery was not painless and required vibratory actuation to minimise pain on insertion [231]. Davis *et al.* reported that needle geometry was essential in minimising insertion pain and optimising efficacy while Sivamani *et al.* reported that volunteers felt pressure, not pain, upon microneedle insertion [232, 233].

Needles vary considerably in their design. Initial microneedle systems contain solid needles which, once they had pierced the skin, were removed and a topical formulation applied over the same site. More recent designs have included porous needles though which the drug can diffuse once the skin is pierced or needles which are constructed from composite materials, including drugs, and which dissolve *in situ*, releasing the drug in a controlled manner.

Silicone microneedles were employed by Coulman *et al.* for non-viral gene delivery [234]. Scanning electron microscopy was used to visualise the microconduits created in human epidermal tissue following the administration of microfabricated silicon microneedles. Diffusion of fluorescent polystyrene nanospheres and lipid:polycation:pDNA (LPD) nonviral gene therapy vectors was determined *in vitro* via Franz-type diffusion cells employing human epidermal sheets. It was observed that the diffusion of 100 nm diameter fluorescent polystyrene nanospheres and LPD complexes was significantly enhanced following membrane treatment with microneedles. Cell culture studies confirmed that LPD complexes mediated efficient reporter gene expression in human keratinocytes. Most recently, Martin *et al.* reported the fabrication of biodegradable sugar glass microneedles for the transdermal drug delivery. Solid amorphous biodegradable sugar glasses containing low residual quantities of water were created by dehydration of trehalose and sucrose sugar combination solutions. These microneedles demonstrated that they were able to facilitate transdermal delivery of a wide range of molecules [235].

Donnelly *et al.* manufactured microneedle devices using hydrogel-forming microarrays [236]. Further, they demonstrated the ability of microneedle pre-treatment to facilitate the delivery of a polylactic-*co*-glycolic acid nanoencapsulated dye across porcine skin, observing a significant improvement in microneedle pre-treated skin [237]. They have also demonstrated microneedle-mediated transdermal bacteriophage delivery [238] and, more recently, the development of hydrogel-forming and dissolving microneedles [239–241].

References

[1] Woolfson, A.D. and McCafferty, D.F. (1993) *Percutaneous Local Anaesthesia*, Ellis Horwood, London.
[2] Wilkes, G.L., Brown, I.A. and Wildnauer, R.H. (1973) The biomechanicalproperties of skin. *CRC Crit. Rev. Bioeng.*, **1**, 453–495.
[3] Cross, S.E. and Roberts, M.S. (1993) Subcutaneous absorption kinetics of interferon and other solutes. *J. Pharm. Pharmacol.*, **45**, 606–609.
[4] Spearman, R.I. (1973) *The Integument: A Textbook of Skin Biology*, Cambridge University Press, London.
[5] Williams, A.C. (2003) *Transdermal and Topical Drug Delivery*, The Pharmaceutical Press, London.
[6] Bronaugh, R.L. and Maibach, H.I. (1999) *Percutaneous Absorption*, 3rd edn, Marcel Dekker, Inc./CRC Press, New York.
[7] Weddell, G. (1941) The pattern of cutaneous innervation in relation to cutaneous sensibility. *J. Anat.*, **75**, 346–367.
[8] Barlow, H.B. and Mallon, J.D. (eds) (1982) *The Senses*, Cambridge University Press, London.
[9] Brodal, A. (1981) *Neurological Anatomy in Relation to Clinical Medicine*, Oxford University Press, London.
[10] Sinclair, D. (1981) *The Mechanisms of Cutaneous Sensations*, 2nd edn, Oxford University Press, London.
[11] Montagna, W. and Yun, J.S. (1964) The skin of the domestic pig. *J. Invest. Dermatol.*, **42**, 11–21.
[12] Elliott, H.C. (1969) *Textbook of Neuroanatomy*, Lippincott, Philadelphia.
[13] Melzack, R. and Wall, P.D. (1965) Pain mechanisms: a new theory. *Science*, **150**, 971–975.
[14] Perl, E.R. (1968) Myelinated afferent fibres innervating the primate skin and their response to noxious stimuli. *J. Physiol.*, **197**, 593–615.
[15] Schmidt, R.F. (1981) Somarovisceral sensibility, in *Fundamentals of Sensory Physiology*, 2nd edn (ed R.F. Schmidt), Springer and Verlag, Berlin.
[16] Scheuplein, R.J. and Blank, I.H. (1971) Permeability of the skin. *Physiol. Rev.*, **51**, 702–747.
[17] Anderson, R.L. and Cassidy, J.M. (1973) Variations in physical dimensions and chemical composition of human *stratum corneum*. *J. Invest. Dermatol.*, **61**, 30–32.
[18] MacKensie, I.C. and Linder, J.C. (1973) An examination of cellular organization within the *stratum corneum* by a silver staining method. *J. Invest. Dermatol.*, **61**, 254–260.
[19] Matoltsy, A.G. (1976) Keratinisation. *J. Invest. Dermatol.*, **67**, 20–25.
[20] Michaelis, A.S., Chandrasekaran, S.K. and Shaw, J.E. (1975) Drug permeation through human skin: theory and *in vitro* experimental measurement. *AIChE*, **21**, 985–996.
[21] Elias, P.M., Cooper, E.R., Korc, A. and Brown, B.E. (1981) Percutaneous transport in relation to *stratum corneum* structure and lipid composition. *J. Invest. Dermatol.*, **76**, 297–301.

[22] Harding, C.R. (2004) The stratum corneum: structure and function in health and disease. *Dermatol. Ther.*, **17S**, 6–15.

[23] Rawlings, A.V. (2010) Recent advances in skin 'barrier' research. *J. Pharm. Pharmacol.*, **62**, 671–677.

[24] Moss, G.P., Wilkinson, S.C. and Sun, Y. (2012) Mathematical modelling of percutaneous absorption. *Curr. Opin. Colloid Interface Sci.*, **17**, 166–172.

[25] Rawlings, A.V. (2003) Trends in stratum corneum research and the management of dry skin conditions. *Int. J. Cosmet. Sci.*, **25**, 63–95.

[26] Norlen, L. (2006) Stratum corneum keratin structure, function and formation – a comprehensive review. *Int. J. Cosmet. Sci.*, **28**, 397–425.

[27] Norlen, L. (2007) Nanostructure of the stratum corneum extracellular lipid matrix as observed by cryo-electron microscopy of vitreous skin sections. *Int. J. Cosmet. Sci.*, **29**, 335–52.

[28] Norlen, L., Plasencia, I. and Bagatolli, L. (2008) Stratum corneum lipid organization as observed by atomic force, confocal and two-photon excitation fluorescence microscopy. *Int. J. Cosmet. Sci.*, **30**, 391–411.

[29] Long, S., Banks, J., Watkinson, A. *et al.* (1996) Desmocollin 1: a key marker for desmosome processing in the stratum corneum. *J. Invest. Dermatol.*, **106**, 397.

[30] Michel, S., Schmidt, R., Shroot, B. and Reichert, U. (1988) Morphological and biochemical characterization of the cornified envelopes from human epidermal keratinocytes of different origin. *J. Invest. Dermatol.*, **91**, 11–15.

[31] Harding, C.R., Long, S., Richardson, J. *et al.* (2003) The cornified cell envelope: an important marker of stratum corneum maturation in healthy and dry skin. *Int. J. Cosmet. Sci.*, **25**, 157–167.

[32] Hirao, T., Terui, T., Takeuchi, I. *et al.* (2003) Ratio of immature cornified envelopes does not correlate with parakeratosis in inflammatory skin disorders. *Exp. Dermatol.*, **12**, 591–601.

[33] Bouwstra, J.A., Gooris, G.S., van der Spek, J.A. *et al.* (1994) The lipid and protein structure of mouse stratum corneum: a wide and small angle diffraction study. *Biochim. Biophys. Acta*, **1212**, 183–192.

[34] Norlen, L. (2003) Skin barrier structure, function and formation – learning from cryo-electron microscopy of vitreous, fully hydrated native human epidermis. *Int. J. Cosmet. Sci.*, **25**, 209–226.

[35] Masukawa, Y., Narita, H., Shimizu, E. *et al.* (2008) Characterization of overall ceramide species in human stratum corneum. *J. Lipid Res.*, **49**, 1466–1476.

[36] Masukawa, Y., Narita, H., Sato, H. *et al.* (2009) Comprehensive quantification of ceramide species in human stratum corneum. *J. Lipid Res.*, **50**, 1708–1719.

[37] Uchida, Y., Hama, H., Alderson, N.L. *et al.* (2007) Fatty acid 2-hydroxylase, encoded by FA2H, accounts for differentiation-associated increase in 2-OH ceramides during keratinocyte differentiation. *J. Biol. Chem.*, **282**, 13211–13219.

[38] Uchida, Y. and Holleran, W.M. (2008) Omega-O-acylceramide, a lipid essential for mammalian survival. *J. Dermatol. Sci.*, **51**, 77–87.

[39] Madison, K.C., Swartzendruber, D.C., Wertz, P.W. and Downing, D.T. (1987) Presence of intact intercellular lipid lamellae in the upper layers of the stratum corneum. *J. Invest. Dermatol.*, **88**, 714–718.

[40] Harding, C.R., Watkinson, A., Rawlings, A.V. and Scott, I.R. (2000) Dry skin, moisturization and corneodesmolysis. *Int. J. Cosmet. Sci.*, **22**, 21–52.

[41] Rawlings, A.V., Watkinson, A., Rogers, J. *et al.* (1994) Abnormalities in stratum corneum structure, lipid composition and desmosome degradation in soap-induced winter xerosis. *J. Soc. Cosmet. Chem.*, **45**, 203–220.

[42] Berry, N., Charmeil, C., Goujon, C. *et al.* (1999) A clinical, biometrological and ultrastructural study of xerotic skin. *Int. J. Cosmet. Sci.*, **21**, 241–252.

[43] Bouwstra, J.A., Gooris, G.S., Dubbelaar, F.E. *et al.* (1998) pH, cholesterol sulfate, and fatty acids affect the stratum corneum lipid organization. *J. Investig. Dermatol. Symp. Proc.*, **3**, 69–74.

[44] Bouwstra, J.A., Gooris, G.S. and Ponec, M. (2008) Skin lipid organization, composition and barrier function. *Int. J. Cosmet. Sci.*, **30**, 388.

[45] Pilgram, G.S., Engelsma-van Pelt, A.M., Bouwstra, J.A. and Koerten, H.K. (1999) Electron diffraction provides new information on human stratum corneum lipid organization studied in relation to depth and temperature. *J. Invest. Dermatol.*, **113**, 403–409.

[46] Pilgram, G.S., van der Meulen, J., Gooris, G.S. *et al.* (2001) The influence of two azones and sebaceous lipids on the lateral organization of lipids isolated from human stratum corneum. *Biochim. Biophys. Acta*, **511**, 244–254.

[47] Elias, P.M. (1988) Structure and function of the *stratum corneum* permeability barrier. *Drug Dev. Res.*, **13**, 97–105.

[48] Treherne, J.E. (1956) Premeability of skin to some non-electrolytes. *J. Physiol.*, **133**, 171–180.

[49] Blank, I.H. (1964) Penetration of low molecular weight alcohols into the skin. I. Effect of concentration of alcohol and type of vehicle. *J. Invest. Dermatol.*, **43**, 415–420.

[50] Scheuplein, R.J. (1965) Mechanism of percutaneous absorption. I. Routes of penetration and the influence of solubility. *J. Invest. Dermatol.*, **45**, 334–346.

[51] Scheuplein, R.J. (1967) Mechanism of percutaneous absorption. II. Transient diffusion and the relative importance of various routes of skin penetration. *J. Invest. Dermatol.*, **48**, 79–88.

[52] Scheuplein, R.J., Blank, I.H., Brauner, G.I. and MacFarlane, D.J. (1969) Percutaneous absorption of steroids. *J. Invest. Dermatol.*, **52**, 63–70.

[53] Elias, P.M. (1983) Epidermal layers, barrier function and desquamation. *J. Invest. Dermatol.*, **80S**, 44–49.

[54] Barry, B.W. (1983) *Dermatological Formulations: Percutaneous Absorption*, Marcel Dekker, New York.

[55] Heuber, F., Besnard, M., Schaefer, H. and Wepierre, J. (1994) Percutaneous absorption of estradiol and progesterone in normal and appendage-free skin of the hairless rat – lack of importance of nutritional blood flow. *Skin Pharmacol.*, **7**, 245–256.

[56] Barr, M. (1962) Percutaneous absorption. *J. Pharm. Sci.*, **51**, 395–409.

[57] Heuber, F., Wepierre, J. and Schaefer, H. (1992) Role of transepidermal and transfollicular routes in percutaneous absorption of hydrocortisone and testosterone – *In vivo* study in the hairless rat. *Skin Pharmacol.*, **5**, 99–107.

[58] Lauer, A.C., Ramachandran, C., Lieb, I.M. *et al.* (1996) Targeted drug delivery to the pilosebaceous unit via liposomes. *Adv. Drug Del. Rev.*, **18**, 311–324.

[59] Martin, A., Swarbrick, J. and Cammarata, A. (1983) *Physical Pharmacy*, 3rd edn, Lea & Febinger, Philadelphia.

[60] Moss, G.P., Dearden, J.C., Patel, H. and Cronin, M.T. (2002) Quantitative structure-permeability relationships (QSPRs) for percutaneous absorption. *Toxicol. In Vitro*, **16**, 299–317.

[61] Poulsen, B.J. (1973) Design of topical drug products: biopharmaceutics, in *Drug Design*, vol. **4** (ed E.J. Ariens), Academic Press, New York, pp. 149–192.

[62] Ostrenga, J., Steinmetz, C. and Poulsen, B. (1971) Significance of vehicle composition I: relationship between topical vehicle composition, skin penetrability and clinical efficacy. *J. Pharm. Sci.*, **60**, 1175–1179.

[63] Ostrenga, J., Steinmetz, C., Poulsen, B. and Yett, S. (1971) Significance of vehicle composition II: prediction of optimal vehicle composition. *J. Pharm. Sci.*, **60**, 1180–1183.

[64] Barry, B.W. (1987) Mode of action of penetration enhancers in human skin. *J. Control. Rel.*, **6**, 85–97.

[65] Sinko, P.J. (2005) *Martin's Physical Pharmacy and Pharmaceutical Sciences*, 5th revised edn, Lippincott, Williams and Wilkins, Baltimore.

[66] Singh, S. and Singh, J. (1993) Transdermal drug delivery by passive diffusion and iontophoresis: a review. *Med. Res. Rev.*, **13**, 569–621.

[67] Bronaugh, R.L. and Congdon, E.R. (1984) Percutaneous absorption of hair dyes: correlation with partition coefficients. *J. Invest. Dermatol.*, **83**, 124–127.

[68] Le, V.H. and Lippold, B.C. (1995) Influence of physicochemical properties of homologous esters of nicotinic acid on skin permeability and maximum flux. *Int. J. Pharm.*, **124**, 285–292.

[69] Higo, N., Sato, S., Irie, T. and Uekama, K. (1995) Percutaneous penetration and metabolism of salicyclic acid derivatives across hairless mouse skin in diffusion cell *in vitro*. *STP Pharma Sci.*, **5**, 302–308.

[70] Mitragotri, S., Anissimov, Y.G., Bunge, A.L. *et al.* (2011) Mathematical models of skin permeability: an overview. *Int. J. Pharm.*, **418**, 115–129.

[71] Idson, B. (1975) Percutaneous absorption. *J. Pharm. Sci.*, **64**, 901–924.

[72] Crank, J. (1975) *The Mathematics of Diffusion*, 2nd edn, Clarendon Press, Oxford.

[73] Maibach, H.I. and Feldman, R.J. (1969) Effect of applied concentration on percutaneous absorption in man. *J. Invest. Dermatol.*, **52**, 382.

[74] Woolfson, A.D., McCafferty, D.F. and McGowan, K.E. (1992) Percutaneous penetration characteristics of amethocaine through porcine and human skin. *Int. J. Pharm.*, **78**, 209–216.

[75] Wester, R.C. and Noonan, P.K. (1980) Relevance of animal models for percutaneous absorption. *Int. J. Pharm.*, **7**, 99–110.

[76] Sved, S., McClean, W.M. and McGilvernay, I.J. (1981) Influence of the method of application on the pharmacokinetics of nitroglycerin from ointments in humans. *J. Pharm. Sci.*, **70**, 1368–1369.

[77] Wester, R.C., Noonan, P.K., Cole, M.P. and Maibach, H.I. (1977) Frequency of application on the percutaneous absorption of hydrocortisone. *Arch. Dermatol. Res.*, **113**, 620–622.

[78] Wilson, J.S. and Holland, L.M. (1982) The effect of application frequency on epidermal carcinogenesis assays. *Toxicology*, **24**, 45–54.

[79] Howes, D. and Black, J.G. (1976) Percutaneous absorption of triclocarban in rat and man. *Toxicology*, **6**, 67–76.

[80] Nakaue, H.S. and Buhler, D.R. (1976) Percutaneous absorption of hexachlorophene in the rat. *Toxicol. Appl. Pharmacol.*, **35**, 381–391.

[81] Katz, M. and Shaikh, Z.I. (1965) Percutaneous corticosteroid absorption correlated to partition coefficient. *J. Pharm. Sci.*, **54**, 591–594.

[82] Idson, B. (1971) Biophysical factors in skin penetration. *J. Soc. Cosmet. Chem.*, **22**, 615–634.

[83] Katz, M. and Poulsen, B.J. (1972) Corticoid, vehicle and skin interactions in percutaneous absorption. *J. Soc. Cosmet. Chem.*, **23**, 565–590.

[84] Hawkins, G.S. and Reifenrath, W.G. (1986) Influence of skin source, penetration cell fluid, and partition coefficient on *in vitro* skin penetration. *J. Pharm. Sci.*, **75**, 378–381.

[85] Tojo, K. (1987) Random brick model for drug transport across *stratum corneum*. *J. Pharm. Sci.*, **76**, 889–891.

[86] Shore, P.A., Brodie, B.B. and Hogben, C.A.M. (1957) The gastric secretion of drugs: a pH partition hypothesis. *J. Pharmacol. Exp. Ther.*, **119**, 361–369.

[87] Swarbrick, J., Lee, G., Brom, J. and Gensmantel, N.P. (1984) Drug permeation through the human skin II: permeability of ionized compounds. *J. Pharm. Sci.*, **73**, 1352–1355.

[88] Parry, G.E., Bunge, A.L., Silcox, G.D. *et al.* (1990) Percutaneous absorption of benzoic acid across human skin. 1. *In vitro* experiments and mathematical modelling. *Pharm. Res.*, **7**, 230–236.

[89] Roy, S.D. and Flynn, G.L. (1990) Transdermal delivery of narcotic analgesics – pH, anatomical and subject influences on cutaneous permeability of fentanyl and sufentanil. *Pharm. Res.*, **7**, 842–847.

[90] Barker, N. and Hadgraft, J. (1981) Facilitated percutaneous absorption: a model system. *Int. J. Pharm.*, **8**, 193–202.

[91] Siddiqui, O., Roberts, M.S. and Polack, A.E. (1985) Topical absorption of methotrexate: role of dermal transport. *Int. J. Pharm.*, **27**, 193–203.

[92] Santi, P., Catellani, P.L., Colombo, P. *et al.* (1991) Partition and transport of verapamil and nicotine through artificial membranes. *Int. J. Pharm.*, **68**, 43–49.

[93] Obata, Y., Takayama, K. and Maitani, Y. (1993) Effect of ethanol on skin permeation of nonionized and ionized diclofenac. *Int. J. Pharm.*, **89**, 191–198.

[94] Treager, R.T. (1966) *Physical Functions of the Skin*, The Academic Press, London.

[95] Lee, G., Swarbrick, J., Kiyohara, G. and Payling, D.W. (1985) Drug permeation through human skin III. Effect of pH on the partitioning behaviour of a chromone-2-carboxylic acid. *Int. J. Pharm.*, **23**, 43–54.

[96] Green, P.G. and Hadgraft, J. (1987) Facilitated transfer of cationic drugs across a lipoidal membrane by oleic acid and lauric acid. *Int. J. Pharm.*, **37**, 251–255.

[97] Oakely, D.M. and Swarbrick, J. (1987) Effects of ionization on the percutaneous absorption of drugs: partitioning of nicotine into organic liquids and the hydrated *stratum corneum*. *J. Pharm. Sci.*, **76**, 866–871.

[98] Siddiqui, O., Ying, S., Liu, J.C. and Chien, Y.W. (1987) Facilitated transdermal transport of insulin. *J. Pharm. Sci.*, **76**, 341–345.

[99]   Siddiqui, O., Roberts, M.S. and Polack, A.E. (1989) Iontophoretic transport of weak electrolytes through the excised human *stratum corneum. J. Phar. Pharmacol.*, **41**, 430–432.

[100] Woolfson, A.D., McCafferty, D.F. and Moss, G.P. (1998) Development and characterisation of a moisture-activated bioadhesive drug delivery system for percutaneous local anaesthesia. *Int. J. Pharm.*, **169**, 83–94.

[101] Moss, G.P., Woolfson, A.D., Gullick, D.R. and McCafferty, D.F. (2006) Mechanical characterisation and drug permeation properties of tetracaine-loaded bioadhesive films for percutaneous local anaesthesia. *Drug Dev. Ind. Pharm.*, **32**, 163–174.

[102] Barry, B.W. (1975) Medicaments for topical application – biopharmaceutics of dermatological preparations. *Pharm. J.*, **215**, 322–325.

[103] Flynn, G.L. (1985) Mechanism of percutaneous absorption from physicochemical evidence, in *Percutaneous Penetration* (eds R.I. Bronaugh and H.I. Maibach), Dekker, London, pp. 17–42.

[104] Imokawa, G., Kuno, H. and Kawai, M. (1991) *Stratum corneum* lipids act as a bound water modulator. *J. Invest. Dermatol.*, **96**, 845–851.

[105] Wiedmann, T.S. (1988) Influence of hydration on epdiermal tissue. *J. Pharm. Sci.*, **77**, 1037–1041.

[106] Auriol, F., Vaillant, L., Machet, L. *et al.* (1993) Effects of short time hydration on skin extensibility. *Acta Dermato-Venerelogica*, **73**, 344–347.

[107] Ohlsen, L.O. and Jemec, G.B.E. (1993) The influence of water, glycerin, paraffin oil and ethanol on skin mechanics. *Acta Dermato-Venerelogica*, **73**, 404–406.

[108] Fritsch, W.C. and Stoughton, R.B. (1963) The effect of temperature and humidity on the penetration of ^{14}C-acetylsalicylic acid in excised human skin. *J. Invest. Dermatol.*, **41**, 307–311.

[109] Zhai, H.B. and Maibach, H.I. (2001) Effects of skin occlusion on percutaneous absorption: an overview. *Skin Pharmacol. Appl. Skin Physiol.*, **14**, 1–10.

[110] Shelmire, J.B. (1960) Factors determining the skin-drug-vehicle relationship. *Arch. Dermatol.*, **82**, 24–31.

[111] Coldman, M.F., Lockerbie, L. and Laws, E.A. (1971) The evaluation of several topical preparations in the blanching test. *Br. J. Dermatol.*, **85**, 381–387.

[112] Scholtz, J.R. and Calif, P. (1961) Topical therapy of psoriasis with flucinolone acetonide. *Arch. Dermatol.*, **84**, 1029–1030.

[113] Sulzberger, M.B. and Witten, V.H. (1961) Thin pliable plastic films in topical dermatological therapy. *Arch. Dermatol.*, **84**, 1027–1029.

[114] Feldman, R.J. and Maibach, H.I. (1967) Regional variation in percutaneous absorption of ^{14}C-cortisol in man. *J. Invest. Dermatol.*, **48**, 181–183.

[115] Vickers, C.F.H. (1963) Existence of reservoir in the *stratum corneum*: experimental proof. *Arch. Dermatol.*, **88**, 72–75.

[116] Edwardson, P.A.D., Walker, M. and Breheny, C. (1993) Quantitative FT-IR determination of skin hydration following occulsion with hydrocolloid containing adhesive dressings. *Int. J. Pharm.*, **91**, 51–57.

[117] Treffel, P., Muret, P., Muretdaniello, P. *et al.* (1992) Effect of occlusion on *in vitro* percutaneous absorption of two compounds with different physicochemical properties. *Skin Pharmacol.*, **5**, 108–113.

[118] Stinchomb, A.L., Pirot, F. and Touraille, G.D. (1999) Chemical uptake into human stratum corneum in vivo from volatile and non-volatile solvents. *Pharm. Res.*, **16**, 1288–1293.

[119] Taylor, L.J., Lee, R.S., Long, M. *et al.* (2002) Effect of occlusion on the percutaneous penetration of linoleic acid and glycerol. *Int. J. Pharm.*, **249**, 157–164.

[120] Barrett, D.A. and Rutter, N. (1994) Percutaneous lignocaine absorption in newborn infants. *Arch. Dis. Child. Fetal Neonatal Ed.*, **71**, 122–124.

[121] Plunkett, L.M., Turnbull, D. and Rodricks, J.V. (1992) Differences between adults and children affecting exposure assessment, in *Similarities and Differences Between Children and Adults.*

Implications for Risk Assessment (eds P.S. Guzelian, C.J. Henry and S.S. Olin), ILSI Press, Washington, pp. 79–94.

[122] Christophers, E. and Kligman, A.M. (1964) Percutaneous absorption in aged skin, in *Advances in the Biology of the Skin* (ed W. Montagna), Permagon, New York, p. 163.

[123] Potts, R.O., Buras, E.M. and Chrisman, D.A. (1984) Changes with age in moisture content of human skin. *J. Invest. Dermatol.*, **82**, 97–100.

[124] Rougier, A., Lotte, C., Corcuff, P. *et al.* (1988) Relationship between skin permeability and corneocyte size according to anatomical age, site and sex in man. *J. Soc. Cosmet. Chem.*, **39**, 15–26.

[125] Wilhelm, K., Surber, C. and Maibach, H.I. (1991) Effect of sodium lauryl sulphate-induced skin irritation on in vivo percutaneous absorption of four drugs. *J. Invest. Dermatol.*, **97**, 927–932.

[126] Wester, R.C. and Maibach, H.I. (1999) Regional variations in percutaneous absorption, in *Percutaneous Absorption*, 2nd edn (eds R.L. Bronaugh and H.I. Maibach), Marcel Dekker, New York, pp. 111–120.

[127] Marzulli, F.N. (1969) Barriers to skin penetration. *J. Invest. Dermatol.*, **39**, 387–393.

[128] Southwell, D., Barry, B.W. and Woodford, R. (1984) Variations in permeability of human skin within and between specimens. *Int. J. Pharm.*, **18**, 299–309.

[129] Wester, R.C. and Maibach, H.I. (1999) Regional variation in percutaneous absorption, in *Percutaneous Absorption; Drugs – Cosmetics – Mechanisms – Methodology*, 3rd edn (eds R.L. Bronaugh and H.I. Maibach), Marcel Dekker, New York, pp. 107–116.

[130] Lotte, C., Wester, R.C., Rougier, A. and Maibach, H.I. (1993) Racial differences in the in vivo percutaneous absorption of some organic compounds: a comparison between black, Caucasian and Asian subjects. *Arch. Dermatol. Res.*, **284**, 456–459.

[131] Kompaore, F. and Tsuruta, H. (1993) In vivo differences between Asian, black and white in the stratum corneum barrier function. *Int. Arch. Occup. Environ. Health*, **65**, 223–225.

[132] Berardesca, E., de Rigal, J., Leveque, J.L. and Maibach, H.I. (1991) In vivo biophysical characterisation of skin physiological differences in races. *Dermatologica*, **182**, 89–93.

[133] Malkinson, F.D. and Gehlmann, L. (1977) Factors affecting percutaneous absorption, in *Cutaneous Toxicity* (eds V.A. Drill and P. Lazar), Academic Press, London, pp. 63–81.

[134] van der Merwe, E., Ackermann, C. and van Wyk, C.J. (1988) Factors affecting the permeability of urea and water through nude mouse skin *in vitro*. I. Temperature and time of hydration. *Int. J. Pharm.*, **44**, 71–74.

[135] Woolfson, A.D. and McCafferty, D.F. (1993) Percutaneous local anaesthesia: drug release characteristics of the amethocaine phase-change system. *Int. J. Pharm.*, **94**, 75–80.

[136] Allenby, A.C., Creasey, N.H., Edginton, J.A.G. *et al.* (1969) Mechanism of action of accelerants on skin penetration. *Br. J. Dermatol.*, **81**, 47–55.

[137] Bronaugh, R.L. and Franz, T.J. (1986) Vehicle effects on percutaneous absorption: *in vivo* and *in vitro* comparisons with human skin. *Br. J. Dermatol.*, **115**, 1–11.

[138] Shahi, V. and Zatz, J.L. (1978) Effect of formulation factors on penetration of hydrocortisone through mouse skin. *J. Pharm. Sci.*, **67**, 789–792.

[139] Berner, B., Mazzenga, G.C., Otte, J.H. *et al.* (1989) Ethanol: water mutually enhanced transdermal therapeutic system III: skin permeation of ethanol and nitroglycerin. *J. Pharm. Sci.*, **78**, 402–407.

[140] Idson, B. (1983) Vehicle effects in percutaneous absorption. *Drug Met. Rev.*, **14**, 207–222.

[141] Davis, A.F. and Hadgraft, J. (1993) Supersaturated solutions as topical drug delivery systems, in *Pharmaceutical Skin Penetration Enhancement* (eds K.A. Walters and J. Hadgraft), Marcel Dekker, New York, pp. 243–267.

[142] Megrab, N.A., Williams, A.C. and Barry, B.W. (1995) Oestradiol permeation across human skin, silastic and snake skin membranes: the effects of ethanol/water co-solvent systems. *Int. J. Pharm.*, **116**, 101–112.

[143] Pellett, M.A., Watkinson, A.C., Hadgraft, J. and Brain, K.R. (1997) Comparison of permeability data from traditional diffusion cells and ATR-FTIR spectroscopy. Part 1: Synthetic membranes. *Int. J. Pharm.*, **154**, 205–215.

[144] Pellett, M.A., Watkinson, A.C., Hadgraft, J. and Brain, K.R. (1997) Comparison of permeability data from traditional diffusion cells and ATR-FTIR spectroscopy. Part 2: determination of diffusional pathlengths in synthetic membranes and human stratum corneum. *Int. J. Pharm.*, **154**, 217–227.

[145] Iervolino, M., Cappello, B., Raghavan, S.L. and Hadgraft, J. (2001) Penetration enhancement of ibuprofen from supersaturated solutions through human skin. *Int. J. Pharm.*, **212**, 131–141.

[146] Hori, M., Satoh, S. and Maibach, H.I. (1990) Classification of percutaneous penetration enhancers: a conceptual diagram. *J. Pharm. Phrmacol.*, **42**, 71–72.

[147] Moss, G.P., Shah, A.J., Adams, R.G. *et al.* (2012) The application of discriminant analysis and Machine Learning methods as tools to identify and classify compounds with potential as transdermal enhancers. *Eur. J. Pharm. Sci.*, **45**, 116–127.

[148] Turraha, L. and Ali-Fossi, N. (1987) Influence of propylene glycol on the release of hydrocortisone and its acetate ester from Carbopol® hydrogels. *Acta Pharm. Fenn.*, **96**, 15–21.

[149] Baker, H.J. (1968) The effects of dimethylsulfoxide, dimethylformamide and dimethylacetamide on the cutaneous barrier to water in human skin. *J. Invest. Dermatol.*, **50**, 283–288.

[150] Vickers, C.F.H. (1969) Percutaneous absorption of sodium fusidate and fusidic acid. *Br. J. Dermatol.*, **81**, 902–908.

[151] Sams, W.M., Caroll, N.V. and Cratz, K. (1966) Effect of dimethylsulfoxide on isolated-innervated skeletal, smooth and cardiac muscle. *Proc. Soc. Exp. Biol. Med.*, **122**, 103–107.

[152] Brechner, V.L., Cohen, D.D. and Pretsky, I. (1967) Dermal anaesthesia by topical application of tetracaine base dissolved in dimethylsulfoxide. *Ann. N.Y. Acad. Sci.*, **141**, 524–531.

[153] Embery, G. and Duggard, P.H. (1971) The isolation of dimethylsulfoxide soluble components from human epidermal preparations: a possible mechanism of action of dimethylsulphoxide in effecting percutaneous migration phenomena. *J. Invest. Dermatol.*, **57**, 308–311.

[154] Creasey, N.H., Battensby, J. and Fletcher, J.A. (1978) Factors affecting the permeability of skin. *Curr. Probl. Dermatol.*, **7**, 95–105.

[155] Sharata, H.H. and Burnette, R.R. (1988) Effect of dipolar aprotic permeability enhancers on the basal stratum corneum. *J. Pharm. Sci.*, **77**, 27–32.

[156] Stoughton, R.B. (1976) Composition and method for topical administration of griseofulvin. US Patent 3,932,653.

[157] Schroer, R.A. (1976) Topical anti-inflammatory composition and method of use. US Patent 3,957,994.

[158] Akhter, S.A. and Barry, B.W. (1985) Absorption through human skin of ibuprofen and flurbiprofen: effect of dose variations, deposited drug films, occlusion and the penetration enhancer N-methyl-2-pyrrolidone. *J. Pharm. Pharmacol.*, **37**, 27–37.

[159] Southwell, D. and Barry, B.W. (1984) Penetration enhancement in human skin: effect of 2-pyrrolidone, dimethylformamide and increased hydration of finite dose permeation of aspirin and caffeine. *Int. J. Pharm.*, **22**, 291–298.

[160] Naik, A., Pechtold, L., Potts, R.O. *et al.* (1995) Mechanism of oleic acid induced skin penetration enhancement *in vivo* in humans. *J. Control. Rel.*, **37**, 299–306.

[161] Tanojo, H., BosvanGeest, A., Bouwstra, J.A. *et al.* (1997) In vitro human skin barrier perturbation by oleic acid: thermal analysis and freeze fracture electron microscopy studies. *Thermochim. Acta.*, **293**, 77–85.

[162] Williams, A.C. and Barry, B.W. (1989) Urea analogues in propylene glycol as penetration enhancers in human skin. *Int. J. Pharm.*, **56**, 43–50.

[163] Hellgren, L. and Larsson, K. (1979) On the effect of urea on human epidermis. *Dermatologica*, **149**, 289–293.

[164] Beastall, J.C., Hadgraft, J. and Washington, C. (1988) Mechanism of action of Azone® as a percutaneous penetration enhancer: lipid bilayer fluidity and transition temperature effects. *Int. J. Pharm.*, **43**, 207–213.

[165] Bettley, F.R. (1965) The influence of detergents and surfactants on epidermal permeability. *Br. J. Dermatol.*, **77**, 98–100.

[166] Scheuplein, R.J. and Ross, L. (1970) Effect of surfactants and solvents on the permeability of the epidermis. *J. Soc. Cosmet. Chem.*, **21**, 853–873.

[167] Gillian, G.M.N. and Florence, A.T. (1973) The influence of the non-ionic surfactant type on the transport of a drug across a biological membrane. *J. Pharm. Pharmacol.*, **25** (Supplement), 136.

[168] Cork, M.J. and Danby, S. (2011) Aqueous cream damages the skin barrier. *Br. J. Dermatol.*, **164**, 1179–1180.

[169] Danby, S.G., Al-Enezi, T., Sultan, A. *et al.* (2011) The effect of aqueous cream BP on the ski barrier in volunteers with a previous history of atopic dermatitis. *Br. J. Dermatol.*, **165**, 329–334.

[170] Cork, M.J. (2012) Skin barrier breakdown in atopic dermatitis: the effect of topical products. *Int. J. Cosmet. Sci.*, **34**, 363–364.

[171] Stoughton, R.B. and McClure, W.O. (1983) Azone®: a new non-toxic enhancer of cutaneous penetration. *Drug Del. Ind. Pharm.*, **9**, 725–744.

[172] Lewis, D. and Hadgraft, J. (1990) Mixed monolayers of dipalmitoylphosphatidylcholine with Azone® or oleic acid at the air-water interface. *Int. J. Pharm.*, **65**, 211–218.

[173] Wiechers, J.W., Drenth, B.F.H., Jonkman, J.H.G. *et al.* (1990) Percutaneous absorption of triaminolone acetonide from creams with and without Azone® in humans *in vivo. Int. J. Pharm.*, **66**, 53–62.

[174] Ogiso, T., Iwaki, M., Bechako, K. and Tsutsumi, Y. (1992) Enhancement of percutaneous absorption by laurocapram. *J. Pharm. Sci.*, **81**, 762–767.

[175] Monti, D., Saettone, M.F., Giannaccini, B. and Galli-Angeli, D. (1995) Enhancement of trans-dermal penetration of dapiprazole through hairless mouse skin. *J. Control. Rel.*, **33**, 71–77.

[176] Baker, E.J. and Hadgraft, J. (1995) *In vitro* percutaneous absorption of Arildone, a highly lipophilic drug, and the apparent no effect of the penetration enhancer Azone® in excised human skin. *Pharm. Res.*, **12**, 993–997.

[177] Williams, A.C. and Barry, B.W. (1991) The enhancement index concept applied to terpene penetration enhancers for human skin and model lipophilic (oestradiol) and hydrophilic (5-fluorouracil) drugs. *Int. J. Pharm.*, **74**, 157–168.

[178] Williams, A.C. and Barry, B.W. (1991) Terpenes and the lipid-protein-partitioning theory of skin penetration enhancers. *Pharm. Res.*, **8**, 17–24.

[179] Kato, A., Ishibashi, Y. and Miyake, Y. (1987) Effect of egg yolk lecithin on transdermal delivery of bunazosin hydrochloride. *J. Pharm. Pharmacol.*, **39**, 399–400.

[180] Adachi, H., Irie, T., Uekama, K. *et al.* (1993) Combination effects of *o*-carboxymethyl-*o*-ethyl-ß-cyclodextrin and penetration enhancer HPE-101 on transdermal delivery of prostaglandin-E(1) in hairless mice. *Eur. J. Pharm. Sci.*, **1**, 117–123.

[181] Zaya, M.J., Shull, K.L., Brunden, M.N. *et al.* (1994) SEPA®, a novel penetration enhancer, elicits earlier and significantly greater minoxidil-induced hair growth compared to Rogaine® topical solution (TS) in the balding stumptail Macaque. *J. Invest. Dermatol.*, **102**, 599.

[182] Diani, A.R., Shull, K.L., Zaya, M.J. and Brunden, M.N. (1995) The penetration enhancer SEPA® augments stimulation of scalp hair-growth by topical minoxidil in the balding stumptail Macaque. *Skin Pharmacol.*, **8**, 221–228.

[183] Harris, W.T., Tenjarla, S.N., Holbrook, J.M. *et al.* (1995) N-pentyl-N-acetylprolinate – a new skin penetration enhancer. *J. Pharm. Sci.*, **84**, 640–642.

[184] Rautio, J., Nevalainen, T., Taipale, H. *et al.* (2000) Synthesis and in vitro evaluation of novel morpholinyl- and methylpiperazinylacyloalkyl prodrugs of 2-(6-methoxy-2-naphthyl)propionic acid (naproxen) for topical drug delivery. *J. Med. Chem.*, **43**, 1489–1494.

[185] Bonino, F.P., Montenegro, L., Decapraris, P. *et al.* (1991) 1-Alkylazacycloalkan-2-one esters as prodrugs of indomethacin for improved delivery through human skin. *Int. J. Pharm.*, **77**, 21–29.

[186] Sintov, A.C., Behar-Canetti, C., Friedman, Y. and Tamarkin, D. (2002) Percutaneous penetration and skin metabolisms of ethylsalicylate-containing agent, TU-2100: in vitro and in vivo evaluation in guinea pigs. *J. Control. Rel.*, **79**, 113–122.

[187] Liu, P.C. and Bergstrom, T.K. (1996) Quantitative evaluation of aqueous isopropyl alcohol enhancement on skin flux of terbutaline (sulphate). 2. Permeability contributions of equilibrated drug species across human skin in vitro. *J. Pharm. Sci.*, **85**, 320–325.

[188] Valenta, C., Siman, U., Kratzel, M. and Hargraft, J. (2000) The dermal delivery of lignocaine: influence of ion pairing. *Int. J. Pharm.*, **197**, 77–85.

[189] Takahashi, K. and Rytting, J.H. (2001) Novel approach to improve ondansetron across shed snake skin as a model membrane. *J. Pharm. Pharmacol.*, **53**, 789–794.

[190] Stott, P.W., Williams, A.C. and Barry, B.W. (2001) Mechanistic study into the enhanced transdermal permeation of a model β-blocker, propranolol, by fatty acids; a melting point depression effect. *Int. J. Pharm.*, **219**, 161–176.

[191] McCafferty, D.F., Woolfson, A.D., Handley, J. and Allen, G. (1997) Effect of percutaneous local anaesthetics on pain reduction during pulse dye laser treatment of Port Wine Stains. *Br. J. Anaesth.*, **78**, 286–289.

[192] McCafferty, D.F. and Woolfson, A.D. (1993) New patch delivery system for percutaneous local anaesthesia. *Br. J. Anaesth.*, **71**, 370–374.

[193] McCafferty, D.F., Woolfson, A.D. and Moss, G.P. (2000) Novel bioadhesive delivery system for percutaneous local anaesthesia. *Br. J. Anaesth.*, **84**, 456–458.

[194] Woolfson, A.D., McCafferty, D.F., McGowan, K.E. and Boston, V. (1989) Non-invasive monitoring of percutaneous local anaesthesia using Laser Doppler Velocimetry. *Int. J. Pharm.*, **51**, 183–187.

[195] Woolfson, A.D., McCafferty, D.F. and Boston, V. (1990) Clinical experiences with a novel percutaneous amethocaine preparation – prevention of pain due to venepuncture in children. *Br. J. Clin. Pharmacol.*, **30**, 273–279.

[196] Kondo, S., Yamasaki-Konishi, H. and Sugimoto, I. (1987) Enhancement of transdermal delivery by superfluous thermodynamic potential. II. In vitro-in vivo correlation of percutaneous nifedipine transport. *J. Pharmacobio-Dyn.*, **10**, 743–749.

[197] Henmi, T., Fujii, M., Kikuchi, K. *et al.* (1994) Application of an oily gel formed by hydrogenated soybean phospholipids as a percutaneous absorption-type ointment base. *Chem. Pharm. Bull.*, **42**, 651–655.

[198] Pellett, M.A., Davies, A.F. and Hadgraft, J. (1994) Effect of supersaturation on membrane transport. 2. Piroxicam. *Int. J. Pharm.*, **111**, 1–6.

[199] Patel, A., Cevc, G. and Bachhawat, B.K. (1995) Transdermal immunization with large proteins by means of ultradeformable carriers *Eur. J. Immunol.*, **25**, 3521–3524.

[200] Honzak, H.E.J., van der Geest, R., Bodde, H.E. *et al.* (1994) Estradiol permeation from non-ionic surfactant vesicles through human stratum corneum *in vitro*. *Pharm. Res.*, **11**, 659–664.

[201] Cevc, G. and Blume, G. (2001) New, highly efficient formulation of diclofenac for the topical, transdermal administration in ultradeformable drug carriers, Transfersomes. *Biochim. Biophys. Acta*, **1514**, 191–205.

[202] Filipe, P., Silva, J.N., Silva, R. *et al.* (2009) Stratum corneum is an effective barrier to TiO2 and ZnO nanoparticle percutaneous absorption. *Skin Pharmacol. Physiol.*, **22**, 266–275.

[203] Kuo, T.R., Wu, C.L., Hsu, C.T. *et al.* (2009) Chemical enhancer induced changes in the mechanisms of transdermal delivery of zinc oxide nanoparticles. *Biomaterials*, **30**, 3002–3008.

[204] Szikszai, Z., Kertesz, Z., Bodnar, E. *et al.* (2010) Nuclear microprobe investigation of the penetration of ultrafine zinc oxide into intact and tape-stripped human skin. *Nucl. Instrum. Methods. Phys. Res. B*, **268**, 2160–2163.

[205] Jung, S., Otberg, N., Thiede, G. *et al.* (2006) Innovative liposomes as a transfollicular drug delivery system: penetration into porcine hair follicles. *J. Invest. Dermatol.*, **126**, 1728–1732.

[206] Nohynek, G.J., Lademann, J., Ribaud, C. and Roberts, M.S. (2007) Grey goo on the skin? Nanotechnology, cosmetic and sunscreen safety. *Crit. Rev. Toxicol.*, **37**, 251–277.

[207] Lademann, J., Patzelt, A., Richter, H. *et al.* (2009) Comparisons of two in vitro models for the analysis of follicular penetration and its prevention by barrier emulsions. *Eur. J. Pharm. Biopharm.*, **72**, 600–604.

[208] Luengo, J., Weiss, B., Schneider, M. *et al.* (2006) Influence of nanoencapsulation on human skin transport of flufenamic acid. *Skin Pharmacol. Physiol.*, **19**, 190–197.

[209] NCT01125215, Capsaicin Nanoparticle in Patient With Painful Diabetic Neuropathy (2009) http://clinicaltrials.gov/ct2/show/NCT01125215 (accessed 16 May 2014).

[210] Erdogan, S. (2009) Liposomal nanocarriers for tumor imaging. *J. Biomed. Nanotechnol.*, **5**, 141–150.

[211] Liang, C.H. and Chou, T.H. (2009) Effect of chain length on physicochemical properties and cytotoxicity of cationic vesicles composed of phosphatidylcholines and dialkyldimethylammonium bromides. *Chem. Phys. Lipids*, **158**, 81–90.

[212] Muller, R.H., Petersen, R.D., Hommoss, A. and Pardeike, J. (2007) Nanostructured lipid carriers (NLC) in cosmetic dermal products. *Adv. Drug Del. Rev.*, **59**, 522–530.

[213] Schneider, M., Stracke, F., Hansen, S. and Schaefer, U.F. (2009) Nanoparticles and their interactions with the dermal barrier. *Dermatoendocrinology*, **1**, 197–206.

[214] Dubina, M. and Goldenberg, G. (2009) Viral-associated nonmelanoma skin cancers: a review. *Am. J. Dermatopathol.*, **31**, 561–573.

[215] Carmona-Ribeiro, A.M. (2010) Biomimetic nanoparticles: preparation, characterization and biomedical applications. *Int. J. Nanomed.*, **5**, 249–259.

[216] Bryan, J.T. and Brown, D.R. (2001) Transmission of human papillomavirus type 11 infection by desquamated cornified cells. *Virology*, **281**, 35–42.

[217] Wolf, J. (1939) Die innere strucktur der zellen des stratum desqamans der menschlichen epidermis. *Anat. Forsch.*, **46**, 170–202.

[218] Jacques, S.L., McAuliffe, D.J., Blank, I.H. and Parrish, J.A. (1987) Controlled removal of human stratum corneum by pulsed laser. *J. Invest. Dermatol.*, **88**, 88–93.

[219] Banga, A.K. and Chien, Y.W. (1988) Iontophoretic delivery of drugs: fundamentals, developments and biomedical applications. *J. Control. Rel.*, **7**, 1–24.

[220] Chien, Y.W., Lelawongs, P., Siddiqui, O. *et al.* (1990) Facilitated transdermal delivery of therapeutic peptides and proteins by iontophoretic delivery devices. *J. Control. Rel.*, **13**, 263–278.

[221] Singh, P. and Maibach, H.I. (1994) Iontophoresis in drug delivery: basic principles and applications. *Crit. Rev. Ther. Drug Carrier Syst.*, **11**, 161–213.

[222] Papa, C.M. and Kligman, A.M. (1966) Mechanisms of eccrine anidrosis. I. High level blockade. *J. Invest. Dermatol.*, **47**, 1–9.

[223] Burnette, R.R. and Marrero, D. (1986) Comparison between the iontophoretic and passive transport of thyrotropin releasing hormone across nude mouse skin. *J. Pharm. Sci.*, **75**, 738–743.

[224] Burnette, R.R. and Ongpipattankul, B. (1987) Characterisation of the permselective properties of human skin during iontophoresis. *J. Pharm. Sci.*, **76**, 765–773.

[225] Praissman, M., Miller, I.F. and Berkowitz, J.M. (1973) Ion-mediated water flow. I: electroosmosis. *J. Membr. Biol.*, **11**, 139–51.

[226] Gangarosa, L.P., Park, N.H., Wiggins, C.A. and Hill, J.M. (1980) Increased penetration of non-electrolytes into mouse skin during iontophoretic water transport (iontohydrokinesis). *J. Pharmacol. Exp. Ther.*, **212**, 377–381.

[227] McElnay, J.C., Matthews, M.P., Harland, R. and McCafferty, D.F. (1985) The effect of ultrasound on the percutaneous absorption of lignocaine. *Br. J. Clin. Pharmacol.*, **20**, 421–424.

[228] Krishnan, G.J., Edwards, J., Chen, Y. *et al.* (2010) Enhanced skin permeation of naltrexone by pulsed electromagnetic fields in human skin in vitro. *J. Pharm. Sci.*, **99**, 2724–2731.

[229] Krishnan, G., Grice, J.E., Roberts, M.S. *et al.* (2013) Enhanced sonophoretic delivery of 5-aminolevulinic acid: preliminary human ex vivo permeation data. *Skin Res. Technol.*, **19**, 283–289.

[230] Henry, S., McAllister, D.V., Allen, M.G. and Prausnitz, M.R. (1998) Microfabricated microneedles: a novel approach to transdermal drug delivery. *J. Pharm. Sci.*, **87**, 922–925.

[231] Kaushik, S., Hord, A.H., Denson, D.D. *et al.* (2001) Lack of pain associated with microfabricated microneedles. *Anesth. Analg.*, **92**, 502–504.

[232] Yang, M. and Zahn, J.D. (2004) Microneedle insertion force reduction using vibratory actuation. *Biomed. Microdevices*, **6**, 177–182.

[233] Davis, S.P., Benjamin, J.L., Adams, Z.H. *et al.* (2004) Insertion of microneedles into skin: measurement and prediction of insertion force and needle fracture force. *J. Biomechanics*, **37**, 1155–1163.

[234] Sivamani, R.K., Stoeber, B., Wu, G.C. *et al.* (2005) Clinical microneedle injection of methyl nicotinate: stratum corneum penetration. *Skin Res. Technol.*, **11**, 152–156.

[235] Coulman, S.A., Barrow, D., Anstey, A. *et al.* (2006) Minimally invasive cutaneous delivery of macromolecules and plasmid DNA via microneedles. *Curr. Drug Del.*, **3**, 65–75.

[236] Martin, C.J., Allender, C.J., Brain, K.R. *et al.* (2012) Low temperature fabrication of biodegradable sugar glass microneedles for transdermal drug delivery applications. *J. Control. Rel.*, **158**, 93–101.

[237] Donnelly, R.F., Singh, T.R.R., Garland, M.J. *et al.* (2012) Hydrogel-forming microneedle arrays for enhanced transdermal drug delivery. *Adv. Funct. Mater.*, **22**, 4879–4890.

[238] Gomaa, Y.A., Garland, M.J., McInnes, F. *et al.* (2012) Laser-engineered dissolving microneedles for active transdermal delivery of nadroparin calcium. *Eur. J. Pharm. Biopharm.*, **82**, 299–307.

[239] Ryan, E., Garland, M.J., Singh, T.R.R. *et al.* (2012) Microneedle-mediated transdermal bacteriophage delivery. *Eur. J. Pharm. Sci.*, **47**, 297–304.

[240] McCrudden, M.T.C., Alkilani, A.Z., McCrudden, C.M. *et al.* (2014) Design and physicochemical characterisation of novel dissolving polymeric microneedle arrays for transdermal delivery of high dose, low molecular weight drugs. *J. Control. Rel.*, **180**, 71–80.

[241] Donnelly, R.F., Mooney, K., McCrudden, M.T.C. *et al.* (2014) Hydrogel-forming microneedles increase in volume during swelling in skin, but skin barrier function recovery is unaffected. *J. Pharm. Sci.*, **103**, 1478–1486.

[242] Donnelly, R.F., Morrow, D.I.J., McCrudden, M.T.C. *et al.* (2014) Hydrogel-forming and dissolving microneedles for enhanced delivery of photosensitizers and precursors. *Photochem. Photobiol.*, **90**, 641–647.

2

Application of Spectroscopic Techniques to Interrogate Skin

Jonathan Hadgraft, Rita Mateus and Majella E. Lane

Department of Pharmaceutics, UCL School of Pharmacy, London, UK

2.1 Introduction

It is surprising that in the twenty-first century, it is still difficult to treat skin diseases effectively and this has a huge impact on the quality of life of a large number of people. It also puts a considerable burden on health providers. The reasons for the lack of effective treatment result partly from the fact that APIs to treat skin diseases are not designed, *ab initio*, for this purpose and also that the mechanisms of skin penetration have not been fully elucidated. This is particularly true when investigating the way in which excipients interact with various components of the skin. As a result of increased sophistication, sensitivity and specificity of spectroscopic techniques, some of the questions concerning mechanisms of skin permeation are being answered. The skin is the largest organ of the human body, and represents 10% of the total body mass in adults, with an average total surface area of $2\,m^2$. It is a complex organ with a diverse cellular population and a range of physiological activities. The main function of the skin is the protection of internal organs from the external environment by preventing the egress of water and the ingress of toxins. Despite this barrier role the skin is also an organ which is exploited for drug administration, both local and to a lesser degree, transdermal. The outer layer of the skin, the stratum corneum, provides the major barrier with a structure that has been likened to a brick wall. The 'bricks' are the corneocytes comprising keratin, and the 'mortar' is a complex lipid comprising ceramides, cholesterol, cholesterol sulphate and free fatty acids. Our current

Novel Delivery Systems for Transdermal and Intradermal Drug Delivery, First Edition.
Ryan F. Donnelly and Thakur Raghu Raj Singh.

understanding is that the predominant route of permeation for drugs across the stratum corneum is via the intercellular lipid pathway. Because the lipids are structured into bilayers, a permeating AP has to pass from lipophilic to hydrophilic domains, and only a few substances have the optimum properties to cross the skin. Typically these molecules are small, with balanced partition characteristics and good solubility in both water and oils, for example nicotine or nitroglycerin. A range of spectroscopic techniques have been used to interrogate and understand the skin barrier. The methods which have been investigated over the years have ranged from X-rays through ultra-violet (UV) to infrared (IR), mass spectroscopy, sound waves and radio waves, in the form of nuclear magnetic resonance (NMR). These will be discussed in further detail in the next sections along with the newer approaches which have come to prominence in recent years.

2.2   Vibrational Spectroscopic Methods

The features that are most commonly examined in skin using IR are the CH symmetric and asymmetric stretch (Figure 2.1) at around 2850 and 2920 cm^{-1}. The carbonyl stretch, amide 1 and amide 2 bands in the fingerprint region are of interest because they provide information on lipid and protein conformation.

The use of attenuated total reflectance IR spectroscopy for studying skin was first suggested by Puttnam [1]. The use of IR in an understanding of the role of the skin lipids in permeation was further advanced in the mid 1980s [2, 3]. Polymorphism of the skin lipids [4] and lipid organisation were also investigated using spectroscopic approaches [5, 6]. Most studies have been conducted on the forearm, but recently Gorcea and colleagues reported

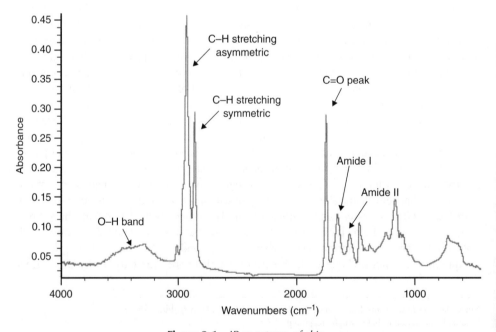

Figure 2.1   *IR spectrum of skin.*

the use of attenuated total reflectance–Fourier transform IR (ATR FTIR) complemented with transepidermal water loss (TEWL) and tape stripping techniques to monitor changes in the composition and organisation of SC lipids and bound water in facial skin *in vivo* [7].

The influence of penetration enhancers on skin lipids, namely octadecanoic acids and oleic acid, has been investigated by a number of research groups [8, 9]. Using perdeuterated oleic acid, Ongpipattanakul and co-workers [10] demonstrated the ability of this molecule to form phase separated pools within the stratum corneum. This is in contrast to Azone® which appears to be distributed uniformly throughout the lipids [11]. Evidence for the intercellular route as the primary permeation pathway was provided by the investigation of water diffusion through porcine skin; water permeability was followed as a function of temperature with the –CH stretch being monitored simultaneously [12]. A high correlation between water permeability and the –CH stretch frequency suggests that water movement reflects its passage through the intercellular lipids. The occlusive effects of hydrocolloid dressings have also been studied using FTIR [13], but even so the precise reasons why hydrated skin is more permeable than normal skin are unclear.

IR data suggest that increased hydration does not affect lipid fluidity although skin hydration has been reported to enhance skin permeation of actives [14]. Ethanol has been used as a skin penetration enhancer in topical and transdermal dosage forms for many years. ATR-FTIR reveals no influence of this molecule on lipid ordering, indicating that the mechanism of penetration is likely related to solubility effects [15].

The effects of different enhancers on skin permeation *in vitro* (Azone and Transcutol) have also been deconvoluted using ATR-FTIR [16]. *In vivo* tape stripping studies combined with ATR-FTIR have also advanced our understanding of the effects of penetration enhancers on drug disposition in skin [17, 18]. Dias *et al.* applied perdeuterated alkanols to human subjects and demonstrated that the skin lipid disorder was directly related to the uptake of the alkanols [19]. Using *in vitro* ATR-FTIR, diffusion experiments it was possible to follow the permeation from a commercial ibuprofen formulation through model membranes and skin; chemometric approaches were used to separate the signals of the drug from excipients [20]. Data were deconvoluted successfully for the commercial formulation as it permeated through skin and the different components of the formulation identified except for benzyl alcohol. This may reflect the low concentrations of benzyl alcohol used in topical formulations.

Imaging using IR (with a resolution down to ~2 µm) has also been used with advanced chemometrics to obtain spatial resolution of both active and excipients [21–23]. The conformation of the lipids has been imaged [24], and the interaction of dimyristoylphosphatidylcholine with skin lipids has also been studied using this approach [25]. Andanson *et al.* [22] used spectroscopic imaging to measure, *in situ*, the diffusion of benzyl nicotinate (BN) through the outer layer of human SC. Image analysis demonstrated a strong correlation between the distribution of lipids and BN. We have also recently investigated the effects of three fatty acid esters on skin permeation using ATR FTIR [26]. Propylene glycol dipelargonate (DPPG), isopropyl myristate (IPM) and isostearyl isostearate (ISIS) were selected as pharmaceutically relevant solvents with a range of lipophilicities; cyanophenol (CNP) was used as a model API. The diffusion coefficients of CNP across epidermis in the different solvents were not significantly different. Using chemometric data analysis diffusion profiles for the solvents were deconvoluted from that of the skin and modelled. Each of these solvents was found to diffuse at a faster rate across the skin than CNP. DPPG considerably increased the concentration of CNP in the stratum corneum in comparison with the other solvents indicating strong penetration enhancer potential. By contrast IPM produced a similar CNP concentration in the stratum corneum to water as the solvent, with ISIS resulting in a lower CNP concentration suggesting negligible enhancement and penetration retardation effects for these two solvents respectively. This would suggest that the solvents have various functions in the permeation process and the degree to which they are taken into the stratum corneum and their residence times therein are important determinants.

There is also evidence that IR can be used in diagnosis of skin disorders [27]. Unpublished data from our own laboratory show a difference in the spectral signature from psoriatic skin pre- and post-treatment with UV radiation (Figure 2.2). More recently Janssens *et al.* [28] have used ATR FTIR to obtain information on the skin lipid organisation of patients with atopic eczema (AE) compared with control subjects. Two types of vibrations were monitored: the CH_2 symmetric stretching vibrations and the CH_2 scissoring vibrations. The mean value of the position of the CH_2 symmetric stretching vibrations of AE patients shows a small but significant shift to higher values compared with control subjects ($2849.2\,cm^{-1}$ vs. $2848.8\,cm^{-1}$, respectively; $P=0.0013$). In addition, the variance in the group of AE patients is larger than in control subjects. To distinguish between an orthorhombic (dense) and hexagonal (less dense) lateral organisation, the bandwidth of the CH_2 scissoring vibrations was monitored. The average bandwidth of the scissoring vibrations was significantly lower in AE patients compared with control subjects ($10.6\,cm^{-1}$ vs. $11.6\,cm^{-1}$, respectively; $P=0.010$), demonstrating a reduction of lipids in an orthorhombic organisation and thus a less dense lipid organisation.

Near infrared (NIR) has been used to monitor water profiles in atopy [29], but the results were inconclusive; however, the study did indicate the uses of NIR in the study of skin. Bodén and co-workers [30] have reported the characterisation of healthy skin using NIR and skin impedance (IMP). Measurements were performed *in vivo* on healthy skin at five anatomic body sites on eight female subjects. Inter-individually the NIR model gave 100% correct classification, while the IMP model provided 92%. Intra-individually the NIR model gave 88% correct classification, whereas the IMP model did not provide any useful classification. NIR was used to follow the interactions of isopropyl myristate and polyethylene glycol on water distribution in normal and atopic skin and differences noted [31]. In the spectral range from 1100 to 2060 nm, NIR provides analytical data that, after appropriate statistical analysis, enable normal and atopic subjects to be distinguished on the basis of skin changes in response to each of the three compounds tested. The compounds caused marked changes in the spectral responses of the skin. Irrespective of the compound applied and the class of subject, there was a considerable change in the skin spectrum after exposure to the chemical agent for 10 min. Egawa [32] used NIR to study urea and water content in the heel of human subjects after a 2 h application of a cream. The estimated water content tended to increase 1 and 2 h after the treatment compared with before the treatment, and subsequently gradually decreased with time. The estimated urea content significantly increased 1 h after the treatment. Significant changes in peak wavelength to shorter wavelengths 1, 2 and 4 h after the treatment were observed, suggesting that the water mobility increased.

Raman spectroscopy is a vibrational spectroscopy, similar to IR spectroscopy but based on inelastic light scattering rather than absorption of light. The spectral profiles produced

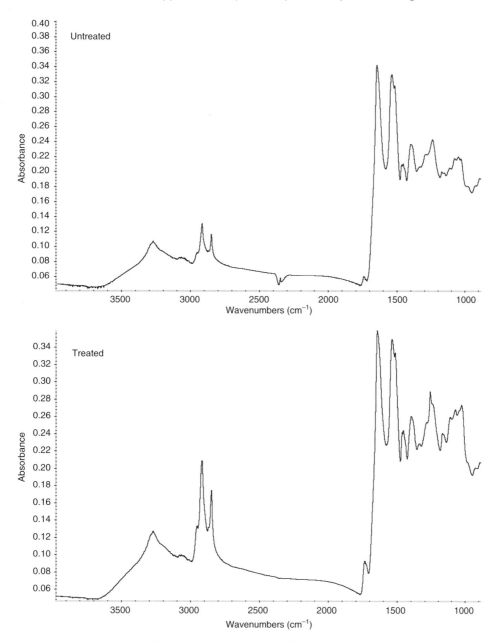

Figure 2.2 *IR spectra from psoriatic skin pre- and post-treatment with UV radiation.*

are similar with good intra- and inter-variability from different skin samples [33, 34]. Williams and co-workers reported spectral signatures from a skin sample over 5000 years old which are comparable to those found today [35]. The use of confocal Raman spectroscopy (CRS) *in vivo* to depth profile the skin [36–38] has been reported, and the hydration

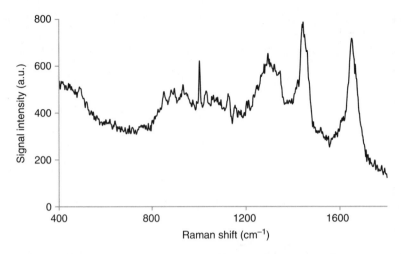

Figure 2.3 *Raman spectrum of human skin (untreated).*

level of the skin as well as SC thickness can be followed [39]. There is an indication from our own *in vivo* studies on the –CH stretch frequency from Confocal Raman that the fluidity of the lipids is higher the deeper into the stratum corneum (J. Hadgraft, P. Matts, M. Lane and J. Crowther, In vivo depth profiling the lipids of the skin using Laser Confocal Raman Spectroscopy. Unpublished data, 2009) which is consistent with previous findings. Mélot *et al.* reported the simultaneous monitoring of an active (retinol) and an excipient (propylene glycol) *in vivo* [40] using CRS; the technique has also been used to monitor the penetration of DMSO [41]. Stamatas and Boireau-Adamezyk [42] have reported the measurement of concentration profiles of topically applied caffeine in skin *in vivo* using CRS. We have also demonstrated the usefulness of CRS in monitoring the permeation of ibuprofen in skin *in vivo* (Figure 2.3); the profiles compared very well with previously published tape stripping data [43].

Most recently a very good correlation was demonstrated between *in vitro* Franz cell studies and *in vivo* human studies which monitored niacinamide skin penetration from a range of simple vehicles [44]. High wave-number CRS (2800–3125 cm^{-1}) has been evaluated for its ability to provide sufficient information for non-invasive discrimination between basal cell carcinoma (BCC) and non-involved skin [45]. Raman spectra were obtained from 19 BCC biopsy specimens and 9 biopsy specimens of perilesional skin using a fibre optic probe. A linear discriminant analysis (LDA)-based tissue classification model was developed, which discriminated between BCC and noninvolved skin with high accuracy.

The application of stimulated Raman scattering (SRS) for skin imaging has been pioneered by Xie and colleagues. SRS has been applied to image the skin permeation of trans-retinol (vitamin A) both *in vitro* and *in vivo*; the distribution of DMSO in skin was also studied [46].

Depth profiling of skin can also be conducted with photoacoustic FTIR spectroscopy and the diffusion of cyanodecane was reported to be 1.6 times higher in the inner regions of the stratum corneum than the outer [47]. This is in agreement with the CRS results suggesting higher lipid fluidity in deeper regions of the stratum corneum. *In vivo* non-invasive monitoring of glucose concentration in human epidermis by mid-infrared pulsed photoacoustic spectroscopy has been reported by Pleitez *et al.* [48]. Similar glucose concentrations were reported for blood sampling and PF FTIR measurements in healthy subjects and diabetics, demonstrating the potential of the technique for non-invasive glucose monitoring.

2.3 Electronic Spectroscopic Methods

2.3.1 UV and Fluorescence

Although diffuse reflectance UV spectroscopy has been used to examine the surface of the skin, there is little penetration into the stratum corneum at these wavelengths. Kim *et al.* [49] reported the use of ultraviolet/visible (UV/vis) diffusive reflectance spectroscopy combined with fibre-optics to measure the penetration time of aminolevulinic acid when applied to the human forearm.

The fluorescent properties of skin have been reviewed by Kollias and co-authors [50]. Most studies have used this technique to study the fate of fluorescent probes, labelled vesicles or particles when applied to skin. Garrison *et al.* [51] used phase modulated fluorescence spectroscopy and the fluorophore 1,6-diphenyl-1,3,5-hexatriene to study the effects of oleic acid on human epidermis. Fluorescence lifetime and limiting anisotropy of a membrane fluorophore 1,6-diphenyl-1,3,5-hexatriene were measured in human epidermis treated with donor formulations containing 0–5% oleic acid. Transport of benzoic acid across human epidermis was also measured under these conditions. The flux enhancement, changes in the spectral parameters, and uptake of oleic acid into epidermis showed a saturation effect with increasing OA concentration in the formulation. Alvarez-Román *et al.* [52] have reviewed the applications of fluorescence spectroscopy to monitor fluorescently tagged vesicles and particles in skin; the effects of solid lipid particles on drug delivery have also been investigated with this approach [53]. Since biological processes modulate epidermal and dermal fluorescence signals in a predictable manner, *in vivo* fluorescence imaging has also been used in diagnosis and the study of skin lesions [54].

Multiphoton laser microscopy (MLM) is based on autofluorescence and fluorescence lifetime imaging and has been used to study skin at the cellular and subcellular level as well as the penetration of drugs and particulates within the skin. Deeper tissues such as collagen can also be visualised [55, 56]. While MLM is based on the fluorescence intensity emitted by epidermal and dermal fluorophores and by the extra-cellular matrix, fluorescence lifetime imaging (FLIM) is generated by the fluorescence decay rate. Seidenari and colleagues have recently investigated the combination of MPT with FLIM in the diagnosis of malignant melanoma [57]. The MPT/FLIM results compared very well with conventional dermoscopy although it should be noted that the studies were carried out on excised lesions.

2.3.2 Nuclear Magnetic Resonance

Packer and Sellwood [58, 59] used NMR to study the hydration of the SC and water mobility using samples of guinea pig skin. Foreman *et al.* [60] noted the presence of 'free' and 'bound' water in human SC. The self-diffusion of water in the stratum corneum and viable epidermis has been reported by McDonald *et al.* [61]. Mixtures of endogenous skin lipids have been examined using deuterium NMR, and thermally induced polymorphism has been detected as well as lamellar structures [62–64]. Water as well as a number of cosmetic ingredients have been investigated using stray field NMR [65–67]. Water loss from samples of skin *in vitro* can be monitored as well as water distribution using stray field NMR [66]. Although most studies have been conducted on *in vitro* skin samples, *in vivo* studies are possible, although the resolution of the technique does not allow precise monitoring of actives in the skin [67]. A novel NMR mobile universal surface explorer (MOUSE) has been developed which allows rapid analysis and which is compact and portable. However, Kornetka *et al.* [68] have observed that precise measurements *in vivo* are compromised by poor reproducibility, long acquisition times, and incompatibility between the geometries of the sensitive area of the instrument and the non-planar structure of the skin.

^{19}F NMR has been used *in vivo* to follow the permeation of flurbiprofen through human and hairless rat skin [69]. Schwarz *et al.* [70] have also used ^{19}F NMR to monitor the skin penetration routes of flufenamic acid and fluconazole using tape stripping experiments on porcine skin. Zemtsov *et al.* [71] assessed the effectiveness of using phosphorus ^{31}P NMR spectroscopy to monitor non-invasively metabolism in psoriasis. In patients with severe psoriasis, in comparison with a control group, elevations in phosphomonoester concentrations and in the phosphomonoester/phosphodiester ratio were observed. Although the approach appears to be a sensitive, non-invasive technique to monitor disease activity in psoriasis it has not been progressed.

2.4   Miscellaneous Spectroscopic Methods

2.4.1   Opto-Thermal Transient Emission Radiometry

Opto-thermal transient emission radiometry (OTTER) uses pulsed laser excitation to induce temperature jumps of the order of a few degree celsius in the outer skin surface. These temperature jumps decay over a time scale of microseconds and do not increase significantly the average substrate temperature. A high-speed infrared detector sensitive to the heat radiation emitted by the surface is used for signal collection. For biological tissue, this radiation is strongest in the mid-infrared 6–13 μm band of wavelengths. The measurement captures the decay dynamics of this transient component of the heat radiation and relates it to the physical properties of the near-surface layers through mathematical models [72]. The technique has been used to measure skin water content *in vivo* [73], the hydration profile within the stratum corneum *in vivo*, and how this profile changes in the presence of petroleum jelly and DMSO [74]. The application of the technique to depth profile ethylene glycol after application to the volar forearm has also been reported [75].

2.4.2   Electron Spin Resonance

Paramagnetic molecules have unpaired electrons and will absorb energy at particular values of an applied magnetic field. Electron spin resonance or ESR (also known as electron paramagnetic resonance) is based on the same principles as NMR however microwave rather than radiowave frequencies are used and the spin transitions of the unpaired electrons rather than nuclei are recorded. The technique typically involves the incorporation of a paramagnetic probe molecule (usually a nitroxide free radical) into the system under investigation. Analysis of the probe's ESR spectrum provides information about the motional character of the probe in that particular environment. Gay and co-workers [76] showed that *n*-decylmethyl sulphoxide increased the degree of disorder in the lipid bilayer in the stratum corneum *in vitro* using 5-doxyl stearic acid as the probe. The diffusion of oxygen through human stratum corneum *in vitro* was reported by Hatcher and Plachy [77]. Permeation of oxygen increased by 200% after application of dimethyl sulphoxide (DMSO) and by 100% after application of decylmethyl sulphoxide (DEMSO) but was unaffected by oleic acid. Decreased mobility of the ESR probe 5-doxyl stearic acid in human stratum corneum was reported after treatment with sodium lauryl sulphate; this was suggested to reflect removal of fluid lipids from the stratum corneum [78]. ESR techniques were used for the determination of localisation and distribution of the spin label 3-carboxy-2,2,5,5-tetramethyl-1-pyrrolidinyloxy (PCA) when incorporated in flexible vesicles [79]. PCA penetration was enhanced 2.5-fold for core multishell nanotransporters and 1.9-fold for invasomes compared to PCA solution. Investigation of penetration depth by step-wise removal of the stratum corneum by tape stripping revealed deepest PCA penetration for invasomes.

2.4.3   Impedance Spectroscopy

Impedance spectroscopy (IS) is based on the analysis of the impedance (resistance of alternating current) of an observed system subject to the applied frequency and exciting signal. Kalia and Guy characterised the effects of iontophoresis on human skin impedance *in vivo* and also showed that IS and transepidermal water loss are complementary techniques for measuring skin permeability *in vivo* [80]. IS was used to investigate the residence time of local anaesthetic formulations in skin *in vivo* by Woolfson *et al.* [81]. Formulation effects on, and uptake into, the SC were evaluated with IS by Curdy *et al.* Lipophilic formulations increased impedance values consistent with the uptake of hydrocarbon ointment base components into the skin which was also substantiated by TEWL and ATR-FTIR measurements [82]. Bunge and co-workers recently reported the application of IS to characterise mechanical and chemical damage to cadaver skin [83, 84]. The application of IS for diagnosis of melanoma has also been investigated [85].

2.4.4   Laser-Induced Breakdown Spectroscopy

Laser-induced breakdown spectroscopy (LIBS) systems typically use a neodymium-doped yttrium aluminium garnet (Nd:YAG) solid-state laser and a spectrometer with a wide spectral range, to generate energy in the near infrared region of the electromagnetic spectrum, with a wavelength of 1064 nm. Excimer (Excited dimer)-type lasers have also been used which generate energy in the visible and ultraviolet regions. Sun and co-workers demonstrated the feasibility of using LIBS to analyse trace elemental concentrations of zinc in the stratum corneum of human skin *in vivo* after application of zinc chloride solution and zinc oxide paste. In a related study [86, 87], LIBS was used to show the effectiveness of various commercial barrier creams to zinc absorption *in vivo*.

2.4.5 Photoacoustic Spectroscopy

The basic principles of PAS involve the irradiation of a sample with light and the measurement of the induced heat response from the material. This technique allows the control, through several parameters, of the depth of detection inside the material. The maximum depth is given by the optical absorption coefficient, and the thermal properties of the medium (heat propagation). The two types of PAS are distinguished by the mode of light irradiation: modulated or pulsed form. Puccetti *et al.* [88] reported the use of pulsed PAS to study the penetration of chemical and physical sunscreens in excised human skin samples. In a later study the same group used pulsed PAS to study the diffusion of emulsion formulation when applied to skin *in vivo* [89]. Discrimination between the different formulations was possible.

2.4.6 Mass Spectrometry Imaging

Performing mass spectrometric imaging experiments on biological tissue using matrix-assisted laser desorption ionisation mass spectrometry (MALDI-MS) is a technique developed by the group of Caprioli *et al.* [90]. In the most common implementation of the technique, the sample is imaged by moving it by set increments under a stationary laser. At each position, the laser is fired for a pre-selected time or number of shots and a mass spectrum is acquired. Images are obtained by plotting the spatial dimensions of *x* and *y* versus the abundance of a selected ion or ions, which is represented as a grey or colour scale. Bunch *et al.* [91] reported the use of MALDI (MALDI) quadrupole time-of-flight mass spectrometry (Q-TOFMS) to detect and image the distribution of ketoconazole. Porcine epidermal tissue was treated with a medicated shampoo containing ketoconazole. Ion images demonstrating the permeation of the applied compound into the skin were achieved by imaging a cross-sectional imprint of treated tissue; a quantitative profile of drug in skin was also obtained. Superimposing the mass spectrometric and histological images appeared to indicate drug permeation into the dermal tissue layer although the tissue was pretreated with the formulation for 1 h at 37°C. The same group reported the absorption of the tricyclic antidepressant imipramine into an artificial model of the human epidermis [92]. The presence of imipramine could be clearly discerned in treated samples by imaging the distribution of the protonated molecule in samples taken 2 and 8 h after treatment with an ethanolic solution of imipramine.

The use of MALDI-MS, MS/MS and MS imaging methods for analysing lipids within cross-sections of *ex vivo* human skin has also been reported by this group [93]. Tentative identification of lipid species has been achieved via accurate mass measurement MALDI-MS, and the identity of a number of these species via MALDI-MS/MS was also confirmed. The main lipid species detected included glycerophospholipids and sphingolipids.

2.5 Conclusions and Future

Considerable developments in the instrumentation and data analysis associated with the major spectroscopic methods have taken place in recent years. The types of features probed, at a molecular level, are the bilayer packing, molecular interactions, diffusion rates and the spatial location of actives and formulation components with the stratum corneum and below. The most significant achievements have been the transference of *in vitro* techniques to *in vivo* studies, and it is anticipated that these will continue with enhanced resolution and sensitivity. The real-time monitoring of actives and excipients in the stratum corneum is now possible, but the limitations of the various techniques for *in vivo* studies remain to be elucidated fully. Some techniques have clear potential to examine the nature of skin and to determine any abnormalities and may have applications in the diagnosis of skin disease and skin cancers.

References

[1] Puttnam, N.A. (1972) Attenuated total reflectance studies of the skin. *J. Soc. Cosmet. Chem.*, **23**, 209–226.

[2] Knutson, K., Potts, R.O., Guzek, D.B. *et al.* (1985) Macro- and molecular physical-chemical considerations in understanding drug transport in the stratum corneum. *J. Control. Release*, **2**, 67–87.

[3] Potts, R.O. and Francoeur, M.L. (1989) Biophysics of stratum corneum barrier function. *Skin Pharmacol.*, **2** (1), 51.

[4] Ongpipattanakul, B., Francoeur, M.L. and Potts, R.O. (1994) Polymorphism in stratum corneum lipids. *Biochim. Biophys. Acta*, **1190** (1), 115–122.

[5] Moore, D.J. and Rerek, M.E. (1998) Biophysics of skin barrier lipid organization. *J. Dermatol. Sci.*, **16** (Suppl. 1), S203.

[6] Moore, D.J., Rerek, M.E. and Mendelsohn, R. (1997) Lipid domains and orthorhombic phases in model stratum corneum: evidence from Fourier transform infrared spectroscopy studies. *Biochem. Biophys. Res. Commun.*, **231** (3), 797–801.

[7] Gorcea, M., Hadgraft, J., Moore, D.J. and Lane, M.E. (2013) *In vivo* barrier challenge and initial recovery in human facial skin. *Skin Res. Technol.*, **19**, e375–e382.

[8] Knutson, K., Krill, S.L. and Zhang, J. (1990) Solvent-mediated alterations of the stratum corneum. *J. Control. Release*, **11** (1–3), 93–103.

[9] Mak, V.H.W., Potts, R.O. and Guy, R.H. (1990) Oleic acid concentration and effect in human stratum corneum: non-invasive determination by attenuated total reflectance infrared spectroscopy *in vivo*. *J. Control. Release*, **12** (1), 67–75.

[10] Ongpipattanakul, B., Burnette, R.R., Potts, R.O. and Francoeur, M.L. (1991) Evidence that oleic acid exists in a separate phase within stratum corneum lipids. *Pharm. Res.*, **8** (3), 350–354.

[11] Harrison, J.E., Groundwater, P.W., Brain, K.R. and Hadgraft, J. (1996) Azone induced fluidity in human stratum corneum: a Fourier transform infrared spectroscopy investigation using the perdeuterated analogue. *J. Control. Release*, **41** (3), 283–290.

[12] Potts, R.O. and Francoeur, M.L. (1990) Lipid biophysics of water loss through the skin. *Proc. Natl. Acad. Sci. U. S. A.*, **87**, 3871–3873.

[13] Edwardson, P.A., Walker, M. and Breheny, C. (1993) Quantitative FT-IR determination of skin hydration following occlusion with hydrocolloid containing adhesive dressings. *Int. J. Pharm.*, **91** (1), 51–57.

[14] Mak, V.H., Potts, R.O. and Guy, R.H. (1991) Does hydration affect intercellular lipid organization in the stratum corneum? *Pharm. Res.*, **8** (8), 1064–1065.

[15] Bommannan, D., Potts, R.O. and Guy, R.H. (1991) Examination of the effect of ethanol on human stratum corneum in vivo using infrared spectroscopy. *J. Control. Release*, **16**, 299–304.

[16] Harrison, J.E., Watkinson, A.C., Green, D.M. *et al.* (1996) The relative effect of Azone and Transcutol on permeant diffusivity and solubility in human stratum corneum. *Pharm. Res.*, **13** (4), 542–546.

[17] Alberti, I., Kalia, Y.N., Naik, A. *et al.* (2001) Effect of ethanol and isopropyl myristate on the availability of topical terbinafine in human stratum corneum, *in vivo*. *Int. J. Pharm.*, **219** (1–2), 11–19.

[18] Alberti, I., Kalia, Y.N., Naik, A. *et al.* (2001) In vivo assessment of enhanced topical delivery of terbinafine to human stratum corneum. *J. Control. Release*, **71** (3), 319–327.

[19] Dias, M., Naik, A., Guy, R.H. *et al.* (2008) In vivo infrared spectroscopy studies of alkanol effects on human skin. *Eur. J. Pharm. Biopharm.*, **69** (3), 1171–1175.

[20] Russeau, W., Mitchell, J., Tetteh, J. *et al.* (2009) Investigation of the permeation of model formulations and a commercial ibuprofen formulation in Carbosil® and human skin using ATR-FTIR and multivariate spectral analysis. *Int. J. Pharm.*, **374** (1–2), 17–25.

[21] Boncheva, M., Tay, F.H. and Kazarian, S.G. (2008) Application of attenuated total reflection Fourier transform infrared imaging and tape-stripping to investigate the three-dimensional distribution of exogenous chemicals and the molecular organization in Stratum corneum. *J. Biomed. Opt.*, **13** (6), 064009.

[22] Andanson, J.M., Hadgraft, J. and Kazarian, S.G. (2009) In situ permeation study of drug through the stratum corneum using attenuated total reflection Fourier transform infrared spectroscopic imaging. *J. Biomed. Opt.*, **14** (3), 34011.

[23] Tetteh, J., Mader, K.T., Andanson, J.M. *et al.* (2009) Local examination of skin diffusion using FTIR spectroscopic imaging and multivariate target factor analysis. *Anal. Chim. Acta*, **642** (1–2), 246–256.

[24] Mendelsohn, R., Flach, C.R. and Moore, D.J. (2006) Determination of molecular conformation and permeation in skin via IR spectroscopy, microscopy, and imaging. *Biochim. Biophys. Acta Biomembr.*, **1758** (7), 923–933.

[25] Xiao, C., Moore, D.J., Flach, C.R. and Mendelsohn, R. (2005) Permeation of dimyristoylphosphatidylcholine into skin – structural and spatial information from IR and Raman microscopic imaging. *Vib. Spectrosc.*, **38** (1–2), 151–158.

[26] McAuley, W.J., Chavda-Sitaram, S., Mader, K.T. *et al.* (2013) The effects of esterified solvents on the diffusion of a model compound across human skin: an ATR-FTIR spectroscopic study. *Int. J. Pharm.*, **447** (1–2), 1–6.

[27] Jackson, M., Kim, K., Tetteh, J. *et al.* (1998) Cancer diagnosis by infrared spectroscopy: methodological aspects. *Proc. SPIE*, **3257**, 24–34.

[28] Janssens, M., Van Smeden, J., Gooris, G.S. *et al.* (2012) Increase in short-chain ceramides correlates with an altered lipid organization and decreased barrier function in atopic eczema patients. *J. Lipid Res.*, **53**, 2755–2766.

[29] Dreassi, E., Ceramelli, G., Fabbri, L. *et al.* (1997) Application of near-infrared reflectance spectrometry in the study of atopy. Part 1. Investigation of skin spectra. *Analyst*, **122** (8), 767–770.

[30] Bodén, I., Nilsson, D., Naredi, P. and Lindholm-Sethson, B. (2008) Characterization of healthy skin using near infrared spectroscopy and skin impedance. *Med. Biol. Eng. Comput.*, **46**, 985–995.

[31] Dreassi, E., Ceramelli, G., Mura, P. *et al.* (1997) Near-infrared reflectance spectrometry in the study of atopy. Part 2. Interactions between the skin and polyethylene glycol 400, isopropyl myristate and hydrogel. *Analyst*, **122** (8), 771–776.

[32] Egawa, M. (2009) *In vivo* simultaneous measurement of urea and water in the human *stratum corneum* by diffuse-reflectance near-infrared spectroscopy. *Skin Res. Technol.*, **15**, 195–199.

[33] Barry, B.W., Edwards, H.G.M. and Williams, A.C. (1992) Fourier-transform Raman and infrared vibrational study of human skin – assignment of spectral bands. *J. Raman Spectrosc*, **23** (11), 641–645.

[34] Williams, A.C., Barry, B.W., Edwards, H.G. and Farwell, D.W. (1993) A critical comparison of some Raman spectroscopic techniques for studies of human stratum corneum. *Pharm. Res.*, **10** (11), 1642–1647.

[35] Williams, A.C., Edwards, H.G.M. and Barry, B.W. (1995) The [`]Iceman': molecular structure of 5200-year-old skin characterised by raman spectroscopy and electron microscopy. *Biochim. Biophys. Acta Protein Struct. Mol. Enzymol.*, **1246** (1), 98–105.

[36] Caspers, P.J., Lucassen, G.W., Bruining, H.A. and Puppels, G.J. (2000) Automated depth-scanning confocal Raman microspectrometer for rapid in vivo determination of water concentration profiles in human skin. *J. Raman Spectrosc.*, **31** (8–9), 813–818.

[37] Caspers, P.J., Lucassen, G.W., Carter, E.A. *et al.* (2001) In vivo confocal Raman microspectroscopy of the skin: noninvasive determination of molecular concentration profiles. *J. Invest. Dermatol.*, **116** (3), 434–442.

[38] Egawa, M. and Kajikawa, T. (2009) Changes in the depth profile of water in the stratum corneum treated with water. *Skin Res. Technol.*, **15** (2), 242–249.

[39] Crowther, J.M., Sieg, A., Blenkiron, P. *et al.* (2008) Measuring the effects of topical moisturizers on changes in stratum corneum thickness, water gradients and hydration in vivo. *Br. J. Dermatol.*, **159** (3), 567–577.

[40] Mélot, M., Pudney, P.D., Williamson, A.M. *et al.* (2009) Studying the effectiveness of penetration enhancers to deliver retinol through the stratum corneum by in vivo confocal Raman spectroscopy. *J. Control. Release*, **138** (1), 32–39.

[41] Caspers, P.J., Williams, A.C., Carter, E.A. *et al.* (2002) Monitoring the penetration enhancer dimethyl sulfoxide in human stratum corneum in vivo by confocal Raman spectroscopy. *Pharm. Res.*, **19** (10), 1577–1580.

[42] Stamatas, G.N. and Boireau-Adamezyk, E. *Development of a Non-Invasive Optical Method for Assessment of Skin Barrier to External Penetration.* Biomedical Optics and 3-D Imaging, 28 April 2012, Miami, FL. Optical Society of America, JM3A.42.

[43] Mateus, R., Abdalghafor, H., Oliveira, G. *et al.* (2013) A new paradigm in dermatopharmacokinetics – confocal Raman spectroscopy. *Int. J. Pharm.*, **444**, 106–108.

[44] Mohammed, D., Matts, P.J., Hadgraft, J. and Lane, M.E. (2013) *In vitro–in vivo* correlation in skin permeation. *Pharm. Res.*, **31**, 394–400.

[45] Nijssen, A., Maquelin, K., Santos, L.F. *et al.* (2007) Discriminating basal cell carcinoma from perilesional skin using high wave-number Raman spectroscopy. *J. Biomed. Opt.*, **12** (3), 034004.

[46] Saar, B.G., Freudiger, C.W., Reichman, J. *et al.* (2010) Video-rate molecular imaging in vivo with stimulated Raman scattering. *Science*, **330** (6009), 1368–1370.

[47] Hanh, B.D., Neubert, R.H.H., Wartewig, S. and Lasch, J. (2001) Penetration of compounds through human stratum corneum as studied by Fourier transform infrared photoacoustic spectroscopy. *J. Control. Release*, **70** (3), 393–398.

[48] Pleitez, M.A., Lieblein, T., Bauer, A. *et al.* (2013) In vivo noninvasive monitoring of glucose concentration in human epidermis by mid-infrared pulsed photoacoustic spectroscopy. *Anal. Chem.*, **85** (2), 1013–1020.

[49] Kim, K.-H., Jheon, S. and Kim, J.-K. (2007) In vivo skin absorption dynamics of topically applied pharmaceuticals monitored by fiber-optic diffuse reflectance spectroscopy. *Spectrochim. Acta A Mol. Biomol. Spectrosc.*, **66** (3), 768–772.

[50] Kollias, N., Zonios, G. and Stamatas, G.N. (2002) Fluorescence spectroscopy of skin. *Vib. Spectrosc.*, **28** (1), 17–23.

[51] Garrison, M.D., Doh, L.M., Potts, R.O. and Abraham, W. (1994) Effect of oleic acid on human epidermis: fluorescence spectroscopic investigation. *J. Control. Release*, **31** (3), 263–269.

[52] Alvarez-Román, R., Naik, A., Kalia, Y.N. *et al.* (2004) Visualization of skin penetration using confocal laser scanning microscopy. *Eur. J. Pharm. Biopharm.*, **58** (2), 301–316.

[53] Lombardi Borgia, S., Regehly, M., Sivaramakrishnan, R. *et al.* (2005) Lipid nanoparticles for skin penetration enhancement – correlation to drug localization within the particle matrix as determined by fluorescence and parelectric spectroscopy. *J. Control. Release*, **110** (1), 151–163.

[54] Fischer, F., Gudgin Dickson, E.F. and Pottier, R.H. (2002) In vivo fluorescence imaging using two excitation and/or emission wavelengths for image contrast enhancement. *Vib. Spectrosc.*, **30** (2), 131–137.

[55] Konig, K., Ehlers, A., Stracke, F. and Riemann, I. (2006) In vivo drug screening in human skin using femtosecond laser multiphoton tomography. *Skin Pharmacol. Physiol.*, **19** (2), 78–88.

[56] Stracke, F., Weiss, B., Lehr, C.M. *et al.* (2006) Multiphoton microscopy for the investigation of dermal penetration of nanoparticle-borne drugs. *J. Invest. Dermatol.*, **126** (10), 2224–2233.

[57] Seidenari, S., Arginelli, F., Dunsby, C. *et al.* (2013) Multiphoton laser tomography and fluorescence lifetime imaging of melanoma: morphologic features and quantitative data for sensitive and specific non-invasive diagnostics. *PLoS One*, **8** (7), e70682.

[58] Packer, K.J. and Sellwood, T.C. (1978) Proton magnetic resonance studies of hydrated stratum corneum part 1. Spin-lattice and transverse relaxation. *J. Chem. Soc. (Great Br.) Faraday Trans. II*, **74**, 1579–1591.

[59] Packer, K.J. and Sellwood, T.C. (1978) Proton magnetic resonance studies of hydrated stratum corneum part 2. Self-diffusion. *J. Chem. Soc. (Great Br.) Faraday Trans. II*, **74**, 1592–1606.

[60] Foreman, M.I., Bladon, P. and Pelling, P. (1979) Proton NMR studies of human stratum corneum. *Bioeng. Skin*, **2**, 48–58.

[61] McDonald, P.J., Akhmerov, A., Backhouse, L.J. and Pitts, S. (2005) Magnetic resonance profiling of human skin in vivo using GARField magnets. *J. Pharm. Sci.*, **94** (8), 1850–1860.

[62] Abraham, W. and Downing, D.T. (1991) Deuterium NMR investigation of polymorphism in stratum corneum lipids. *Biochim. Biophys. Acta*, **1068** (2), 189–194.

[63] Abraham, W. and Downing, D.T. (1992) Lamellar structures formed by stratum corneum lipids in vitro: a deuterium nuclear magnetic resonance (NMR) study. *Pharm. Res.*, **9** (11), 1415–1421.

[64] Kitson, N., Thewalt, J., Lafleur, M. and Bloom, M. (1994) A model membrane approach to the epidermal permeability barrier. *Biochemistry*, **33** (21), 6707–6715.

[65] Dias, M., Hadgraft, J., Glover, P.M. and McDonald, P.J. (2003) Stray field magnetic resonance imaging: a preliminary study of skin hydration. *J. Phys. D-Appl. Phys.*, **36** (4), 364–368.

[66] Backhouse, L., Dias, M., Gorce, J.P. *et al.* (2004) GARField magnetic resonance profiling of the ingress of model skin-care product ingredients into human skin in vitro. *J. Pharm. Sci.*, **93** (9), 2274–2283.

[67] Ciampi, E., Van Ginkel, M., McDonald, P.J. *et al.* (2011) Dynamic *in vivo* mapping of model moisturiser ingress into human skin by GARfield MRI. *NMR Biomed.*, **24**, 135–144.

[68] Kornetka, D., Trammer, M. and Zange, J. (2012) Evaluation of a mobile NMR sensor for determining skin layers and locally estimating the T(2eff) relaxation time in the lower arm. *MAGMA*, **25** (6), 455–466.

[69] Koch, R.L., Micali, G., Burt, C.T. *et al.* (1993) Measurement of flurbiprofen absorption in vivo through human skin using ^{19}fluorine – nuclear magnetic resonance. *J. Invest. Dermatol.*, **100** (4), 594.

[70] Schwarz, J.C., Hoppel, M., Kählig, H. and Valenta, C. (2013) Application of quantitative (19) F nuclear magnetic resonance spectroscopy in tape-stripping experiments with natural microemulsions. *J. Pharm. Sci.*, **102** (8), 2699–2706.

[71] Zemtsov, A., Dixon, L. and Cameron, G. (1994) Human in vivo phosphorus 31 magnetic resonance spectroscopy of psoriasis: a noninvasive tool to monitor response to treatment and to study pathophysiology of the disease. *J. Am. Acad. Dermatol.*, **30** (6), 959–965.

[72] Imhof, R.E., Zhang, B. and Birch, D.J.S. (1994) Progress in photothermal and photoacoustic science and technology, in *Photothermal Radiometry for NDE*, vol. **II**, 2nd edn (ed A. Mandelis), PTR Prentice Hall, Upper Saddle River, pp. 185–236.

[73] Bindra, R.M.S., Imhof, R.E., Mochan, A. and Eccleston, G.M. (1994) Optothermal technique for in-vivo stratum-corneum hydration measurement. *J. Phys. IV*, **4** (C7), 465–468.

[74] Xiao, P. and Imhof, R.E. (1996) *Opto-Thermal Skin Water Concentration Gradient Measurement.* Proceedings of SPIE 2681, Laser-Tissue Interaction VII, 31, 7 May 1996, doi:10.1117/12.239588.

[75] Xiao, P., Cowen, J.A. and Imhof, R.E. (2001) In-vivo transdermal drug diffusion depth profiling – a new approach to opto-thermal signal analysis. *Anal. Sci.*, **17**, S349–S352.

[76] Gay, C.L., Murphy, T.M., Hadgraft, J. *et al.* (1989) An electron spin resonance study of skin penetration enhancers. *Int. J. Pharm.*, **49**, 39–45.

[77] Hatcher, M.E. and Plachy, W.Z. (1993) Dioxygen diffusion in the stratum corneum: an EPR spin label study. *Biochim. Biophys. Acta*, **1149** (1), 73–78.

[78] Mukherjee, S., Margosiak, M., Prowell, S. *et al.* (1994) In vitro spectroscopic study of surfactant-stratum corneum interactions. *J. Invest. Dermatol.*, **102** (4), 606.

[79] Haag, S.F., Fleige, E., Chen, M. *et al.* (2011) Skin penetration enhancement of core–multishell nanotransporters and invasomes measured by electron paramagnetic resonance spectroscopy. *Int. J. Pharm.*, **416**, 223–228.

[80] Kalia, Y.N. and Guy, R.H. (1995) The electrical characteristics of human skin in vivo. *Pharm. Res.*, **12** (11), 1605–1613.

[81] Woolfson, A.D., Moss, G.P., McCafferty, D.F. *et al.* (1999) Changes in skin A.C. impedance parameters in vivo during the percutaneous absorption of local anesthetics. *Pharm. Res.*, **16** (3), 459–462.

[82] Curdy, C., Naik, A., Kalia, Y.N. *et al.* (2004) Non-invasive assessment of the effect of formulation excipients on stratum corneum barrier function in vivo. *Int. J. Pharm.*, **271** (1–2), 251–256.

[83] White, E.A., Orazem, M.E. and Bunge, A.L. (2013) Characterization of damaged skin by impedance spectroscopy: chemical damage by dimethyl sulfoxide. *Pharm. Res.*, **30** (10), 2607–2624.

[84] White, E.A., Orazem, M.E. and Bunge, A.L. (2013) Characterization of damaged skin by impedance spectroscopy: mechanical damage. *Pharm. Res.*, **30** (8), 2036–2049.

[85] Aberg, P., Birgersson, U., Elsner, P. *et al.* (2011) Electrical impedance spectroscopy and the diagnostic accuracy for malignant melanoma. *Exp. Dermatol.*, **20** (8), 648–652.

[86] Sun, Q., Tran, M., Smith, B.W. and Winefordner, J.D. (2000) Zinc analysis in human skin by laser induced-breakdown spectroscopy. *Talanta*, **52** (2), 293–300.

[87] Sun, Q., Tran, M., Smith, B. and Winefordner, J.D. (2000) In-situ evaluation of barrier-cream performance on human skin using laser-induced breakdown spectroscopy. *Contact Dermatitis*, **43** (5), 259–263.

[88] Puccetti, G., Lahjomri, F. and Leblanc, R.M. (1997) Pulsed photoacoustic spectroscopy applied to the diffusion of sunscreen chromophores in human skin: the weakly absorbent regime. *J. Photochem. Photobiol. B*, **39** (2), 110–120.

[89] Lahjomri, F., Benamar, N., Chatri, E. and Leblanc, R.M. (2003) Study of the diffusion of some emulsions in the human skin by pulsed photoacoustic spectroscopy. *Phys. Med. Biol.*, **48** (16), 2729–2738.

[90] Caprioli, R.M., Farmer, T.B. and Gile, J. (1997) Molecular imaging of biological samples: localization of peptides and proteins using MALDI-TOFMS. *Anal. Chem.*, **69**, 4751–4760.

[91] Bunch, J., Clench, M.R. and Richards, D.S. (2004) Determination of pharmaceutical compounds in skin by imaging matrix-assisted laser desorption/ionisation mass spectrometry. *Rapid Commun. Mass Spectrom.*, **18** (24), 3051–3060.

[92] Avery, J.L., McEwen, A., Flinders, B. *et al.* (2011) Matrix-assisted laser desorption mass spectrometry imaging for the examination of imipramine absorption by Straticell-RHE-EPI/001 an artificial model of the human epidermis. *Xenobiotica*, **41** (8), 735–742.

[93] Hart, P.J., Francese, S., Claude, E. *et al.* (2011) MALDI-MS imaging of lipids in ex vivo human skin. *Anal. Bioanal. Chem.*, **401** (1), 115–125.

3

Analysis of the Native Structure of the Skin Barrier by Cryo-TEM Combined with EM-Simulation

Lars Norlén

Department of Cell and Molecular Biology (CMB), Karolinska Institute, Stockholm, Sweden
Dermatology Clinic, Karolinska University Hospital, Stockholm, Sweden

3.1 Introduction

The horny layer represents the main barrier towards delivery of drugs to and through skin. It comprises keratin-filled cells embedded in an extracellular fat matrix that is largely impermeable to both lipophilic and hydrophilic compounds.

The horny layer's molecular organisation remains an outstanding problem despite its crucial role for skin permeability.

Recently, cryo transmission electron microscopy of vitreous sections (CEMOVIS) combined with electron microscopy simulation (EM simulation) has been used to analyse the near-native molecular organisation of the extracellular fat matrix.

The fat matrix is organised in an arrangement not previously described in a biological system – stacked bilayers of fully extended ceramides with cholesterol molecules associated with the ceramide sphingoid moiety.

Efforts are now directed towards building an atomic 3D model of the horny layer including both the keratin-filled cells and the extracellular fat matrix. Such a model may contribute with atomic-level information about the horny layer's physical and chemical

Novel Delivery Systems for Transdermal and Intradermal Drug Delivery, First Edition.
Ryan F. Donnelly and Thakur Raghu Raj Singh.
© 2015 John Wiley & Sons, Ltd. Published 2015 by John Wiley & Sons, Ltd.

properties, and in the prolongation support optimisation of transdermal and intradermal drug delivery.

We here briefly review biomolecular structure determination by means of CEMOVIS combined with EM simulation as well as our current understanding of skin barrier structure and function.

3.2   Our Approach: *In Situ* Biomolecular Structure Determination in Near-Native Skin

In situ biomolecular structure determination by means of CEMOVIS combined with EM-simulation was for the first time presented in 2012 [1], and used to analyse the near-native molecular organisation of the extracellular lipid matrix of the horny layer (Figure 3.1).

The procedure involves the following four steps:

1. CEMOVIS to yield high (sub-nanometer) resolution 2D electron micrographs of the targeted biomolecules in their native state (Figure 3.2).
2. Construction of systematic series of 3D molecular (atomic) models for the targeted biomolecules' organisation (Figure 3.3d).
3. Simulation of 2D electron micrographs, at different tilt-angles and underfocusses, for each of the proposed 3D molecular models (Figure 3.3g).
4. Confrontation of the simulated 2D electron micrographs with those observed experimentally (Figure 3.4) to select the 3D model that optimises the fit between observed and simulated 2D data (Figure 3.5a–c).

At a first glance, the idea of building up an atomic 3D model of a specimen from 2D electron microscopy images seems futile. First, the atomic model is 3D, whereas imaging data is 2D. Second, resolution in the 3D model should allow for atomic modeling, whereas 2D data has a resolution limited to the molecular level. Third, 2D electron microscopy data are noisy.

The proposed approach is possible only if one can make use of additional a priori information about the specimen that leaves detectable traces in the experimental data (i.e. the 2D CEMOVIS images). The idea is to use multi-scale geometric models for macromolecular systems to account for such a priori information. The starting point is the atomic-resolution arrangement of basic molecules, given as 3D structures from X-ray crystallography. The atomic-level geometric arrangement is further folded into a mesoscopic geometric arrangement, which in turn is part of a microscopic arrangement that makes up the specimen. At each scale there is a good way to parameterise the conformational freedom for reasonable models. As an example, at the atomic level the 3D X-ray crystallographic structure corresponds to the ground state of the molecule, so alternative flexible states can be computationally calculated as low energy conformations from the ground state. In summary, the 3D model-building problem is recast to the problem of inferring the geometry parameters, each representing the arrangement at a specific scale, from experimental data.

The above inference problem is only feasible if there are reasons to believe that small changes in most of the geometry parameters result in visible contrast changes in the 2D CEMOVIS images. We use a validated and accurate simulator for electron microscopy [3] for assessing which of the geometry parameters that leave such detectable traces. In fact,

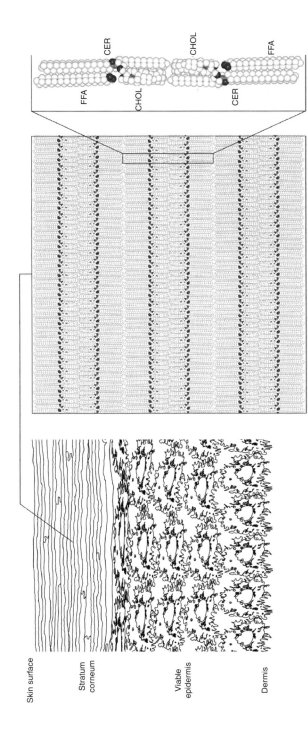

Figure 3.1 Schematic drawing of skin. Left part: schematic cellular-scale drawing of epidermis. Midpart: molecular-scale drawing of the lamellar lipid matrix occupying the space between the cells of the stratum corneum. Right part: atomic model of the lipid matrix repeating unit, composed of two mirrored subunits, each composed of one fully extended ceramide molecule (CER), one cholesterol molecule (CHOL) and one free fatty acid (FFA) molecule. Reproduced with permission from Ref. [2]. With kind permission from Springer Science and Business Media.

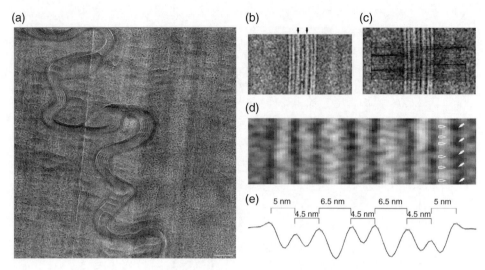

Figure 3.2 *Step 1: Cryo-electron microscopy of vitreous sections (CEMOVIS). (a) Medium magnification CEMOVIS micrograph of the interface between two cells in the midpart of stratum corneum. Note that in CEMOVIS the tissue is unstained, and that the pixel intensity is directly related to the local electron density of the sample. The stacked lamellar pattern represents the extracellular lipid matrix. Dark approximately 10 nm dots represent keratin intermediate filaments filling out the intracellular space. (b) High magnification CEMOVIS micrograph of the extracellular space in the midpart of stratum corneum. The averaged intensity profile of the lipid matrix was obtained by fuzzy distance-based image analysis. The stars in (b) represent the manually chosen start and end points for fuzzy distance based path growing. (c) The vertical line in the centre represents the traced out path. Stacked lines mark extracted intensity profiles. (d) Enlarged area of the central part of (b). (e) Reversed averaged pixel intensity profile obtained from the extracted area in (c). Peaks in (e) correspond to dark bands and valleys to lucent bands in (d). Black arrows in (b) denote electron lucent narrow bands at the centre of the 6.5 nm bands. Section thickness approximately 50 nm (a–d). Scale bar (a): 100 nm. Pixel size in (a–d): 6.02 Å. Reproduced with permission from Ref. [1]. Nature Publishing Group.*

for the corneocyte keratin filament matrix like for the horny layer extracellular fat matrix, most of these geometry parameters are detectable from 2D CEMOVIS images. The reason is the high level of order in the geometric arrangement, albeit not as ordered as in a crystal. Due to this ordering, even small changes in geometry parameters result in visible contrast changes in simulated 2D CEMOVIS images (Figure 3.5a–c).

3.2.1   Step 1: Cryo-Electron Microscopy of Vitreous Sections

In CEMOVIS the native, hydrated tissue is preserved down to the molecular level, and the image pixel intensity is directly related to the local electron density of the biomolecules themselves (Figure 3.2). The procedure involves only freezing, so there is no chemical fixation, staining and dehydration. Hence, scientists can use CEMOVIS to study the native organisation of biomolecules inside tissues at high (sub-nanometre) resolution (Figure 3.4g–i). Next,

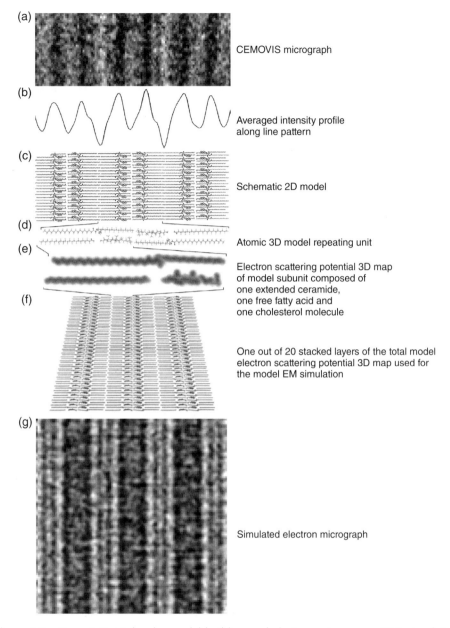

(a) CEMOVIS micrograph

(b) Averaged intensity profile along line pattern

(c) Schematic 2D model

(d) Atomic 3D model repeating unit

(e) Electron scattering potential 3D map of model subunit composed of one extended ceramide, one free fatty acid and one cholesterol molecule

(f) One out of 20 stacked layers of the total model electron scattering potential 3D map used for the model EM simulation

(g) Simulated electron micrograph

Figure 3.3 Steps 2–3: Molecular model building and electron microscopy (EM) simulation. (a) High-magnification CEMOVIS micrograph of the extracellular space in the midpart of stratum corneum. (b) Corresponding averaged intensity profile obtained by fuzzy distance based path growing (cf. Figure 2b and c). (c) Schematic 2D illustration of ceramides (tetracosanylphytosphingosine (C24:0)) in fully extended conformation with cholesterol associated with the ceramide sphingoid part and free fatty acids (lignoceric acid (C24:0)) associated with the ceramide fatty acid part. (d) Atomic 3D model of the repeating unit composed of two mirrored subunits, each composed of one fully extended ceramide molecule, one cholesterol molecule and one free fatty acid molecule. (e) Calculated electron scattering potential of one model subunit. (f) Calculated electron scattering potential 3D maps of the topmost layer out of 20 superimposed layers used to generate the simulated electron micrograph (g). Defocus (a, g): −2.5 μm. Pixel size in (a and g): 3.31 Å. Reproduced with permission from Ref. [1]. Nature Publishing Group.

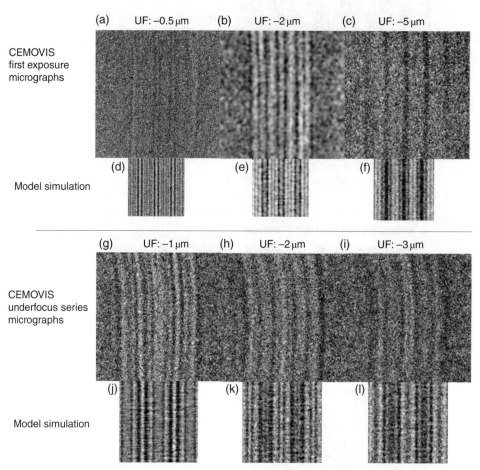

Figure 3.4 *Step 4: Confrontation of observed data with simulated data. Electron microscopy simulation of alternating fully extended ceramides with selective localisation of cholesterol to the ceramide sphingoid part. (a–c) High-magnification CEMOVIS micrographs (first exposition images) of the extracellular space in the midpart of stratum corneum obtained at −0.5 μm (a), −2 μm (b) and −5 μm (c) defocus. (d–f) represents corresponding atomic 3D model (cf. Figure 3.1, right part) electron microscopy simulation images recorded at −0.5 μm (d), −2 μm (e) and −5 μm (f) defocus. (g–i) Sequential CEMOVIS micrograph defocus-series obtained at very high magnification (1.88 Å pixel-size). Note the fine changes in interference patterns caused by gradually increasing the microscope's defocus during repeated image acquisition at a fixed position. (j–l) represents corresponding atomic 3D model (see Figure 3.1, right part) electron microscopy simulation images recorded at −1 μm (j), −2 μm (k) and −3 μm (l) defocus. It is shown that the atomic 3D model in Figure 3.1 accurately accounts not only for the major features of the CEMOVIS micrographs (a–f) but also for the interference intensity pattern changes observed upon varying the microscope's defocus during image acquisition at very high magnification (g–l). Pixel size in (c and f): 3.31 Å, in (b and e): 6.02 Å, and in (a and d and g–l): 1.88 Å. Reproduced with permission from Ref. [1]. Nature Publishing Group.*

(a)

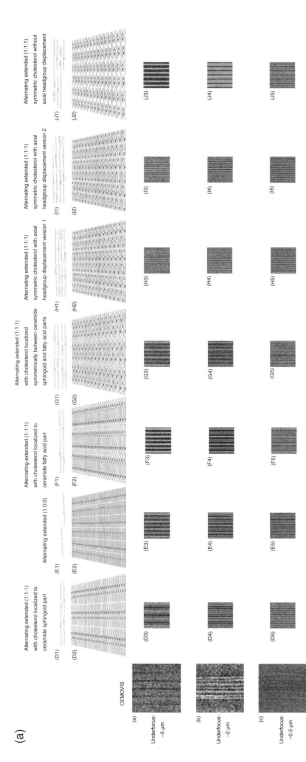

Figure 3.5 (a) Electron microscopy simulation results from seven fully extended ceramide bilayer models with varying cholesterol distribution. (A–C) CEMOVIS micrographs of the stratum corneum extracellular lipid matrix acquired at −5 µm (a), −2 µm (B) and −0.5 µm (C) defocus. (D3–J5) Corresponding simulated electron micrographs obtained from seven fully extended ceramide models. (D1–J1) Repeating units for each simulated model. (D2–J2) Calculated electron scattering potential 3D maps of the topmost layer out of 20 superimposed layers used to generate each individual simulated micrograph (D3–J5). In model (D), cholesterol is selectively localised to the ceramide sphingoid part. In model (E), cholesterol has been removed to evaluate whether the simulation method could discriminate the presence (D) or absence (E) of cholesterol. In model (F), cholesterol is selectively localised to the ceramide fatty acid part. In models (G–J), cholesterol is homogenously distributed between the ceramide sphingoid and fatty acid parts. Contrary to models (G and I), models (H and I) express axial headgroup displacement of cholesterol and free fatty acids. Models (H and I) differ in that model (H) expresses a pair-wise lateral distribution of ceramides, while model (I) expresses a homogeneous lateral distribution of ceramides. Note that except for the position of the lipid headgroups, the localisation of cholesterol within the fully extended ceramide structure largely determines the electron scattering properties of the models.

(b)

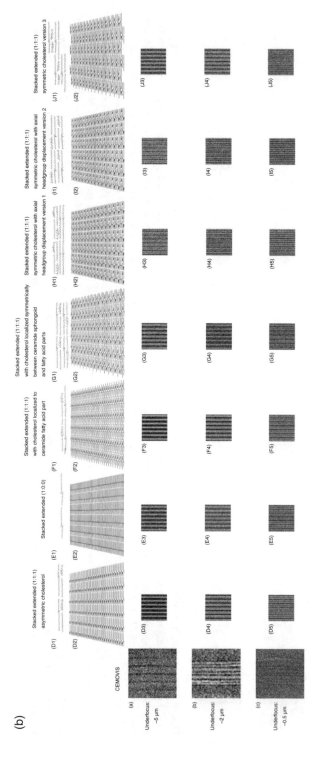

Figure 3.5 *(continued) (b) Electron microscopy simulation results from seven fully extended ceramide stacked monolayer models with varying cholesterol distribution. (A–C) CEMOVIS micrographs of the stratum corneum extracellular lipid matrix acquired at −5 μm (A), −2 μm (B) and −0.5 μm (C) defocus. (D3–J5) Corresponding simulated electron micrographs obtained from seven stacked fully extended ceramide models. (D1–J1) Two repeating units for each simulated model. (D2–J2) Calculated electron scattering potential 3D maps of the topmost layer out of 20 superimposed layers used to generate each individual simulated micrograph (D3–J5). In model (D) cholesterol is selectively localised to the ceramide sphingoid part. In model (E) cholesterol is selectively localised to the ceramide fatty acid part. In models (G–J), cholesterol is distributed homogenously between the ceramide sphingoid and fatty acid parts. Contrary to models (G and I), models (H and I) express axial headgroup displacement of cholesterol and free fatty acids. Models (H and I) differs in that model (H) expresses a pair-wise lateral distribution of ceramides while model (I) expresses a homogeneous lateral distribution of ceramides. Note that except for the position of the lipid headgroups, the localisation of cholesterol within the fully extended ceramide structure largely determines the electron scattering properties of the models.*

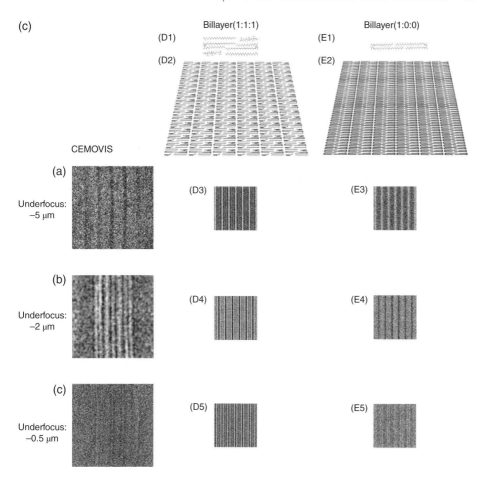

Figure 3.5 *(continued) (c) Electron microscopy simulation results from two-folded ceramide bilayer models with and without the presence of cholesterol. (A–C) CEMOVIS micrographs of the stratum corneum extracellular lipid matrix acquired at –5 μm (A), –2 μm (B) and –0.5 μm (C) defocus. (D3–E5) Corresponding simulated electron micrographs obtained from two-folded ceramide models. (D1–E1) Repeating units for each simulated model. (D2–E2) Calculated electron scattering potential 3D maps of the topmost layer out of 20 superimposed layers used to generate each individual simulated micrograph (D3–E5). In model (D), cholesterol is present. In model (E), cholesterol has been removed to ascertain if the simulation method could distinguish the presence (D) or absence (E) of cholesterol. Note that the presence of cholesterol within the folded ceramide structure largely determines the electron scattering properties of the models. Reproduced with permission from Ref. [1]. Nature Publishing Group.*

CEMOVIS data can be combined with molecular 3D model building and electron microscopy simulation; thereby enabling *in situ* 3D molecular structure determination [1].

The CEMOVIS procedure is as follows: small tissue biopsies (~1 × 1 mm) are collected and immediately (within 30 s) vitrified using a Leica EMPACT2 high-pressure freezer. The vitreous tissue samples are sectioned with a diamond knife at −140°C with a nominal section thickness of 20–30 nm. The vitreous sections are transferred to pre-cooled copper grids and inserted into the cryo-electron microscope. During the whole procedure sample temperature stays below −140°C. For more detailed accounts of CEMOVIS, see Al-Amoudi *et al.* [4] and Norlén *et al.* [5].

3.2.2   Steps 2–3: Molecular Model Building and Electron Microscopy Simulation

Our approach employs multiscale geometric models for macromolecular systems to account for *a priori* information suitable for the horny layer's extracellular fat matrix and corneocyte keratin matrix (Figure 3.3d and g). Next, simulated cryo-electron micrographs from generated molecular models are created with a simulator program recently developed by Ozan Öktem and co-workers [3]. The first part of the program is a phantom generator that can read one or more atomic models in the RCSB Protein Data Bank (PDB) format and construct a model scenario with molecules at defined positions. An electron scattering potential map is then generated with a background structure and potential corresponding to that of vitrified water. The second part of the program simulates the interaction between the potential map and the electron beam, the optical transformation effect of the lens system of the microscope and the image formation on the detector. Parameters defining optical properties of the microscope, for example acceleration voltage, aberration constants and defocus, and the point spread function of the detector, are set to mimic the conditions in the real experiment. The software is freely available at http://tem-simulator. sourceforge.net/.

For a more detailed account of the mathematical procedures employed during molecular model building and electron microscopy simulation, see Rullgård *et al.* [3] and Norlén *et al.* [6].

3.2.3   Step 4: Confrontation of Observed Data with Simulated Data

The match between the model data and the observed data is usually guided by least squares minimisation, which coincides with the maximum likelihood method when noise in data is independent and Gaussian (Figures 3.4 and 3.5a–c). The least squares method attempts to estimate those parameter values that minimise the sum of the deviations between the observed data and the simulated data. Another method for measuring image fidelity over multiple scales is the structural similarity (SSIM) index [7, 8]. The SSIM approach was originally motivated by the observation that natural image signals are highly structured. It measures the similarities of three elements of the image patches: the similarity of the local patch intensity values, the similarity of the local patch contrasts and the similarity of the local patch structures. These local similarities are expressed using simple statistics and combined together to form a single similarity measure.

The data modelling approach outlined earlier is highly successful when the number of data points is larger than the number of parameters. It is applied widely and underpins, amongst other application domains, the whole of crystallography – where the interest is similar to that at hand, that is the determination of the molecular structure and packing. X-ray or neutron diffraction or scattering data does not directly reveal atomic positions; rather the diffraction/scattering arising from a proposed molecular structure/packing is modelled, and the model structure manipulated until it best describes the data. The assumption then is that the modelled structure characterises the structure of the sample.

3.3   Molecular Organisation of the Horny Layer's Fat Matrix

The skin lipids consist of a heterogeneous mixture of saturated, long-chain ceramides (CERs), free fatty acids (FFAs) and cholesterol (CHOL) in a roughly 1:1:1 molar ratio [9]. More than 300 different species have been identified in the ceramide fraction alone [10].

The most characteristic features of the stratum corneum lipid composition [9] are (i) extensive compositional heterogeneity with broad, but invariable, chain length distributions (20–32C; peaking at 24C) in the ceramide fatty acid and FFA fractions, (ii) almost complete dominance of saturated very long hydrocarbon chains (C20:0–C32:0) and (iii) large relative amounts of cholesterol (about 30 mol%) [9].

According to our CEMOVIS defocus series/EM simulation-based structure determination [1], the skin lipids' basic molecular organisation is that of stacked bilayers of fully extended ceramides with the sphingoid moieties interfacing. Both cholesterol and the FFAs are selectively distributed: cholesterol at the ceramide sphingoid end and the FFAs at the ceramide fatty acid end (Figure 3.1).

The major physiological consequence of this molecular organisation is that the fat matrix will be largely impermeable to both hydrophilic and lipophilic substances, because of the condensed state and the presence of alternating lipophilic (alkyl chain) and hydrophilic (headgroup) regions. It will be resistant towards both hydration and dehydration because of the absence of exchangeable water between lipid leaflets. Macroscopically, the structure allows for horny layer cell cohesion and simultaneously for lateral displacement of horny layer cells to accommodate skin bending. At the molecular level the individual extended bilayers are free to slide with respect to each other, making the fat matrix pliable. The fat matrix thus meets the barrier needs of skin by being simultaneously impermeable and robust.

It is possible that the horny layer's unique fat organisation possesses unique physico-chemical properties that could be exploited for transdermal drug delivery. By studying more in detail different physicochemical properties of the fully extended ceramide bilayer system, one may perhaps propose new venues for drug delivery through the fat matrix. One approach is to use computational molecular modeling combined with *in vitro* model experiments.

3.4   Molecular Organisation of the Horny Layer's Keratin
Filament Matrix

The elucidation of the molecular organisation of the horny layer's fat matrix is but a first step towards building a complete atomic 3D model of the horny layer that also includes the corneocytes. The outstanding question in this respect is the molecular-level structure of the corneocyte keratin filament network.

The presently dominating model of horny layer keratin cytoskeletal network organisation is based on a chiral body centred cubic rod-packing (the periodic Σ+ rod packing with twisted rods) symmetry [11, 12].

The physical properties of keratin intermediate filament networks are essentially controlled by interplay between filament hydration [13] and mesoscale geometry of the filament network [12]. The balance of solvation-free energy and elasticity induces swelling of the corneocyte keratin cytoskeletal network with complete reversibility [12]. In the non-swollen state a keratin filament network packed according to the periodic Σ+ rod packing with twisted filaments in contact would occupy greater than 68% [14] of the cell volume, not counting the keratin head and tail domains that may protrude from the filament cores into the water-enriched interfilament space. For hydrophilic compounds, keratin binding is likely to be an important determinant of stratum corneum permeability [15–17].

3.5   Final Remark

Simulation is now firmly established as one of the major pillars of modern science and engineering, complementing experimental and theoretical approaches. It is routinely used in many parts of science and industry. It can have a dramatic effect on the discovery process and increase the speed with which new discoveries can be made. We believe simulation is set to play an increasingly important role in biomedical research, from understanding physiological processes and linking biomolecules' structure to function, to developing novel delivery systems for transdermal and intradermal drug delivery.

References

[1] Iwai, I., Han, H., den Hollander, L. *et al.* (2012) The human skin barrier is organized as stacked bilayers of fully-extended ceramides with cholesterol molecules associated with the ceramide sphingoid moiety. *J. Invest. Dermatol.*, **132**, 2215–2225.

[2] Norlén, L. (2012) Skin lipids, in *Encyclopedia of Biophysics*, vol. **5** (eds C. Gordon and K. Roberts), Springer-Verlag, Berlin Heidelberg, pp. 2368–2373.

[3] Rullgård, H., Öfverstedt, L.-G., Masich, S. *et al.* (2011) Simulation of transmission electron microscope images of biological specimens. *J. Microsc.*, **243** (3), 234–256.

[4] Al-Amoudi, A., Norlén, L. and Dubochet, J. (2004) Cryo-electron microscopy of vitreous sections of native biological cells and tissues. *J Struct. Biol.*, **148** (1), 131–135.

[5] Norlén, L., Öktem, O. and Skoglund, U. (2009) Molecular cryo-electron tomography of vitreous tissue sections: current challenges. *J. Microsc.*, **235**, 293–307.

[6] Norlén, L., Anwar, J. and Öktem, O. (2014) Accessing the molecular organization of the stratum corneum using high resolution electron microscopy and computer simulation, in *Computational Biophysics of the Skin* (ed B. Querleux), Pan Stanford Publishing, Singapore, pp. 289–331.

[7] Wang, Z., Bovik, A.C., Sheikh, H.R. and Simoncelli, E.P. (2004) Image quality assessment: from error visibility to structural similarity. *IEEE Trans. Image Process.*, **13** (4), 600–612.

[8] Wang, Z. and Bovik, A.C. (2009) Mean squared error: love it or leave it? A new look at signal fidelity measures. *IEEE Signal Process. Mag.*, **2009**, 98–117.

[9] Wertz, P. and Norlén, L. (2003) 'Confidence intervals' for the 'true' lipid composition of the human skin barrier, in *Skin, Hair and Nails – Structure and Function* (eds B. Forslind and M. Lindberg), Marcel Dekker, New York, pp. 85–106.

[10] Masukawa, Y., Narita, H., Sato, H. *et al.* (2009) Comprehensive quantification of ceramide species in human stratum corneum. *J. Lipid Res.*, **50**, 1708–1719.

[11] Norlén, L. and Al-Amoudi, A. (2004) Stratum corneum keratin structure, function, and formation: the cubic rod-packing and membrane templating model. *J. Invest Dermatol.*, **123** (4), 715–732.

[12] Evans, M. and Roth, R. (2014) Shaping the skin: the interplay of mesoscale geometry and corneocyte swelling. *Phys. Rev. Lett.*, **112** (038102), 1–5.

[13] Greenberg, D.A. and Fudge, D.S. (2013) Regulation of hard alpha-keratin mechanics via control of intermediate filament hydration: matrix squeeze revisited. *Proc. R. Soc. B*, **280** (1750), 20122158.

[14] Andersson, S. and O'Keeffe, M. (1977) Body-centred cubic cylinder packing and the garnet structure. *Nature*, **267**, 605–606.

[15] Barbero, A.M. and Frasch, H.F. (2006) Transcellular route of diffusion through stratum corneum: results from finite element models. *J. Pharm. Sci.*, **95** (10), 2186–2194.

[16] Hansen, S., Selzer, D., Schaeffer, U.F. and Kasting, G. (2011) An extended database of keratin binding. *J. Pharm. Sci.*, **100** (5), 1712–1726.

[17] Akinshina, A., Jambon-Puillet, E., Warren, P. and Noro, M. (2013) Self-consistent field theory for the interactions between keratin intermediate filaments. *BMC Biophys.*, **6**, 1–16.

4

Intradermal Vaccination

Marija Zaric and Adrien Kissenpfennig

The Centre for Infection & Immunity, Queen's University Belfast, Belfast, UK

4.1 Vaccination

Vaccination is one of the most effective achievements of modern medicine. A vaccine can be defined as a biological preparation used to improve immunity against a particular pathogen. Vaccination is the process of exposing the immune system to a vaccine, in order to generate immunological memory. Modern vaccination began in the eighteenth century when Edward Jenner discovered a method to successfully vaccinate against smallpox. Since then, vaccination has dramatically improved global public health. Before the era of vaccination, over 400 000 lives were claimed by smallpox every year. Eventually, in 1979, a very successful public health campaign resulted in a declaration by the United Nations and World Health Organization (WHO) that smallpox had been completely eradicated [1]. Furthermore, other life-threatening infectious diseases that were ordinary just a few generations ago have now been mostly eliminated as a result of effective vaccination programs. The WHO estimates that over 2.5 million child deaths are prevented each year worldwide due to successful vaccination.

Several types of vaccines have been successfully used in medicine:

1. Live-attenuated vaccines are designed on a principle that uses a live, closely related, but less pathogenic organism for the immunisation than the target pathogen, while preserving its antigenicity. Usually, less dangerous organism, with the common antigens as the disease causing pathogen is exposed to the immune system in order to generate immune

Novel Delivery Systems for Transdermal and Intradermal Drug Delivery, First Edition.
Ryan F. Donnelly and Thakur Raghu Raj Singh.
© 2015 John Wiley & Sons, Ltd. Published 2015 by John Wiley & Sons, Ltd.

protection. Vaccines used against viral pathogens such are viruses causing measles, rubella and mumps are examples of the live attenuated vaccines.

2. Killed or inactivated vaccines function by exposing the immune system to an organism that is killed by using heat, radiation or chemicals prior to being used for immunisation. These types of vaccines are available against viral pathogens such as influenza virus, but also against bacterial pathogens such as *Salmonella typhi* that causes typhoid.

3. In some occasions, vaccine is given against a specific toxin that is produced by a particular pathogen, rather than the target microorganism itself. In this case, a harmless toxoid is first produced by the inactivation of the pathogenic toxin and then administered as a vaccine. The tetanus toxoid vaccine is one example of these types of vaccines.

4. Subunit vaccines are designed by isolating a particular antigen from a pathogenic micro-organism that is subsequently delivered to the patient in order to generate protective immune responses. For instance, the hepatitis B vaccine is a subunit vaccine, when isolated surface proteins of the hepatitis B virus are used for immunisation to generate protective immunity.

Overall, all these agents are designed to mimic the target pathogens and activate the body's immune system upon vaccination to recognise the agent as foreign, destroy it and 'remember' it. This consequently allows the immune system to promptly and efficiently challenge disease-causing pathogens upon their subsequent encounters.

4.1.1   Disadvantages Associated with Conventional Vaccination

Undoubtedly, vaccines represent the most efficient way to protect against threatening epidemics by providing means to save lives and maintain good health and quality of life. The success of vaccination depends not only on the identification of effective vaccines but also on a technology for vaccine manufacturing, presence of safety regulations and organised approaches to vaccine delivery. Despite this legacy, vaccination campaigns face many significant challenges. First, insufficient vaccine supply or limitations of vaccine production have proven problematic in instances when mass vaccination is necessary, for example in developing countries. The Global Alliance for Vaccines and Immunization estimates that every year more than 1.5 million children (3/min) die from vaccine-preventable diseases. Also, most conventional vaccines must be maintained within specific temperature ranges to retain their potency, and therefore the associated expense of maintaining the 'cold chain' is estimated to cost vaccine programs $200–300 million annually globally [2]. In developing countries this can be a real challenge to overcome. For nearly all successful licensed vaccines natural immunity to infection has been shown as the vaccine generates the protective immune responses. However, for some pathogens, like human immunodeficiency virus (HIV), tuberculosis and malaria, it has been difficult to induce efficacious protective immunity. Protection against these pathogens requires a distinct approach to vaccine design, based on an understanding the mechanisms of their immuno-pathogenesis.

Appropriate vaccine administration is the key element to ensure successful vaccination. Typically, most vaccines are administered via the subcutaneous (s.c.) or intramuscular (i.m.) routes. Hypodermic injections are associated with pain and distress that might lead to poor patient compliance and require highly trained personnel for administration. They are also associated with a risk of disease transmission due to the possibility of needle-stick injuries or reuse of contaminated needles [3, 4].

Most vaccines in clinical use today protect by eliciting antibody responses. Antibodies provide protection by neutralising viruses, fixing complement, enabling opsonisation and phagocytosis, or promoting antibody-dependent cellular cytotoxicity [5]. However, for some infections, specific antibodies provide insufficient protection, and antigen-specific CD4+ or CD8+ T cell responses are thought to be required. However, a design of safe and efficacious subunit T cell vaccines has faced many challenges. In order to induce effective CD8+ cytotoxic T cell responses, which are particularly important in eradicating intracellular infections or cancers, activation of the cross-presentation pathway in antigen-presenting cells is required.

Killed whole organisms and subunit vaccines administered alone do not usually elicit protective immunity. For that reason, successful vaccines that elicit robust T-cell immune responses contain not only protective antigen but also an adjuvant, a component that triggers innate immune activation. Adjuvants (from the Latin *adjuvare*, meaning to help) are defined as molecules that augment the adaptive immune response and help to generate immunological memory [6]. Still, the traditional adjuvants such is Alum, commonly used in most approved vaccines, do not stimulate broad T cell responses [7].

Thus, although the successes of vaccines have been encouraging so far, various challenges remain to be addressed in the development of future vaccines.

4.2   Dendritic Cells Immunobiology

In order to increase the immunogenicity of vaccines, new vaccine platforms that can serve as improved delivery systems are being developed. Ideal platforms are thought to target antigen to dendritic cells (DCs), the professional APCs of the immune system and the most responsible for initiating the immune responses.

Dendritic cells represent the most important family of professional APCs specialised to capture, process and present particulate and soluble substances [8]. The main role of DCs is to induce specific immunity against pathogenic organisms and maintain tolerance to self-antigens [9, 10]. Anatomically, DCs can be found in lymphoid and non-lymphoid tissues. The plasmacytoid DCs (pDCs) and the conventional DCs, which have been found in both lymphoid and non-lymphoid tissues, can also be distinguished. All peripheral tissues in the steady-state are found to contain non-lymphoid tissue DCs. DCs found in the outer layer of stratified epithelia are called Langerhans cells (LCs), while DCs occupying connective tissues such as the dermis or *lamina propria* are named interstitial DCs [11]. Despite their heterogeneity, all DCs function as a link between the innate and adaptive immune systems, by processing and presenting pathogenic antigens to T and B lymphocytes. This induces activation of B and T cells, and triggers the adaptive immune responses to the specific antigen. Activation of adaptive immune system is a prerequisite for immunological memory.

The high phenotypic levels of expression of CD11c and major histocompatibility complex class II molecules (MHC-II) allows for DCs to be distinguished from other cell types [12]. Upon contact with a foreign antigen, immature DCs from the periphery uptake and process pathogenic antigen and at the same time migrate to the draining lymph nodes (LNs). DCs simultaneously rapidly mature, which involves up-regulation of major histocompatibility complex (MHC) molecules, as the antigens are presented in the context of these molecules to the antigen-specific T cells in the LNs. Simultaneously, DCs up-regulate the expression of CC-chemokine receptor 7 (CCR7) that allows the migration towards a

CC-chemokine ligand (CCL)19/(CCL)21 gradient, found in the draining LNs [13, 14]. During migration, DCs go through many other changes including loss of receptors involved in endocytosis and phagocytosis and an up-regulation of the expression of co-stimulatory molecules such as CD80, CD40 and CD86 [15, 16]. Increased secretion of cytokines also takes place, as they are required for adequate activation of the cells of the adaptive immune system, the B and T lymphocytes.

Naïve B lymphocytes have the capacity to directly bind the foreign antigens through their B cell receptor. However, to become fully activated and functional, naïve B cells need to interact with antigen peptides presented in the context of MHC-II molecules on DCs. In this way, their morphology changes to become plasma cells, which drive the humoral response through secretion of highly specific antibodies, which are involved in the neutralisation of extracellular pathogens.

In contrast to B cells, naïve T lymphocytes are unable to directly bind to the foreign antigen through their T cell receptors (TCRs), as the receptor can only recognise MHC-bound antigen. If antigenic peptide is presented by DC via MHC-II, CD4$^+$ T cells can be activated into T helper (Th) cells which secret various cytokines assisting adaptive immune responses. If peptide is presented via MHC-I molecule, antigen-specific CD8$^+$ T cells can bind and differentiate into cytotoxic T lymphocytes (CTLs). The cell-mediated response, involves recognition and lysis of infected cells by CTLs and the secretion of a wide range of cytokines by T helper cells which leads to the activation of the appropriate immune cells for efficient pathogen elimination [17].

By contrast, in tolerogenic conditions, DCs are known to induce protective regulatory T cells or cause anergy of antigen-specific T cells upon arrival to the lymphoid organs. In the steady-state setting, which represent a typical condition of the immune system, DCs continuously migrate from peripheral organs *via* the lymph to the draining LNs and present self-antigens in order to maintain peripheral tolerance [18].

4.3   Skin Anatomy and Physiology

The skin is the largest organ in the human body and is designed to carry out a wide range of functions [19, 20]. It has barrier properties to ensure that the underlying organs are protected from physical, chemical or microbial insults (Figure 4.1).

The epidermis is composed of the viable epidermis and the *stratum corneum*. The *stratum corneum* consists of layers of hexagonal-shaped, non-viable cells known as corneocytes. In most areas of the skin, there are 10 ± 30 layers of stacked corneocytes. Each corneocyte is surrounded by a protein envelope and is filled with water-retaining keratin proteins. Surrounding the cells in the extracellular space are stacked layers of lipid bilayers. The resulting structure provides the natural physical and water-retaining barrier of the skin.

The viable epidermis consists of four histologically distinct layers: the *stratum germinativum*, *stratum spinosum*, *stratum granulosum* and *stratum lucidum*. The thickness of the human epidermis varies depending on location, ranging from 60 μm on the eyelids to 800 μm on the palms [21]. The layers of the epidermis are avascular and receive nutrients by diffusion of substances from the underlying dermal capillaries. The dermis (or *corium*) resides atop the subcutaneous fat layer and is approximately 3–5 mm thick [22]. The epidermis is tightly connected to the dermis through a basement membrane.

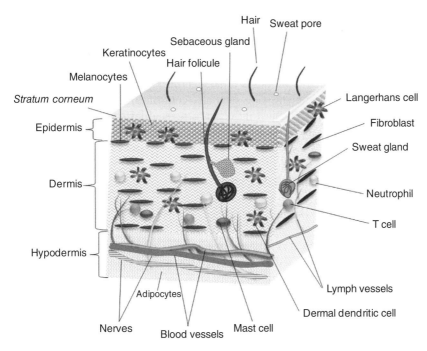

Figure 4.1 *Schematic illustration demonstrating skin structure. The skin contains three main layers – the epidermis, dermis and hypodermis. Different structures present throughout the skin including various cell populations allow for efficient barrier protection against water loss and microbial invasions. The blood and lymph vessels permit the migration of immune cells through the skin during the steady-state and under inflammatory conditions.*

The dermis is composed of several types of cells, including immune cells, but it also contains a network of collagen and elastin fibres embedded in a mucopolysaccharide matrix. Collagen provides the skin with mechanical support [23], whilst the elastic properties of the skin are associated with the presence of elastin [24]. Physiological support to the dermis is provided by a network of blood vessels, lymphatics and nerve endings [25].

The subcutaneous fat layer – sub-cutis, subdermis or hypodermis – lies between the overlying dermis, and the underlying body constituents. Its main functions are to provide physical support to the dermis and epidermis, act as a heat insulator (due to the high content of adipose tissue), and to provide nutritional support [26]. The hypodermis carries the main blood vessels and nerves to the skin, and contains sensory organs.

4.3.1 The Role of Skin in Vaccine Delivery

At present, most vaccines are applied into the subcutaneous tissue or into the muscle underneath the skin. Only few vaccines are deposited into the dermis [27], and even fewer are applied topically onto the skin, also known as transcutaneous [28, 29] or epicutaneous route [30]. Each of these routes of administration relies on the tissue-resident DCs that are able to take up the antigen, process it and present it to T lymphocytes in the draining lymphoid organs. Whereas subcutaneous fat and muscle tissue contain only few DCs, the dermis and the epidermis are populated by a rich network of different DC subsets. Consequently, antigen delivery by hypodermic injection bypasses the skin's immune cells, potentially leading to less-efficient vaccination. For this reason, the skin represents an ideal site for vaccine delivery. In fact, vaccination at this site has been shown to evoke strong immune responses at much lower doses of antigen than intramuscular vaccines [31]. The potential benefits of skin immunisation were also confirmed in a clinical trial in humans, where epidermal vaccination against influenza generated influenza-specific CD8+ T cell response, whereas classically administered intramuscular vaccine did not [32].

4.4   The Skin Dendritic Cell Network

The concept of delivery of vaccines through the skin has been gathering momentum in the past decade, largely due to the increasing recognition that a tight semi-contiguous network of DCs residing in the different skin layers is an ideal target for administration of antigenic agents. Several subsets of DCs that form a rich network of cells present in the different layers of the skin have been described in both mice and humans [33, 34].

4.4.1   Langerhans Cells and the 'Langerhans Cell Paradigm'

Langerhans cells (LC) were first described in the skin in 1868 by Paul Langerhans, and then identified in the secondary lymphoid organs one hundred years later. Only since 1978, LCs have been recognised as the APCs and therefore associated with the immune system [35].

The LC are found in the suprabasal layers of the epidermis and constitute 3–5% of all viable cells in the murine epidermis [36]. They are arranged in a network of cells, protruding their dendrites between neighbouring keratinocytes, the epithelial cells that form the epidermis [37].

Apart from the common DC markers, such as CD45, CD11c and MHCII molecules, murine LCs also express langerin [38, 39], a type II C-type lectin receptor that is involved in binding mannose and correlated sugars [40]. Langerin has also been described as essential for the formulation of the Birbeck granules [41]. Birbeck granules are the rod-shaped organelles and are believed to be a hallmark feature of LCs [37]. LCs also express adhesion molecules such are E-cadherin [42] and epithelial-cell adhesion molecule (EpCAM or gp40) [43]. Another lectin receptor, CD205 (DEC-205), involved in antigen capture and processing [44], has also been identified on the surface of LCs.

LCs migrate to the draining LNs in the steady state [45, 46], and they increase their rate of migration during inflammation [47]. After leaving the epidermis and migration through the dermal lymphatic vessels, LCs localise in the T-cell area of the skin-draining LNs [48, 49]. Migratory LCs up-regulate and redistribute MHC class II molecules on their cell surface [50, 51] and the expression of co-stimulatory molecules such as CD40, CD80, CD86 [52] and CCR7. At the same time, expression of E-cadherin becomes down-regulated, which probably facilitates LC disengagement from surrounding keratinocytes.

Ralph Steinman's work first described the functional characteristics of LCs, starting from the crucial discovery that murine epidermal LCs can develop into mature DCs that have potent immunostimulatory capacity and are able to activate naïve T cells [53].

The work that followed defined the concept of LCs as professional stimulators of T cell immune responses through the term 'LC paradigm' [54].

This concept has delineated three key functions of DCs [55]. First, immature DCs that reside in the periphery during the steady-state constantly sample the local microenvironment and are able to efficiently uptake the antigen. Second, they transport the acquired antigen to the draining LNs that is necessary to ensure interaction with antigen-specific naïve T cells. Third, during their migration, DCs undergo a process of maturation while antigen processing and presentation in the context of MHC-I/MHC-II molecules occur. Simultaneously, DCs up-regulate the expression of co-stimulatory molecules on their surface and produce cytokines which are steering for adequate T cell differentiation. Thus, by the time DCs arrive to the draining LNs, they have expressed the surface phenotype of a functionally mature DCs capable of activating naïve T cells and thereby, initiating an adaptive immune response specifically tailored to fight off antigenic threats.

However, this concept failed to explain the crucial contribution of DCs to the maintenance of peripheral T cell tolerance. Consequently, the concept was adapted such that during their low steady-state migration, immature DCs would mediate self-tolerance, but their full activation and maturation occur in response to inflammation and infection [56]. However, the growing evidence that DCs form a heterogeneous family of APCs, which are phenotypically and functionally distinct, further challenges this concept and questions the proposal that all DC populations follow the life cycle described in the 'LC paradigm' [11].

4.4.2 Dermal Dendritic Cell Network

Apart from LCs, the skin also contains a different type of DCs identified as dermal DCs (dDCs) that are distributed throughout the dermal connective tissue (Figure 4.2). Dermal DCs have been much less studied than LCs due to their poor accessibility and the lack of available specific markers which we are only now beginning to fully understand. DCs in the dermis comprise both dermal resident DCs and migratory LCs on their way to the cutaneous draining LNs [36].

Murine dermal resident DCs were first believed to form a homogenous population that could simply be distinguished from migratory LCs based on the lack of langerin expression [38]. However, several studies in mouse Langerin-reporter models showed that dermal langerin expressing cells include both migratory LCs and a new population of dDCs, known as Langerin$^+$ dDCs [57–59].

Three distinct langerin expressing DC subsets can be identified in steady-state murine skin: epidermal LCs, Langerin$^+$ CD103$^+$ and Langerin$^+$ CD103$^-$ dDCs, in addition to two Langerin$^-$ dDC subsets, characterised by differential expression of CD11b [60]. Further details about the expression of various markers by different murine skin DCs subsets are summarised in Table 4.1.

Langerin$^+$ dDCs are a distinct subset of cutaneous DCs, and are not related to LCs. LCs are radio-resistant, but Langerin$^+$ dDCs are found to be radio-sensitive. This was confirmed in bone marrow radiation chimeras when Langerin$^+$ dDCs were replaced by donor cells [58, 59], whereas LCs remained of recipient origin [61]. Using LC ablation mouse models, it was found that Langerin$^+$ dDCs repopulate the skin differently to LCs. After all langerin-expressing cells disappeared from the skin following diphtheria toxin administration, Langerin$^+$ dDCs started to repopulate the dermis and the skin-draining LNs after few days, while the epidermis was still devoid of LCs [48, 62]. In contrast to epidermal LCs that

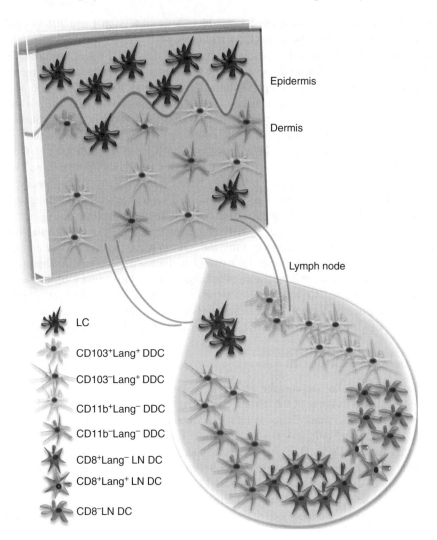

Figure 4.2 *Phenotype of dendritic cell subsets in the skin and cutaneous draining lymph node (cLNs). Based on the expression of langerin, CD11b and CD103, five separate DC subsets can be distinguished in a mouse steady-state skin. In the dermis, two langerin subsets (Langerin CD11b+ dDCs and Langerin CD11b dDCs) coexist with two langerin+ DC subsets (Langerin+CD103 dDCs and Langerin+ CD103+dDCs). The residual MHCII^{high} dermal cells correspond to migratory epidermal LCs on their route to the skin draining LNs. In addition to skin-derived migratory DCs, blood-derived Langerin CD8+ and Langerin+CD8+ DCs, as well as CD8 DC can be identified in the cLNs.*

require transforming growth factor β1 (TGFβ1) for their development [63, 64], the development of Langerin+ dDCs has been shown to depend on the cytokine Flt3 ligand (Flt3L) [65]. This was confirmed in Flt3L knock-out mice, where lack of Langerin+ dDCs was evident, while a practically unaffected population of epidermal LCs was observed [66].

Table 4.1 *Phenotype of the murine cutaneous dendritic cell subsets*

	Langerhans cells	Dermal langerin+ CD103+	Dermal langerin+ CD103-	Dermal langerin- CD11b+	Dermal langerin- CD11b-	Skin macrophages
CD45	+	+	+	+	+	+
CD11c	++	++	++	++	++	-/+
MHCII	+	+	+	+	+	-/+
Langerin	++	+	+	-	-	-
CD103	-	+	-	-	-	-
CD11b	+	-	-	+	-	++
EpCAM	+	-/+	-/+	-	-	-
F4/80	+	-	-	+	+	+

In contrast to other DCs, which are constantly replaced by a circulating pool of bone marrow-derived committed precursors, it was shown that LCs are maintained by local long-lived precursors that seed skin during embryonic development and self-renew under steady-state conditions [61]. It was recently shown that adult LCs have a dual origin [67]. During embryogenesis, the first wave of yolk sac (YS)-derived primitive myeloid progenitors seeds the skin before the onset of foetal liver hematopoiesis. However, YS-derived LC precursors are largely replaced by foetal liver monocytes during late embryogenesis. Consequently, adult LCs derive predominantly from foetal liver monocyte-derived cells with a minor contribution of YS-derived cells.

4.4.3 Dendritic Cell Subsets in the Skin-Draining Lymph Node

Three main populations of DCs can be distinguished in the skin-draining LNs based on the expression on MHCII and CD11c molecules [48]. One population, characterised by strong expression of MHCII and intermediate to high levels of CD11c (MHCII^high, CD11c^inter to high subset) corresponds to skin-derived DCs. This population contains both immigrating LCs and dermal-derived DCs subsets. A second population expressing intermediate levels of MHCII molecules and high levels of CD11c (MHCII^inter, CD11c^high subset) contains blood-derived CD8+ DCs and CD8- DCs. Finally, the third DC population in skin-draining LN expresses intermediate levels of CD11c and low surface levels of MHCII and includes both blood-derived CD11b+ DCs and pDCs.

Three subsets of langerin-expressing DCs can be distinguished in the skin-draining LNs of the mouse: epidermal LCs, Langerin+ dDCs, and blood derived CD8+ DCs. Both Langerin+ dDCs and LCs are found within CD8- fraction of DCs. However, Langerin+ dDCs express low levels of epithelial cell adhesion molecule Ep-CAM, while LCs that have migrated from the epidermis express high levels of Ep-CAM. This selective expression of the EpCAM by LCs allows for LCs to be easily distinguishable from Langerin+ dDCs in the skin and LNs. A LN-resident, blood-derived population of Langerin+ DCs has also been identified in LNs [62, 68]. These DCs can be differentiated from migratory skin-derived Langerin+ cells by their expression of CD8 and higher levels of CD11c, and by the lack of CCR7 expression. However, langerin is expressed on approximately half of CD8[+] blood-derived DCs, and it was found that the levels of langerin expressed by this subset are lower than that of the Langerin[+] skin-derived DC subset [48].

4.4.3.1   The Role of Lymph Node Resident Dendritic Cells and Plasmacytoid Dendritic Cells

Lymph node-resident DCs function by presenting blood borne or lymph borne antigens to provide either immunity or tolerance. Both CD8[+] and CD8[−] CD11b[+] DCs in the LNs have the ability to present antigens. However, CD8[+] DCs are specialised in MHC-I presentation and they appear to be the only LN-resident DC subset dedicated to cross-presentation [69]. In addition, it was shown that CD8[+] DCs phagocytised dying cells as a consequence of bacterial infection and were further involved in MHC class I presentation and cross-presentation of bacterial antigens [70, 71]. Antigen-presentation capability of CD8[−] DCs is less well understood, but they are known to express the components of MHC II presentation machinery [72]. Following Leishmania major infection, it was found that antigen presentation peaked at 24 h after infection and that activation of naïve Leishmania major-specific CD4[+] T cells in draining LNs was mediated by a population of CD8[−] CD11b[+] DCs residing in LNs [73].

Plasmacytoid DCs are described as CD11c[low]CD11b[−]B220[+]Gr1[+]PDCA1[+]SiglecH[+] and develop fully in the bone marrow before entering the blood stream in order to access multiple tissues, including LNs, during steady-state. They were also shown to accumulate in inflamed LNs *via* high endothelial venules [74], but only few pDCs recruited to the peripheral infected tissues seemed to migrate to draining LNs or did so relatively late after onset of infection [75]. The ability of pDCs to secrete large quantities of type I interferons (IFNs), consisting of IFN-α and IFN-β, immediately upon viral infection, has been recognised as a key anti-viral feature of these cells [76]. Exact mechanism of antigen presentation by pDCs has not been understood yet, but they have been described as proficient presenters of endogenous antigens [77]. However, an example of their functional insignificance has been shown in mucosal Herpes simplex virus (HSV) and flu lung infection models where depletion of pDCs did not have an effect on antigen presentation to CD4[+] and CD8[+] T cells [75, 78].

4.4.4   Human Dendritic Cells in the Skin

Similar to murine skin, several distinct DC subsets have been described in healthy human skin. The outer epidermal layer of the homeostatic human skin also contains LCs, constituting around 2% of the total epidermal cell population. Langerhans cells are the only subset in human skin that expresses langerin. The presence of human DC subsets equivalent to the murine Langerin[+] dDCs has been extensively investigated in recent years, but human Langerin[+] dDCs have not yet been identified. Dermal DCs in humans can be additionally divided into a quantitatively minor population expressing CD14 but not CD1a and a larger population characterised by high levels of CD1a but not CD14 expression [79]. Both populations, CD1a[+] and CD14[+] skin DCs have been considered as equivalents of CD11b[+] murine dDCs. Recently, another population of CD141[+] tissue cross-presenting DCs was identified [80]. The complexity of phenotypic characteristics of human skin DC subsets is described in more details in Table 4.2.

Table 4.2 *Phenotype of the human cutaneous dendritic cell subsets*

	LCs	CD1a+ DDCs	CD14+ DDCs	CD141+ DDCs	Skin macrophages
CD45	+	+	+	+	+
CD11c	+	+	+	-/+	-
Langerin	+	-	-	-	-
CD1a	+	+	-	-/+	-
CD14	-	-	+	-	+
CD141	-	-/+	-/+	+	-
E-cadherin	+	-	-	-	-

4.4.5   The Role of Skin Dendritic Cells Subsets in Transdermal Immunisation

Since diverse cutaneous DC subsets have been identified, research interest has been mainly focused on understanding the specific roles of various DC subsets in the skin. However, it has become evident that different skin DCs populations have different physiological functions *in vivo*. Several laboratories have first shown that murine LCs are especially capable of inducing cytotoxic T lymphocytes [81, 82] as opposed to dDCs. Studies in a mouse melanoma model identified both LCs and Langerin+ dDCs as necessary DC subsets in inducing anti-tumour immunity *in vivo,* as the progression of tumour growth was observed when LCs and Langerin+ dDCs were ablated from the skin [83]. In addition, Geijtenbeek and co-workers have shown that LCs are important for protection from HIV infection [84].

Not all DCs have the capacity for cross-presentation and targeting antigen to DC subsets that can induce efficient CTL responses has been a main goal in vaccine development for a long time [9]. There has been strong evidence in recent years that cross-presentation of cutaneous antigen *in vivo* is dependent on CD103+ Langerin+ dDCs or CD8+ LN-resident DCs [85–87]. However, several studies designed to test LC-mediated cross-presentation *in vitro* or *ex vivo* have confirmed their ability to efficiently present exogenously acquired antigen to antigen-specific CD8+ T cells [81, 88]. On the other hand, there has not been evidence to demonstrate LC cross-presentation *in vivo* [89]. Prior to the identification of Langerin+CD103+ dDCs, the requirement of Langerin+ DCs to induce strong CD8+ T cells proliferation and anti-tumour immunity were taken as evidence of cross-presentation by LCs [83].

Recent publications have questioned whether LCs are required for the generation of protective antigen-specific immune responses in different mouse models [90–92]. In addition, several studies with mice when langerin-expressing cells had been selectively eliminated [93] indicated that LCs in fact induced antigen-specific unresponsiveness or tolerance under homeostatic conditions [60, 94, 95]. On the other hand, Langerin+ dDCs appear to be essential for protective immunity in a number of infection and vaccination models. Recent studies in experimental cutaneous leishmaniasis have shown that LCs may have more of a regulatory role. Following the inoculation of the parasite Leishmania major, dDCs induce protective Th1 immunity after antigen presentation, whereas antigen-loaded LCs promote expansion of regulatory T cells, which prevents complete parasite clearance from the host [90]. Moreover, another published study showed that LCs induce deletion of CD4+ T cells even when highly activated by exposure to multiple strong adjuvants. This occurred when LCs were exposed to s.c. antigen or when protein antigen and adjuvant were delivered *via* topical application in aqueous cream [96]. However, it was not examined whether this observation remains true when the structural integrity of the epidermal/dermal barrier is disrupted and an inflammatory signal occurs in the skin. In other immunologic-mediated inflammatory reactions, such as hapten-induced contact hypersensitivity reaction, LCs also seemed to have regulatory function [68]; however, this remains controversial [97].

Peter Hammerl's group used studies involving gene-gun technology to deliver DNA vaccines to the cutaneous DC subsets demonstrate that skin DCs are also important for humoral responses. They identified migratory Langerin$^+$ dDCs as the subset that directly activated CD8$^+$ T cells in LNs, while Langerin$^+$ DCs were also critical for IgG1, but not IgG2a antibody induction, suggesting differential polarisation of CD4$^+$T helper cells by Langerin$^+$ or Langerin$^-$ DCs, respectively [98].

Possible functional differences between the different types of human skin DCs are essentially verified when LCs and CD14$^+$ dDCs were directly isolated from human skin. LCs were shown to be superior in cross-priming CD8$^+$ T cells, while CD14$^+$ dDCs were specialised to prime CD4$^+$ helper T cells that further activated B cells into antibody producing cells [88]. It has also been confirmed that LCs are involved in the initiation of antiviral immunity, as they efficiently stimulate naive CD8$^+$ T cells to differentiate into effector cells with high cytotoxic activity [99]. However, it has recently been observed that human skin also contains CD141$^+$ DCs, which are superior in antigen cross-presentation and functionally might correspond to mouse CD103$^+$ dDCs [80].

Thus, the contribution of different skin DC subsets in mediating the immune responses still remains controversial. Taken together, there has been progress in the characterisation of the different DC populations which may differently direct T cell responses. However, due to many controversial findings, it is still too early to draw final conclusion about specific functional roles of the various skin DC subsets.

4.5   The DTR-DT Depletion System

DCs have been found to form more heterogeneous and phenotypically complex population than it was originally thought [100]. However, it is still unclear whether this phenotypical heterogeneity of DCs impacts their functional characteristics.

In order to study functions and the development of different DC subsets, several *in vivo* depletion systems have been established [101]. One of the first methods was the injection of depleting antibodies, but this approach was proven challenging due to the lack of phenotypic markers characteristic for individual DC subsets. In order to systemically deplete DCs scientists used the injection of liposomal clodronate, which induces apoptosis by inhibiting cell metabolism. However, it was found that application of liposomal clodronate depleted not only DCs but also macrophages in the spleen and liver [102].

In the following years, research has been mainly focused on LCs due to their optimal accessibility and therefore various methods have been used to induce mobilisation of LCs in the epidermis. However, techniques such are UV-irradiation, tape stripping or topical application of sensitisers alter the immunological homeostasis in the skin and can induce inflammation. Therefore, they were used mainly to study the migration and LC precursors *in vivo* [61, 103].

The diphtheria toxin (DT) polypeptide is composed of two subunits: the B subunit, which binds to heparin-binding epidermal growth factor receptor, henceforth referred as diphtheria toxin receptor (DTR) and that is present on the surface of mammalian cells, and the subunit A, which is consequently internalised in the cell where it causes inhibition of protein biosynthesis and induces rapid cell death. The rodent homologue to the mammalian DTR is 10^5 times less sensitive to the DT subunit B attachment, which prevents internalisation of the DT subunit A causing poor responsiveness to DT [104]. Thus, genetically engineered expression of the high-affinity human DTR by a particular cell type is a powerful method of depleting a particular cell type *in vivo*. In addition, depletion in the DT-DTR model is inducible and temporal and more importantly, DT causes cell death in non-dividing, terminally differentiated cells by inducing apoptosis which does not induce pro-inflammatory immune responses [105]. Therefore, depletion using DT can be considered to occur in immunologically steady-state manner. The generation of DT-neutralising antibodies that could potentially have undesirable effect was only seen after repetitive application of DT [106].

4.5.1   Langerin-DTR Mouse Models

As described in the Section 4.4.1, LCs express the C-type lectin-like protein langerin, which is essential for the formulation of the Birbeck granules [41]. Because Birbeck granules were believed to be characteristic feature of LCs, langerin was considered as a specific marker of LCs. This allowed the generation of four different lines of transgenic mice, whose aim was to characterise the function of LCs present in the epidermis of the skin, both in the steady state and under the inflammatory conditions [93, 101]. Basically, these models allow the researcher to track LCs *in vivo*, eliminate langerin expressing cells from a mouse [48, 97] or, instead, have a mouse that never possesses LCs [68, 93].

For the purpose of *in vivo* visualisation of LCs, knock-in mice that express an enhanced green fluorescent protein (eGFP) under the control of the langerin gene were engineered. In the resulting Lang-eGFP mice, the expression of eGFP mimics the expression of the langerin, and permits clear visualisation of LCs *in vivo*, without the need for pre-labelling using antibodies [48].

In order to genetically engineer cell-ablation models, the approach based on the fact that mouse cells are 10^5 times less sensitive to diphtheria toxin as compared with human cells was used. The first group of researchers [48] knocked the high affinity human diphteria toxin receptor (hDTR) into the langerin gene locus. Specifically, the fusion protein IRES-hDTR-eGFP was introduced into the sixth exon in the 3′ untranslated region of the langerin gene. In these Langerin-DTR mice, intraperitoneal administration of the DT leads to the elimination of all langerin expressing cells. In this case, cell ablation is transient. These Lang-hDTR transgenic mice were then subsequently crossed with Lang-eGFP mice to produce F1 mice in which LCs were sensitive to DT and expressed high levels of eGFP (Figure 4.3).

The second group targeted a DTR cassette into the second exon of the langerin gene [97]. Similarly, they reported that LCs were depleted within 24 h following single intraperitoneal (i.p.) injection of DT and repopulated slowly, as patches of LCs stated to reappear at 2–4 weeks post DT injection.

The third group, however, used the alternative approach and generated a bacterial artificial chromosome-transgenic mouse by inserting DT A subunit into the 3′UTR of the

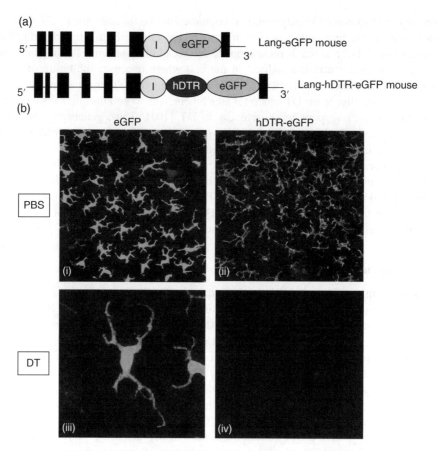

Figure 4.3   *Langerin knock-in mice (a) Schematic representation of different final recombinant langerin genes. (b) Efficient and specific ablation of LCs. Fixed epidermal sheets from Lang-eGFP mouse (i) showing Langerin expressing cells (white) that are not eliminated after DT administration (iii). Epidermal sheets form hDTR-eGFP mouse (ii) showing eGFP⁺ cells corresponding to LCs that are efficiently eliminated 24 h post DT administration (iv). All panels (confocal images) correspond to 206.8 × 206.8 × 10.4 μm. Reproduced with permission from Dr A Kissenpfennig.*

langerin gene contained within human genomic bacterial artificial chromosome DNA [68]. In this model, the absence of LCs is constitutive and persistent from birth in both the epidermis and cutaneous LNs, and it appeared that no other langerin-expressing cells were ablated, most probably due to the use of human langerin promoter.

4.6   Dendritic Cells and the Differentiation of T Lymphocytes

Dendritic cells are key players of the immune system since they orchestrate activation and proliferation of T lymphocytes resulting in appropriate adaptive immune responses. Another critical role of DCs is their ability to orientate the polarisation of the Th cells, depending on the nature and level of the danger identified. They secrete a variety of cytokines which are determined by the type of the pathogen and pattern recognition receptors that have been activated during encounter of that particular pathogen [107, 108]. Those cytokines determine the differentiation of naïve CD4+ T cells into distinct effector CD4+ T cells following interaction between TCR and MHC-II-peptide complexes.

4.6.1   CD8+ T Cell Activation

Unlike CD4+ T cells, priming of CD8+ T cells leads to the differentiation of a unique subset called cytotoxic T lymphocytes. Interleukin (IL)-2 is the major cytokine involved in complete activation of CTLs, but other cytokines such as IL-12, IFN-γ and IL-27 also play an important role [109]. Additionally, activation of Th1 cells was established as an important factor for CTL stimulation [110, 111]. The most specific feature of CTL differentiation is the development of membrane-bound cytoplasmic granules that contain proteins including perforin and granzymes, whose function is to kill target cells. In addition, CTLs are capable of secreting cytokines, mostly IFN-γ, but also lymphotoxin and tumour necrosis factor (TNF), which function to activate phagocytes and induce inflammation.

DCs are known to have superior ability to cross-present exogenous antigens. Cross-presentation is defined as the ability of DCs to deliver exogenously acquired antigens to the MHC class I molecules for specific recognition by CD8+ T cells. Using this mechanism of antigen presentation, DCs can initiate CTL immune responses against tumours or virus infected cells [112, 113]. For that reason, occurrence of antigen cross-presentation is highly desirable when designing anti-tumour vaccines. Thus, vaccine formulations able to facilitate activation of MHC class I antigen processing pathway and subsequently activate anti-tumour CTL responses are being extensively investigated.

4.6.2   CD4+ T Cell Polarisation

Several main effector CD4+ T cell subsets have been characterised. The first two effector CD4+ T cells, Th1 and Th2 cells, were identified through their distinct cytokine profiles [114, 115]. However, new lineages have recently been described, further increasing the complexity of CD4+ T cell population [116]. Specific cytokines in combination with specific transcriptional factors involved in induction of different lineages were then characterised (Figure 4.4).

IFN-γ is the hallmark cytokine of Th1-biased immune responses. T helper1 cells produce high levels of IL-2 and IFN-γ and play a major role in immunity against intracellular pathogens due to their ability to activate macrophages and CTLs. Several studies have identified IL-12 as the key initiating factor for Th1 differentiation [117, 118]. Another cytokine, IL-18, was also found to regulate Th1 development [119], but further studies revealed that both IL-12 and IL-18 synergise to prime Th1 responses [120]. Finally, IFN-γ itself is able to prime Th1 responses in autocrine positive feedback loop [121]. Therefore, IL-12 seems to be important for initiation of IFN-γ secretion, which then can further sustain its own levels. It has been found that following activation, Th1 cells up-regulate the transcription factor T-bet [122], which was later confirmed as a master regulator of Th1 cell differentiation [123]. Production of IFN-γ by Th1 cells promotes clearance of intracellular pathogens in several ways. IFN-γ has an effect in up-regulation of various genes

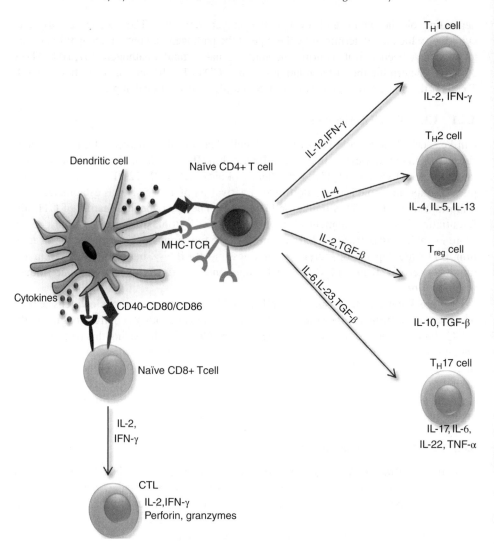

Figure 4.4   *Dendritic cell – T cell activation and polarisation. Naïve CD8⁺ T cells differentiate into effector CTLs, while naïve CD4⁺ T cells, once activated, can differentiate into various CD4⁺ T helper subsets. This differentiation depends on the presence of cytokines implicated in induction of important transcription factors leading to formation of distinct subsets. The different CD4⁺ T cell subsets produce various cytokines and perform effector functions accordingly. Main populations of CD4⁺ T cells, the factors inducing them and their effector profiles are shown.*

involved in the expression and loading of MHC-I and MHC-II molecules, which can favour the activation of CTLs, that consequently can kill infected cells by cytotoxicity [124]. IFN-γ can directly inhibit viral replication through induction of protein kinase R, whose activation leads to the inhibition of protein synthesis, and therefore blocks viral replication

[125]. IFN-γ has been found to also promote macrophage anti-microbial activity and stimulate complement activation [126].

T helper2 phenotype was found to be dependent on the presence of IL-4 [127, 128] and GATA-3 was identified as a master regulator of Th2 differentiation, as it was found to directly control the transcription of Th2 cytokines, while blocking Th1 polarisation [129]. Th2 cells are the major modulator of humoral responses, since they produce IL-4, IL-5 and IL-13 which drive B cells maturation, activation and proliferation [130]. Th2 cytokines also play an important role in promoting eosinophilia and mast cells function [130]. For that reason, the importance of Th2 subset has been recognised in allergic responses.

The recently described lineage of Th17 cells is characterised by the production of pro-inflammatory cytokines IL-17, IL-17F and IL-22, which are known for induction of highly inflammatory responses, often involved in autoimmunity [131, 132]. Nevertheless, they also contribute to the host protection, as the strong responses they promote are required to eliminate certain extracellular pathogens resistant to Th1 and Th2 responses. Th17 cells depend on IL-6, TGF-β and IL-23 for their development and survival, and transcriptional factor RORγt is shown to drive their development [133]. Th17 cells are abundant at mucosal surfaces where they have been shown to help in preventing bacterial and fungal infections [134, 135]. Th17 cytokines are known to promote the secretion of pro-inflammatory cytokines such as IL-6, IL1-β, TNF-β and are also involved in the requirement and proliferation of neutrophils [133].

Regulatory T cells (Tregs) can be defined as the main regulators of immune homeostasis capable of providing extensive protective function. They are found to constitute 5–10% of CD4+ T cells in the periphery. Tregs can be identified by their surface expression of CD25 (IL-2 receptor α-chain) and intra-nuclear forkhead box P3 (Foxp3). They prevent the development of immune reactions directed against "self", modulate the strength of immune responses at the mucosal surfaces to prevent commensal bacteria elimination and can also arrest the immune responses after successful elimination of pathogen to protect the host from extensive inflammation. For that reason, their disregulation is implicated in autoimmune and atopic diseases [136]. Tregs can be categorised into two subsets: natural Tregs, that undergo normal thymic development and are characterised as CD4+CD25+Foxp3+ cells and inducible Tregs (iTregs), that are known to be induced from naïve CD4+CD25−Foxp3− cells in the periphery, under the influence of transforming growth factor β (TGF-β) and IL-2 [137, 138]. iTregs also secrete TGF-β and IL-10, cytokines necessary to mediate their suppressive action [139]. IL-10 is known to induce anergy of T lymphocytes [140] and TGF-β inhibits Th1 and Th2 differentiation and stimulates Foxp3 expression [141].

Another CD4+ T cell subset that aids B cells in specific antibody production are follicular T helper cells (Tfh). However, the presence of Tfh cells, as separate lineage cells is controversial [142, 143]. Recently two other subsets called Th9 and Th22 have also been proposed although their molecular and pathophysiological roles are not clear [144, 145]. In addition, plasticity of T cells also leads to formation of new subsets as Tr1 or Th3. Tr1 cells are known to produce IL-10 under TGF-β and IL-27 cytokine environment, which make them a key subset involved in immunosuppression [146]. Th3 is a subset of CD4+ T cells that secrete TGF-β, and it is likely that most of Th3 cells actually arise from Foxp3+ iTreg [147].

4.7 Summary

Vaccination development remains an important field in both research and pharma, whereby in addition to extending the spectrum of antigens for novel vaccines, developing improved administration strategies to ameliorate vaccine efficacy remains a challenge. The concept of delivery of vaccines through the skin has been gathering momentum in the past decade, largely due to the increasing recognition that a tight semi-contiguous network of immunoregulatory cells reside in the different skin layers with the potential to result in more efficacious vaccines. Dendritic cells, macrophages and neutrophilic granulocytes are the principal phagocytes in the skin, while numerous cells of the adaptive immune system, such are CD8+ T cells and different types of CD4+ T cells can be found in normal skin. DCs play an essential role in regulating the immune responses, and recent research in skin DC immunobiology and the development of new technologies for manipulating the function of these cells highlight their extensive potential for the prevention and treatment of various infectious diseases and cancer. However, DCs are heterogeneous, composed of phenotypically and functionally diverse subsets and for that reason the outcome of immune responses might be determined by the subset of DCs involved. The skin harbours a wide network of these cells, and for that reason is recognised as an attractive target for immunisation. Harnessing the potency of skin DCs as the key regulators of immune responses might play a critical role in the design of modern intradermal vaccines.

References

[1] Aronson, S.M. (2012) Two stealth triumphs in global health. *Med. Health*, **95** (2), 35.

[2] Miller, M.A. and Pisani, E. (1999) The cost of unsafe injections. *Bull. World Health Organ.*, **77** (10), 808–811.

[3] Hegde, N.R., Kaveri, S.V. and Bayry, J. (2011) Recent advances in the administration of vaccines for infectious diseases: microneedles as painless delivery devices for mass vaccination. *Drug Discov. Today*, **16** (23–24), 1061–1068.

[4] Koutsonanos, D.G., del Pilar Martin, M., Zarnitsyn, V.G. *et al.* (2009) Transdermal influenza immunization with vaccine-coated microneedle arrays. *PLoS One*, **4** (3), e4773.

[5] Pulendran, B. and Ahmed, R. (2011) Immunological mechanisms of vaccination. *Nat. Immunol.*, **12** (6), 509–517.

[6] McKee, A.S., Munks, M.W. and Marrack, P. (2007) How do adjuvants work? Important considerations for new generation adjuvants. *Immunity*, **27** (5), 687–690.

[7] Coffman, R.L., Sher, A. and Seder, R.A. (2010) Vaccine adjuvants: putting innate immunity to work. *Immunity*, **33** (4), 492–503.

[8] Steinman, R.M. (2007) Lasker basic medical research award. Dendritic cells: versatile controllers of the immune system. *Nat. Med.*, **13** (10), 1155–1159.

[9] Banchereau, J., Briere, F., Caux, C. *et al.* (2000) Immunobiology of dendritic cells. *Annu. Rev. Immunol.*, **18**, 767–811.

[10] Steinman, R.M., Hawiger, D. and Nussenzweig, M.C. (2003) Tolerogenic dendritic cells. *Annu. Rev. Immunol.*, **21**, 685–711.

[11] Shortman, K. and Naik, S.H. (2007) Steady-state and inflammatory dendritic-cell development. *Nat. Rev. Immunol.*, **7** (1), 19–30.

[12] De Panfilis, G., Soligo, D., Manara, G.C. *et al.* (1989) Adhesion molecules on the plasma membrane of epidermal cells. I. Human resting langerhans cells express two members of the adherence-promoting CD11/CD18 family, namely, H-Mac-1 (CD11b/CD18) and gp 150,95 (CD11c/CD18). *J. Invest. Dermatol.*, **93** (1), 60–69.

[13] Banchereau, J. and Steinman, R.M. (1998) Dendritic cells and the control of immunity. *Nature*, **392** (6673), 245–252.

[14] Ohl, L., Mohaupt, M., Czeloth, N. *et al.* (2004) CCR7 governs skin dendritic cell migration under inflammatory and steady-state conditions. *Immunity*, **21** (2), 279–288.

[15] Sparwasser, T., Koch, E.S., Vabulas, R.M. *et al.* (1998) Bacterial DNA and immunostimulatory CpG oligonucleotides trigger maturation and activation of murine dendritic cells. *Eur. J. Immunol.*, **28** (6), 2045–2054.

[16] Young, J.W., Koulova, L., Soergel, S.A. *et al.* (1992) The B7/BB1 antigen provides one of several costimulatory signals for the activation of CD4+ T lymphocytes by human blood dendritic cells in vitro. *J. Clin. Invest.*, **90** (1), 229–237.

[17] Medzhitov, R. (2007) Recognition of microorganisms and activation of the immune response. *Nature*, **449** (7164), 819–826.

[18] Steinman, R.M. and Hemmi, H. (2006) Dendritic cells: translating innate to adaptive immunity. *Curr. Top. Microbiol. Immunol.*, **311**, 17–58.

[19] Chuong, C.M., Nickoloff, B.J., Elias, P.M. *et al.* (2002) What is the 'true' function of skin? *Exp. Dermatol.*, **11** (2), 159–187.

[20] Wysocki, A.B. (1999) Skin anatomy, physiology, and pathophysiology. *Nurs. Clin. North Am.*, **34** (4), 777–797.

[21] Williams, A.C. and Barry, B.W. (1992) Skin absorption enhancers. *Crit. Rev. Ther. Drug Carrier Syst.*, **9** (3–4), 305–353.

[22] Wiechers, J.W. (1989) The barrier function of the skin in relation to percutaneous absorption of drugs. *Pharm. Weekbl. Sci. Ed.*, **11** (6), 185–198.

[23] Tobin, D.J. (2006) Biochemistry of human skin – our brain on the outside. *Chem. Soc. Rev.*, **35** (1), 52–67.

[24] Asbill, C.S., El-Kattan, A.F. and Michniak, B. (2000) Enhancement of transdermal drug delivery: chemical and physical approaches. *Crit. Rev. Ther. Drug Carrier Syst.*, **17** (6), 621–658.

[25] Menon, G.K. (2002) New insights into skin structure: scratching the surface. *Adv. Drug Deliv. Rev.*, **54** (Suppl. 1), S3–S17.

[26] Siddiqui, O. (1989) Physicochemical, physiological, and mathematical considerations in optimizing percutaneous absorption of drugs. *Crit. Rev. Ther. Drug Carrier Syst.*, **6** (1), 1–38.

[27] Nicolas, J.F. and Guy, B. (2008) Intradermal, epidermal and transcutaneous vaccination: from immunology to clinical practice. *Expert Rev. Vaccines*, **7** (8), 1201–1214.

[28] Stoitzner, P., Sparber, F. and Tripp, C.H. (2010a) Langerhans cells as targets for immunotherapy against skin cancer. *Immunol. Cell Biol.*, **88** (4), 431–437.

[29] Warger, T., Schild, H. and Rechtsteiner, G. (2007) Initiation of adaptive immune responses by transcutaneous immunization. *Immunol. Lett.*, **109** (1), 13–20.

[30] Stoitzner, P., Stingl, G., Merad, M. and Romani, N. (2010b) Langerhans cells at the interface of medicine, science, and industry. *J. Invest. Dermatol.*, **130** (2), 331–335.

[31] Kenney, R.T., Yu, J., Guebre-Xabier, M. *et al.* (2004) Induction of protective immunity against lethal anthrax challenge with a patch. *J. Infect. Dis.*, **190** (4), 774–782.

[32] Combadiere, B., Vogt, A., Mahe, B. *et al.* (2010) Preferential amplification of CD8 effector-T cells after transcutaneous application of an inactivated influenza vaccine: a randomized phase I trial. *PLoS One*, **5** (5), e10818.

[33] Ginhoux, F., Ng, L.G. and Merad, M. (2012) Understanding the murine cutaneous dendritic cell network to improve intradermal vaccination strategies. *Curr. Top. Microbiol. Immunol.*, **351**, 1–24.

[34] Teunissen, M.B., Haniffa, M. and Collin, M.P. (2012) Insight into the immunobiology of human skin and functional specialization of skin dendritic cell subsets to innovate intradermal vaccination design. *Curr. Top. Microbiol. Immunol.*, **351**, 25–76.

[35] Stingl, G., Katz, S.I., Clement, L. *et al.* (1978) Immunologic functions of Ia-bearing epidermal langerhans cells. *J. Immunol.*, **121** (5), 2005–2013.

[36] Merad, M., Ginhoux, F. and Collin, M. (2008) Origin, homeostasis and function of langerhans cells and other langerin-expressing dendritic cells. *Nat. Rev. Immunol.*, **8** (12), 935–947.

[37] Romani, N., Clausen, B.E. and Stoitzner, P. (2010a) Langerhans cells and more: langerin-expressing dendritic cell subsets in the skin. *Immunol. Rev.*, **234** (1), 120–141.

[38] Takahara, K., Omatsu, Y., Yashima, Y. *et al.* (2002) Identification and expression of mouse langerin (CD207) in dendritic cells. *Int. Immunol.*, **14** (5), 433–444.

[39] Valladeau, J., Clair-Moninot, V., Dezutter-Dambuyant, C. *et al.* (2002) Identification of mouse langerin/CD207 in langerhans cells and some dendritic cells of lymphoid tissues. *J. Immunol.*, **168** (2), 782–792.

[40] Valladeau, J. and Saeland, S. (2005) Cutaneous dendritic cells. *Semin. Immunol.*, **17** (4), 273–283.

[41] Valladeau, J., Ravel, O., Dezutter-Dambuyant, C. *et al.* (2000) Langerin, a novel C-type lectin specific to langerhans cells, is an endocytic receptor that induces the formation of Birbeck granules. *Immunity*, **12** (1), 71–81.

[42] Tang, A., Amagai, M., Granger, L.G. *et al.* (1993) Adhesion of epidermal langerhans cells to keratinocytes mediated by E-cadherin. *Nature*, **361** (6407), 82–85.

[43] Borkowski, T.A., Nelson, A.J., Farr, A.G. and Udey, M.C. (1996a) Expression of gp40, the murine homologue of human epithelial cell adhesion molecule (Ep-CAM), by murine dendritic cells. *Eur. J. Immunol.*, **26** (1), 110–114.

[44] Inaba, K., Swiggard, W.J., Inaba, M. *et al.* (1995) Tissue distribution of the DEC-205 protein that is detected by the monoclonal antibody NLDC-145. I. Expression on dendritic cells and other subsets of mouse leukocytes. *Cell. Immunol.*, **163** (1), 148–156.

[45] Kelly, R.H., Balfour, B.M., Armstrong, J.A. and Griffiths, S. (1978) Functional anatomy of lymph nodes. II. Peripheral lymph-borne mononuclear cells. *Anat. Rec.*, **190** (1), 5–21.

[46] Hemmi, H., Yoshino, M., Yamazaki, H. *et al.* (2001) Skin antigens in the steady state are trafficked to regional lymph nodes by transforming growth factor-beta1-dependent cells. *Int. Immunol.*, **13** (5), 695–704.

[47] Stoitzner, P., Tripp, C.H., Douillard, P. *et al.* (2005) Migratory langerhans cells in mouse lymph nodes in steady state and inflammation. *J. Invest. Dermatol.*, **125** (1), 116–125.

[48] Kissenpfennig, A., Henri, S., Dubois, B. *et al.* (2005) Dynamics and function of langerhans cells in vivo: dermal dendritic cells colonize lymph node areas distinct from slower migrating langerhans cells. *Immunity*, **22** (5), 643–654.

[49] Randolph, G.J., Ochando, J. and Partida-Sanchez, S. (2008) Migration of dendritic cell subsets and their precursors. *Annu. Rev. Immunol.*, **26**, 293–316.

[50] Pierre, P., Turley, S.J., Gatti, E. *et al.* (1997) Developmental regulation of MHC class II transport in mouse dendritic cells. *Nature*, **388** (6644), 787–792.

[51] Larsen, C.P., Steinman, R.M., Witmer-Pack, M. *et al.* (1990) Migration and maturation of langerhans cells in skin transplants and explants. *J. Exp. Med.*, **172** (5), 1483–1493.

[52] Ruedl, C., Bachmann, M.F. and Kopf, M. (2000) The antigen dose determines T helper subset development by regulation of CD40 ligand. *Eur. J. Immunol.*, **30** (7), 2056–2064.

[53] Schuler, G. and Steinman, R.M. (1985) Murine epidermal langerhans cells mature into potent immunostimulatory dendritic cells in vitro. *J. Exp. Med.*, **161** (3), 526–546.

[54] Wilson, N.S. and Villadangos, J.A. (2004) Lymphoid organ dendritic cells: beyond the langerhans cells paradigm. *Immunol. Cell Biol.*, **82** (1), 91–98.

[55] Romani, N., Thurnher, M., Idoyaga, J. *et al.* (2010b) Targeting of antigens to skin dendritic cells: possibilities to enhance vaccine efficacy. *Immunol. Cell Biol.*, **88** (4), 424–430.

[56] Steinman, R.M. (2003) Some interfaces of dendritic cell biology. *Acta Pathol. Microbiol. Immunol. Scand.*, **111** (7–8), 675–697.

[57] Bursch, L.S., Wang, L., Igyarto, B. *et al.* (2007) Identification of a novel population of langerin+ dendritic cells. *J. Exp. Med.*, **204** (13), 3147–3156.

[58] Ginhoux, F., Collin, M.P., Bogunovic, M. *et al.* (2007) Blood-derived dermal langerin+ dendritic cells survey the skin in the steady state. *J. Exp. Med.*, **204** (13), 3133–3146.

[59] Poulin, L.F., Henri, S., de Bovis, B. *et al.* (2007) The dermis contains langerin+ dendritic cells that develop and function independently of epidermal Langerhans cells. *J. Exp. Med.*, **204** (13), 3119–3131.

[60] Henri, S., Poulin, L.F., Tamoutounour, S. *et al.* (2010a) CD207+ CD103+ dermal dendritic cells cross-present keratinocyte-derived antigens irrespective of the presence of langerhans cells. *J. Exp. Med.*, **207** (1), 189–206.

[61] Merad, M., Manz, M.G., Karsunky, H. *et al.* (2002) Langerhans cells renew in the skin throughout life under steady-state conditions. *Nat. Immunol.*, **3** (12), 1135–1141.

[62] Henri, S., Guilliams, M., Poulin, L.F. *et al.* (2010b) Disentangling the complexity of the skin dendritic cell network. *Immunol. Cell Biol.*, **88** (4), 366–375.

[63] Borkowski, T.A., Letterio, J.J., Farr, A.G. and Udey, M.C. (1996b) A role for endogenous transforming growth factor beta 1 in Langerhans cell biology: the skin of transforming growth factor beta 1 null mice is devoid of epidermal Langerhans cells. *J. Exp. Med.*, **184** (6), 2417–2422.

[64] Kaplan, D.H., Li, M.O., Jenison, M.C. *et al.* (2007) Autocrine/paracrine TGFbeta1 is required for the development of epidermal Langerhans cells. *J. Exp. Med.*, **204** (10), 2545–2552.

[65] Ginhoux, F., Liu, K., Helft, J. *et al.* (2009) The origin and development of nonlymphoid tissue CD103+ DCs. *J. Exp. Med.*, **206** (6), 3115–3130.

[66] Merad, M. and Manz, M.G. (2009) Dendritic cell homeostasis. *Blood*, **113** (15), 3418–3427.

[67] Hoeffel, G., Wang, Y., Greter, M. *et al.* (2012) Adult langerhans cells derive predominantly from embryonic fetal liver monocytes with a minor contribution of yolk sac-derived macrophages. *J. Exp. Med.*, **209** (6), 1167–1181.

[68] Kaplan, D.H., Jenison, M.C., Saeland, S. *et al.* (2005) Epidermal langerhans cell-deficient mice develop enhanced contact hypersensitivity. *Immunity*, **23** (6), 611–620.

[69] Guilliams, M., Henri, S., Tamoutounour, S. *et al.* (2010) From skin dendritic cells to a simplified classification of human and mouse dendritic cell subsets. *Eur. J. Immunol.*, **40** (8), 2089–2094.

[70] den Haan, J.M., Lehar, S.M. and Bevan, M.J. (2000) CD8(+) but not CD8(−) dendritic cells cross-prime cytotoxic T cells in vivo. *J. Exp. Med.*, **192** (12), 1685–1696.

[71] Iyoda, T., Shimoyama, S., Liu, K. *et al.* (2002) The CD8+ dendritic cell subset selectively endocytoses dying cells in culture and in vivo. *J. Exp. Med.*, **195** (10), 1289–1302.

[72] Dudziak, D., Kamphorst, A.O., Heidkamp, G.F. *et al.* (2007) Differential antigen processing by dendritic cell subsets in vivo. *Science (New York)*, **315** (5808), 107–111.

[73] Iezzi, G., Frohlich, A., Ernst, B. *et al.* (2006) Lymph node resident rather than skin-derived dendritic cells initiate specific T cell responses after Leishmania major infection. *J. Immunol.*, **177** (2), 1250–1256.

[74] Cella, M., Jarrossay, D., Facchetti, F. *et al.* (1999) Plasmacytoid monocytes migrate to inflamed lymph nodes and produce large amounts of type I interferon. *Nat. Med.*, **5** (8), 919–923.

[75] GeurtsvanKessel, C.H., Willart, M.A., van Rijt, L.S. *et al.* (2008) Clearance of influenza virus from the lung depends on migratory langerin+CD11b- but not plasmacytoid dendritic cells. *J. Exp. Med.*, **205** (7), 1621–1634.

[76] Fitzgerald-Bocarsly, P., Dai, J. and Singh, S. (2008) Plasmacytoid dendritic cells and type I IFN: 50 years of convergent history. *Cytokine Growth Factor Rev.*, **19** (1), 3–19.

[77] Sadaka, C., Marloie-Provost, M.A., Soumelis, V. and Benaroch, P. (2009) Developmental regulation of MHC II expression and transport in human plasmacytoid-derived dendritic cells. *Blood*, **113** (10), 2127–2135.

[78] Lund, J.M., Linehan, M.M., Iijima, N. and Iwasaki, A. (2006) Cutting edge: plasmacytoid dendritic cells provide innate immune protection against mucosal viral infection in situ. *J. Immunol.*, **177** (11), 7510–7514.

[79] Angel, C.E., Lala, A., Chen, C.J. *et al.* (2007) CD14+ antigen-presenting cells in human dermis are less mature than their CD1a+ counterparts. *Int. Immunol.*, **19** (11), 1271–1279.

[80] Haniffa, M., Shin, A., Bigley, V. *et al.* (2012) Human tissues contain CD141hi cross-presenting dendritic cells with functional homology to mouse CD103+ nonlymphoid dendritic cells. *Immunity*, **37** (1), 60–73.

[81] Stoitzner, P., Tripp, C.H., Eberhart, A. *et al.* (2006) Langerhans cells cross-present antigen derived from skin. *Proc. Natl. Acad. Sci. U. S. A.*, **103** (20), 7783–7788.

[82] Romani, N., Koide, S., Crowley, M. *et al.* (1989) Presentation of exogenous protein antigens by dendritic cells to T cell clones. Intact protein is presented best by immature, epidermal Langerhans cells. *J. Exp. Med.*, **169** (3), 1169–1178.

[83] Stoitzner, P., Green, L.K., Jung, J.Y. *et al.* (2008) Tumor immunotherapy by epicutaneous immunization requires langerhans cells. *J. Immunol.*, **180** (3), 1991–1998.

[84] Cunningham, A.L., Carbone, F. and Geijtenbeek, T.B. (2008) Langerhans cells and viral immunity. *Eur. J. Immunol.*, **38** (9), 2377–2385.

[85] Olvera-Gomez, I., Hamilton, S.E., Xiao, Z. *et al.* (2012) Cholera toxin activates nonconventional adjuvant pathways that induce protective CD8 T-cell responses after epicutaneous vaccination. *Proc. Natl. Acad. Sci. U. S. A.*, **109** (6), 2072–2077.

[86] Wang, L., Bursch, L.S., Kissenpfennig, A. *et al.* (2008) Langerin expressing cells promote skin immune responses under defined conditions. *J. Immunol.*, **180** (7), 4722–4727.

[87] Bobr, A., Olvera-Gomez, I., Igyarto, B.Z. *et al.* (2010) Acute ablation of langerhans cells enhances skin immune responses. *J. Immunol.*, **185** (8), 4724–4728.

[88] Klechevsky, E., Morita, R., Liu, M. *et al.* (2008) Functional specializations of human epidermal langerhans cells and CD14+ dermal dendritic cells. *Immunity*, **29** (3), 497–510.

[89] Igyarto, B.Z. and Kaplan, D.H. (2013) Antigen presentation by langerhans cells. *Curr. Opin. Immunol.*, **25** (1), 115–119.

[90] Kautz-Neu, K., Noordegraaf, M., Dinges, S. *et al.* (2011) Langerhans cells are negative regulators of the anti-Leishmania response. *J. Exp. Med.*, **208** (5), 885–891.

[91] Ritter, U., Meissner, A., Scheidig, C. and Korner, H. (2004) CD8 alpha- and Langerin-negative dendritic cells, but not Langerhans cells, act as principal antigen-presenting cells in leishmaniasis. *Eur. J. Immunol.*, **34** (6), 1542–1550.

[92] Zhao, X., Deak, E., Soderberg, K. *et al.* (2003) Vaginal submucosal dendritic cells, but not Langerhans cells, induce protective Th1 responses to herpes simplex virus-2. *J. Exp. Med.*, **197** (2), 153–162.

[93] Kaplan, D.H., Kissenpfennig, A. and Clausen, B.E. (2008) Insights into langerhans cell function from langerhans cell ablation models. *Eur. J. Immunol.*, **38** (9), 2369–2376.

[94] Idoyaga, J., Fiorese, C., Zbytnuik, L. *et al.* (2013) Specialized role of migratory dendritic cells in peripheral tolerance induction. *J. Clin. Invest.*, **123** (2), 844–854.

[95] Schwarz, A., Noordegraaf, M., Maeda, A. *et al.* (2010) Langerhans cells are required for UVR-induced immunosuppression. *J. Invest. Dermatol.*, **130** (5), 1419–1427.

[96] Shklovskaya, E., O'Sullivan, B.J., Ng, L.G. *et al.* (2011) Langerhans cells are precommitted to immune tolerance induction. *Proc. Natl. Acad. Sci. U. S. A.*, **108** (44), 18049–18054.

[97] Bennett, C.L., van Rijn, E., Jung, S. *et al.* (2005) Inducible ablation of mouse langerhans cells diminishes but fails to abrogate contact hypersensitivity. *J. Cell Biol.*, **169** (4), 569–576.

[98] Stoecklinger, A., Eticha, T.D., Mesdaghi, M. *et al.* (2011) Langerin+ dermal dendritic cells are critical for CD8+ T cell activation and IgH gamma-1 class switching in response to gene gun vaccines. *J. Immunol.*, **186** (3), 1377–1383.

[99] Van der Aar, A.M., de Groot, R., Sanchez-Hernandez, M. *et al.* (2011) Cutting edge: virus selectively primes human langerhans cells for CD70 expression promoting CD8+ T cell responses. *J. Immunol.*, **187** (7), 3488–3492.

[100] Villadangos, J.A. and Heath, W.R. (2005) Life cycle, migration and antigen presenting functions of spleen and lymph node dendritic cells: limitations of the langerhans cells paradigm. *Semin. Immunol.*, **17** (4), 262–272.

[101] Bennett, C.L. and Clausen, B.E. (2007) DC ablation in mice: promises, pitfalls, and challenges. *Trends Immunol.*, **28** (12), 525–531.

[102] Zhang, Y., Shlomchik, W.D., Joe, G. *et al.* (2002) "APCs in the liver and spleen recruit activated allogeneic CD8+ T cells to elicit hepatic graft-versus-host disease. *J. Immunol.*, **169** (12), 7111–7118.

[103] Holzmann, S., Tripp, C.H., Schmuth, M. *et al.* (2004) A model system using tape stripping for characterization of langerhans cell-precursors in vivo. *J. Invest. Dermatol.*, **122** (5), 1165–1174.

[104] Saito, M., Iwawaki, T., Taya, C. *et al.* (2001) Diphtheria toxin receptor-mediated conditional and targeted cell ablation in transgenic mice. *Nat. Biotechnol.*, **19** (8), 746–750.

[105] Thorburn, J., Frankel, A.E. and Thorburn, A. (2003) Apoptosis by leukemia cell-targeted diphtheria toxin occurs via receptor-independent activation of Fas-associated death domain protein. *Clin. Cancer Res. Off. J. Am. Assoc. Cancer Res.*, **9** (2), 861–865.

[106] Buch, T., Heppner, F.L., Tertilt, C. *et al.* (2005) A Cre-inducible diphtheria toxin receptor mediates cell lineage ablation after toxin administration. *Nat. Methods*, **2** (6), 419–426.

[107] Kapsenberg, M.L. (2003) Dendritic-cell control of pathogen-driven T-cell polarization. *Nat. Rev. Immunol.*, **3** (12), 984–993.

[108] Coquerelle, C. and Moser, M. (2010) DC subsets in positive and negative regulation of immunity. *Immunol. Rev.*, **234** (1), 317–334.

[109] Morishima, N., Mizoguchi, I., Okumura, M. *et al.* (2010) A pivotal role for interleukin-27 in CD8+ T cell functions and generation of cytotoxic T lymphocytes. *J. Biomed. Biotechnol.*, **2010**, 605483.

[110] Bennett, S.R., Carbone, F.R., Karamalis, F. *et al.* (1997) Induction of a CD8+ cytotoxic T lymphocyte response by cross-priming requires cognate CD4+ T cell help. *J. Exp. Med.*, **186** (1), 65–70.

[111] Schoenberger, S.P., Toes, R.E., van der Voort, E.I. *et al.* (1998) T-cell help for cytotoxic T lymphocytes is mediated by CD40-CD40L interactions. *Nature*, **393** (6684), 480–483.

[112] Kurts, C. (2000) Cross-presentation: inducing CD8 T cell immunity and tolerance. *J. Mol. Med.*, **78** (6), 326–332.

[113] Heath, W.R. and Carbone, F.R. (2001) Cross-presentation, dendritic cells, tolerance and immunity. *Annu. Rev. Immunol.*, **19**, 47–64.

[114] Killar, L., MacDonald, G., West, J. *et al.* (1987) Cloned, Ia-restricted T cells that do not produce interleukin 4(IL 4)/B cell stimulatory factor 1(BSF-1) fail to help antigen-specific B cells. *J. Immunol.*, **138** (6), 1674–1679.

[115] Mosmann, T.R., Cherwinski, H., Bond, M.W. *et al.* (1986) Two types of murine helper T cell clone. I. Definition according to profiles of lymphokine activities and secreted proteins. *J. Immunol.*, **136** (7), 2348–2357.

[116] Zhu, J., Yamane, H. and Paul, W.E. (2010) Differentiation of effector CD4 T cell populations (*). *Annu. Rev. Immunol.*, **28**, 445–489.

[117] Hsieh, C.S., Macatonia, S.E., Tripp, C.S. *et al.* (1993) Development of TH1 CD4+ T cells through IL-12 produced by Listeria-induced macrophages. *Science*, **260** (5107), 547–549.

[118] Mattner, F., Magram, J., Ferrante, J. *et al.* (1996) Genetically resistant mice lacking interleukin-12 are susceptible to infection with Leishmania major and mount a polarized Th2 cell response. *Eur. J. Immunol.*, **26** (7), 1553–1559.

[119] Takeda, K., Tsutsui, H., Yoshimoto, T. *et al.* (1998) Defective NK cell activity and Th1 response in IL-18-deficient mice. *Immunity*, **8** (3), 383–390.

[120] Yoshimoto, T., Takeda, K., Tanaka, T. *et al.* (1998) IL-12 up-regulates IL-18 receptor expression on T cells, Th1 cells, and B cells: synergism with IL-18 for IFN-gamma production. *J. Immunol.*, **161** (7), 3400–3407.

[121] Bradley, L.M., Dalton, D.K. and Croft, M. (1996) A direct role for IFN-gamma in regulation of Th1 cell development. *J. Immunol.*, **157** (4), 1350–1358.

[122] Glimcher, L.H. and Murphy, K.M. (2000) Lineage commitment in the immune system: the T helper lymphocyte grows up. *Genes Dev.*, **14** (14), 1693–1711.

[123] Szabo, S.J., Sullivan, B.M., Stemmann, C. *et al.* (2002) Distinct effects of T-bet in TH1 lineage commitment and IFN-gamma production in CD4 and CD8 T cells. *Science*, **295** (5553), 338–342.

[124] Schroder, K., Hertzog, P.J., Ravasi, T. and Hume, D.A. (2004) Interferon-gamma: an overview of signals, mechanisms and functions. *J. Leukoc. Biol.*, **75** (2), 163–189.

[125] Hovanessian, A.G. and Galabru, J. (1987) The double-stranded RNA-dependent protein kinase is also activated by heparin. *Eur. J. Biochem. FEBS*, **167** (3), 467–473.

[126] Strunk, R.C., Cole, F.S., Perlmutter, D.H. and Colten, H.R. (1985) gamma-Interferon increases expression of class III complement genes C2 and factor B in human monocytes and in murine fibroblasts transfected with human C2 and factor B genes. *J. Biol. Chem.*, **260** (28), 15280–15285.

[127] Swain, S.L., Weinberg, A.D., English, M. and Huston, G. (1990) IL-4 directs the development of Th2-like helper effectors. *J. Immunol.*, **145** (11), 3796–3806.

[128] Kopf, M., Le Gros, G., Bachmann, M. *et al.* (1993) Disruption of the murine IL-4 gene blocks Th2 cytokine responses. *Nature*, **362** (6417), 245–248.

[129] Ouyang, W., Ranganath, S.H., Weindel, K. *et al.* (1998) Inhibition of Th1 development mediated by GATA-3 through an IL-4-independent mechanism. *Immunity*, **9** (5), 745–755.

[130] Brightling, C.E., Saha, S. and Hollins, F. (2010) Interleukin-13: prospects for new treatments. *Clin. Exp. Allergy J. Br. Soc. Allergy Clin. Immunol.*, **40** (1), 42–49.

[131] Harrington, L.E., Hatton, R.D., Mangan, P.R. *et al.* (2005) Interleukin 17-producing CD4+ effector T cells develop via a lineage distinct from the T helper type 1 and 2 lineages. *Nat. Immunol.*, **6** (11), 1123–1132.

[132] Park, H., Li, Z., Yang, X.O. *et al.* (2005) A distinct lineage of CD4 T cells regulates tissue inflammation by producing interleukin 17. *Nat. Immunol.*, **6** (11), 1133–1141.

[133] Bettelli, E., Korn, T., Oukka, M. and Kuchroo, V.K. (2008) Induction and effector functions of T(H)17 cells. *Nature*, **453** (7198), 1051–1057.

[134] Huang, W., Na, L., Fidel, P.L. and Schwarzenberger, P. (2004) Requirement of interleukin-17A for systemic anti-Candida albicans host defense in mice. *J. Infect. Dis.*, **190** (3), 624–631.

[135] Yang, X.O., Pappu, B.P., Nurieva, R. *et al.* (2008) T helper 17 lineage differentiation is programmed by orphan nuclear receptors ROR alpha and ROR gamma. *Immunity*, **28** (1), 29–39.

[136] Sakaguchi, S., Sakaguchi, N., Asano, M. *et al.* (1995) Immunologic self-tolerance maintained by activated T cells expressing IL-2 receptor alpha-chains (CD25). Breakdown of a single mechanism of self-tolerance causes various autoimmune diseases. *J. Immunol.*, **155** (3), 1151–1164.

[137] Zheng, S.G., Wang, J., Wang, P. *et al.* (2007) IL-2 is essential for TGF-beta to convert naive CD4+CD25- cells to CD25+Foxp3+ regulatory T cells and for expansion of these cells. *J. Immunol.*, **178** (4), 2018–2027.

[138] Mucida, D., Kutchukhidze, N., Erazo, A. *et al.* (2005) Oral tolerance in the absence of naturally occurring Tregs. *J. Clin. Invest.*, **115** (7), 1923–1933.

[139] DiPaolo, R.J., Brinster, C., Davidson, T.S. *et al.* (2007) Autoantigen-specific TGFbeta-induced Foxp3+ regulatory T cells prevent autoimmunity by inhibiting dendritic cells from activating autoreactive T cells. *J. Immunol.*, **179** (7), 4685–4693.

[140] Akdis, M., Verhagen, J., Taylor, A. *et al.* (2004) Immune responses in healthy and allergic individuals are characterized by a fine balance between allergen-specific T regulatory 1 and T helper 2 cells. *J. Exp. Med.*, **199** (11), 1567–1575.

[141] Marie, J.C., Letterio, J.J., Gavin, M. and Rudensky, A.Y. (2005) TGF-beta1 maintains suppressor function and Foxp3 expression in CD4+CD25+ regulatory T cells. *J. Exp. Med.*, **201** (7), 1061–1067.

[142] Fazilleau, N., Mark, L., McHeyzer-Williams, L.J. and McHeyzer-Williams, M.G. (2009) Follicular helper T cells: lineage and location. *Immunity*, **30** (3), 324–335.

[143] Yu, D., Batten, M., Mackay, C.R. and King, C. (2009) Lineage specification and heterogeneity of T follicular helper cells. *Curr. Opin. Immunol.*, **21** (6), 619–625.

[144] Dardalhon, V., Awasthi, A., Kwon, H. *et al.* (2008) IL-4 inhibits TGF-beta-induced Foxp3+ T cells and, together with TGF-beta, generates IL-9+ IL-10+ Foxp3(-) effector T cells. *Nat. Immunol.*, **9** (12), 1347–1355.

[145] Eyerich, S., Eyerich, K., Pennino, D. *et al.* (2009) Th22 cells represent a distinct human T cell subset involved in epidermal immunity and remodeling. *J. Clin. Invest.*, **119** (12), 3573–3585.

[146] Awasthi, A., Carrier, Y., Peron, J.P. *et al.* (2007) A dominant function for interleukin 27 in generating interleukin 10-producing anti-inflammatory T cells. *Nat. Immunol.*, **8** (12), 1380–1389.

[147] Gagliani, N., Ferraro, A., Roncarolo, M.G. and Battaglia, M. (2009) Autoimmune diabetic patients undergoing allogeneic islet transplantation: are we ready for a regulatory T-cell therapy? *Immunol. Lett.*, **127** (1), 1–7.

5

Film-Forming and Heated Systems

William J. McAuley and Francesco Caserta

Department of Pharmacy, University of Hertfordshire, Hatfield, UK

Two technologies that have generated considerable amounts of attention recently for improving drug delivery across the skin are formulations that dry on the skin to form a drug-containing film or patch and systems that heat up on the skin surface providing a physiologically tolerable temperature and improve drug permeation. Both of these approaches offer opportunities to increase the drug absorption of a range of different drug molecules above that of conventional transdermal/topical systems and improve patient treatment. Even though both strategies have successfully been used in products that have been brought to market, there remains considerable scope to improve understanding of how to best use them to provide optimal therapeutic outcomes. Whilst these approaches do not typically offer the potential to deliver the sorts of molecules, for example proteins that can be delivered via other technologies such as ablative systems, they are generally much cheaper to manufacture and easier for patients to use and may therefore have greater potential to be translated into successful products that improve patient treatment. The advantages of both of these drug penetration enhancement approaches and their potential for the development of transdermal/topical products are reviewed in this chapter.

5.1   Film-Forming Systems

Film-forming systems contain drug and film-forming excipients in a formulation containing volatile solvent(s). On contacting the skin, the solvent evaporates leaving behind the drug in a film of excipients on the skin surface. The residual film may take different forms

Novel Delivery Systems for Transdermal and Intradermal Drug Delivery, First Edition.
Ryan F. Donnelly and Thakur Raghu Raj Singh.
© 2015 John Wiley & Sons, Ltd. Published 2015 by John Wiley & Sons, Ltd.

and, for example, may be a solid polymeric material or a residual liquid film that is rapidly absorbed into the stratum corneum. The actual formulation types that are commonly used for this film-forming purpose are often a spray or solution, but other designs, such as a gel may potentially be used. Examples of film-forming solution products include Novartis's Lamisil Once, and Almirall's Actikerall, and several pharmaceutical companies have licensed spray technology for transdermal and topical drug delivery, indicating the feasibility of this type of approach to produce films for topical delivery purposes. Examples include Medspray (Medpharm, Guilford, UK), Metered Dose Transdermal System (MDTS) (Acrux, Melbourne, Australia), Cortispray (DVM Pharmaceuticals, Saint Joseph, MO, USA) and Cortizone Quickshot Spray (Pfizer Consumer Healthcare).

5.1.1   The Design of Film-Forming Systems

Early examples of improving drug delivery across the skin using a volatile solvent that evaporates from a formulation leaving the drug in residual solvent or film of material date back to the 1960s [1, 2]. The simplest design of these systems is a solution that in which the volatile solvent is the main formulation ingredient and is used as the carrier for the rest of the formulation. Upon application to the skin the volatile solvent evaporates rapidly leaving behind the drug and any film-forming excipients. Although simple in design, such cutaneous solutions have been shown to confer advantages over more conventional semisolid formulation types. The other common approach used to form drug-containing films on the skin surface is to use a spray. These are easier to apply to the skin in comparison to a cutaneous solution and potentially may also be able to further improve drug delivery. Sprays can be separated into two categories, pump sprays and aerosols, which relate to how the spray is actuated. Both pump and aerosol spray systems consist of a canister to hold the formulation and an actuator that allows the release of the formulation. Pump sprays use the mechanical force of pressing on the actuator to pump the spray out of the canister. In contrast, pressurised aerosol sprays use a propellant to force the formulation out of the canister. In contrast to pump-based systems which always deliver a finite amount of spray for each actuation aerosol systems can spray either continuously or in finite amounts as a 'metered dose' per actuation. Continuous sprays are useful for application to large areas but are typically unable to precisely control the applied dose. Pump spray systems often have a similar formulation composition to the cutaneous solutions described earlier, whereas the aerosols require the addition of the propellant' which is generally the major component of the formulation. Currently pump sprays appear to be used more commonly in topical medicines for application to the skin; however, the use of a propellant may have advantages in particular scenarios, for example enabling more viscous formulations to be delivered, the need for organic volatile solvents in the formulation can be reduced or avoided and potentially they can also offer improvements in drug delivery.

The volatile solvents selected for use in cutaneous solutions or pump spray systems are most commonly ethanol and/or isopropyl alcohol, both of which have regulatory approval for topical use. Water which is less volatile may also constitute a significant part of the solvent composition. The relative proportions of these solvents will affect the ability of the solvent system to dissolve the other formulation components, the drying time of the formulation and its ability to deliver a drug. These solutions usually also contain non-volatile solvent(s), which act as a carrier for the drug to ensure absorption into the skin and to prevent it from crystallising out when the volatile solvents evaporate. In addition the non-volatile solvents may be specifically selected to enhance the penetration of the drug. Polymers may be included in the formulation and may be present for different functions. For example they can thicken of a cutaneous solution which can make it easier to apply, or they may be included specifically to form a polymeric residue on the skin surface to increase the residence time of the formulation and improve drug delivery. Both the presence and lack of a residual film have been made a feature of particular spray system designs. For example Medpharm's Medspray system which is often designed to leave film on the skin delivers the drug as a 'Patch in a can®', whereas Acrux have trade marked their MTDS system, which is normally designed to be a liquid film that is rapidly absorbed into the stratum corneum as a 'Patchless patch®'. Depending on the polymer used in the formulation, the residual film may also act to occlude the skin, increasing skin hydration which can be beneficial for particular skin conditions again and improve drug delivery. The use of a polymer in the formulation may necessitate that a plasticiser is also used. Plasticisers reduce the brittleness of the polymeric film on the skin surface helping it to be flexible enough to move with the skin surface, improving patient acceptability.

Aerosols spray systems also contain a propellant. Propellants are gases at room temperature and pressure, but exist as liquids when stored under pressure in the aerosol can. Normally the rest of the excipients will exist in a single solution phase in the canister with the propellant. A number of different propellants are available, including hydrofluoroalkanes (HFAs) and flammable propellants such as, propane, butane and dimethyl ether.

Short chain hydrocarbon propellants such as butane and propane are inexpensive and are relatively common; however, their use is to some extent limited by their flammability. Dimethyl ether is also a flammable propellant that can be used for topical use but is more polar than the hydrocarbons or HFAs. HFA propellants have been used in pressurised metered dose inhalers (pMDIs) since the mid 1990s, and may also be used for aerosols that can be used on the skin. HFAs are highly volatile like hydrocarbons, but are not flammable making their handling easier. Different propellants have different capabilities as solvents which affects their selection in a formulation. The physiochemical properties of HFAs are such that they are poor solvents for a wide range of drug molecules, including both hydrophilic and hydrophobic therapeutic agents. In contrast, short chain alkanes are usually good solvents for hydrophobic drugs. Dimethyl ether shows an ability to be a good solvent and in contrast to the other propellants also allows an appreciable amount of water to be included in the formulation which may be advantageous in particular circumstances. To improve the ability of propellants to solubilise drugs, co-solvents such as ethanol and isopropanol may be used. For these types of formulations that contain both a propellant and a volatile solvent, evaporation occurs as a two-stage process. The propellant initially evaporates upon actuation of the valve followed by the evaporation of the other volatile components once the solution reaches the skin [3]. This process can potentially be used to control the drug delivery rate to suit different drugs and dosage regimens [4].

5.1.2 Advantages of Using Film-Forming Systems for Drug Delivery

One of the major advantages of using film-forming systems is that they can use supersaturation or at least generate high drug thermodynamic activity in the formulation to improve drug permeation across the skin. Supersaturation offers the possibility of increasing drug delivery to the skin without affecting the skin's barrier, therefore avoiding any possible irritancy or side effects that may be associated with other drug delivery enhancement approaches. The concept of supersaturation can be explained with reference to an adapted form of Fick's law. The common form of Fick's first law used to model skin permeation data under steady-state conditions is shown in Equation 5.1. The equation describes the rate of drug permeation (the flux, J), passing through a unit area of skin per unit time, where D is the diffusion coefficient of the permeant in the skin, K is the partition coefficient between the vehicle and the superficial layer of the skin, C_v is the concentration of the permeant of the in the donor vehicle and h is the diffusional pathlength.

$$J = \frac{DKC_v}{h} \tag{5.1}$$

Equation 5.1 suggests that the drug flux across the skin is proportional to the concentration of the drug within the vehicle under steady-state conditions. However this is an adequate description only under particular circumstances when all of the drug concentration is dissolved within the vehicle (i.e. the formulation is not a suspension) and that the different concentrations of drug that are being compared are within the same formulation. Equation 5.2 shows another adaptation of Fick's First Law [5] in which the partition coefficient and drug concentration within the vehicle are replaced with the ratio between the thermodynamic activity of the drug within the formulation (α) and the thermodynamic drug activity within the membrane (γ).

$$J = \frac{\alpha D}{\gamma h} \tag{5.2}$$

The thermodynamic activity of the drug is any formulation is at a maximum stable value (which is given a value of 1) when the drug is saturated in the formulation, that is at its maximum solubility value, regardless of the drug concentration that is required to produce this saturated system. As long as the vehicle components in the different formulations do not alter the membrane properties, that is have different penetration enhancing effects, the resultant drug flux from the different formulations that are equally saturated with drug contain different drug concentrations should be the same. This theory has been demonstrated experimentally, for example Twist and Zatz have shown that for different formulations that did not interact with the model membrane, similar permeant flux values across the membrane were obtained from saturated solutions despite considerable differences in the solubility of the permeant in the different vehicles [6].

This indicates that simply modifying a formulation to allow it to contain an increased concentration of drug through increasing the solubility of the drug in the formulation may not be a suitable strategy to improve skin permeation (unless the solubility of the drug is very low). However increasing drug thermodynamic activity in the formulation beyond that at the saturated solubility limit using supersaturated systems will increase the drug transport across the membrane in a proportional manner assuming all other parameters are constant. In fact supersaturation is typically measured in degrees of saturation (DS) with a saturated solution having a DS of 1 and supersaturated systems having values greater than this which are expected to be directly proportional to the increase in drug flux observed with a saturated system, for example a system with a DS of 2 would be expected to exhibit a twofold increase in drug flux across the skin compared to a saturated system

5.1.3 Production of a Supersaturated State

Producing supersaturated systems to realise their potential to improve drug permeation across skin presents a challenge. Initially some means must be used to take the dissolved drug concentration above the equilibrium solubility in the vehicle. Then, given that the supersaturated state is unstable, the drug will eventually crystallise out of the system so that the drug concentration reverts to the equilibrium value and any improvement in skin permeation is lost. As a result of this, these types of formulation are designed to become supersaturated *in situ* on the skin surface, thereby mitigating problems associated with drug stability that would likely appear on storage over a pharmaceutically relevant shelf-life. Two main methods are most commonly cited in the literature to produce supersaturated systems on the skin: the co-solvent technique and solvent evaporation. Although the co-solvent method is commonly cited in the literature and patents exist describing its use in products for application to skin, it appears to be less suitable for translation into medicines [7]. The method requires two separate formulations to be mixed on the skin surface, one being a saturated solution of a drug in a solvent in which it has a high saturated solubility, with the other being a poor solvent. However it is likely that the requirement for appropriate packaging to enable this to occur and ensuring satisfactory mixing on the skin under clinical conditions has limited its use in commercial products. In contrast, solvent evaporation is commonly used in topical products and can be used to produce supersaturation [8]. If a drug is dissolved in a vehicle that contains a volatile solvent, when this volatile component evaporates following application of the product to the skin, the concentration and thermodynamic activity of the drug in the vehicle will increase and may increase to the extent that the system becomes supersaturated. The extent of supersaturation produced will depend on the initial concentration of the drug and its solubility in the volatile and non-volatile components and as solvent evaporation systems are dynamic, the extent of supersaturation would be expected to vary over time. Aerosol systems in particular which contain a propellant that evaporates on actuation can be used to rapidly generate a highly supersaturated system on the skin surface.

Not only can this approach be used to increase drug absorption across the skin, but it can also be used to reduce the amount of drug required in a formulation to deliver a particular dose into the body. Conventional topical and transdermal formulations are notoriously inefficient, with for example the typical bioavailabilities of topical corticosteroids being estimated to be a few percent [9]. Supersaturation can be used to improve bioavailability such that the same amount of drug can be delivered, with the application of a lower dose [10]. This potentially offers not only the possibility of cost savings through a reduction in the amount of drug required in a formulation but also potential increases in safety in relation to the decreased amount of drug remaining in a product or on the skin surface.

The principle of supersaturation has been shown to produce significant increases in drug transport for various drug molecules *in vitro* across both model membranes and human skin [8, 11–20]. The extent to which drug flux can be increased across the skin has been seen to be considerable. For example Megrab *et al.* reported a 13-fold increase in oestradiol flux across human skin from a supersaturated formulation over a 12 h period [21]. Examples of other drug molecules for which formulations containing the drug at 2–5 degrees of saturation produced consequent improvements in drug flux over extended periods of time, for example 24 h, include ibuprofen, piroxicam, sodium noviamide and a model lavendustin derivative [11, 13, 18, 22]. In contrast other literature reports exist whereby only modest levels of supersaturation could be achieved (<2 degrees of saturation), or where only transient effects were obtainable [8, 16]. Similar variability in data has been reported in *in vivo* studies, for example whilst some improved permeation in animal models has been attributable to drug supersaturation [23, 24], Schwarb *et al.* using the pharmacodynamic corticosteroid skin blanching assay obtained did not find the blanching response to be related to the degree of saturation of the drug in the formulation [25]. The unstable nature of supersaturated systems means that the drug may crystallise from the formulation even if it is generated *in situ* on the skin surface, thus making the drug unavailable for absorption [19]. As a result of these specific formulation ingredients, for example anti-nucleant polymers may be required to delay or retard crystallisation [15].

A range of different polymers have been found to be able to stabilise supersaturated systems, and thereby improve drug permeation including, cellulose derivatives, polyvinylpyrollidone (PVP) and Eudragits. The mechanisms by which these polymers exert their actions are not fully understood making their selection on a rational basis difficult [3]. Different polymers have been shown to have superior stabilising properties for different drug molecules. For example Megrab *et al.* found PVP to be superior to hydroxypropyl cellulose (HPC) in stabilising supersaturated systems of oestradiol, whereas Raghavan *et al.* found that the optimal concentration of PVP for stabilising hydrocortisone acetate was 20 times higher than that of HPC and PVP was not able to stabilise the drug to the same level of supersaturation as HPC [19, 21]. Others have found sodium carboxymethyl cellulose to have superior stabilising properties, and copolymers and cyclodextrin derivatives have also been suggested to be useful antinucleants [3, 26]. The anti-nucleant polymers have been shown to delay crystal nucleation, slow drug crystal growth and alter crystal shape [13, 27–29]. Evidence from X-ray diffraction (XRD) studies link the alteration of crystal shape with a change in crystal habit rather than change in polymorphic form [28, 29]. Crystal shape is known to be determined by the slowest growing crystal faces, if polymers disrupt the growth at particular faces they will alter the crystal habit. This suggests that the polymers can interact or adsorb preferentially to certain crystal faces. In other cases polymers can decrease crystal growth rate without modifying crystal shape indicating less specific adsorption/interactions [27]. Current consensus considers adsorption of polymers to the crystal interface important for the stabilisation of supersaturated states, where the polymer reduces the integration kinetics of the crystal [28, 30, 31]. Within this mechanism, interactions between the polymer and crystal face, most usually hydrogen bonds, are considered to be important for the anti-nucleant action [13, 27].

It is important to note that much of this work on anti-nucleant polymers has been done on infinite dose systems using the co-solvent method to prepare the supersaturated formulation. In these cases the polymer concentration is often relatively low, for example 1% w/w. Translation of the beneficial effects of these polymers for stabilising supersaturation to film-forming systems, where a polymer may make up a considerable proportion of the residual film may not be directly applicable.

5.1.4 Use with Chemical Penetration Enhancers

Supersaturation may be used alongside other penetration enhancement approaches with chemical penetration enhancers commonly forming key constituents of any formulation. For film-forming systems both the volatile and non-volatile solvents can act as penetration enhancers. A wide variety of different types of molecules have been shown to be able to act as chemical penetration enhancers increasing drug permeation across the skin including alcohols, esters and fatty acids [32–34]. They modify the properties of the stratum corneum, typically improving the diffusion coefficient of the drug or improving its partitioning into the skin, increasing drug transport. Both ethanol and isopropanol, the most common volatile solvents in film-forming systems, can act as penetration enhancers. Common non-volatile solvents include small polar molecules which tend to provide good solubility to a large number of drugs and more lipophilic molecules with alkyl chains attached to polar head groups. Examples of the small polar molecules include propylene glycol, ethanol, transcutol and dimethyl isosorbide (DMI), whilst the more lipophilic examples include isopropyl myristate (IPM) and oleic acid [21, 35–37]. Other molecules that have shown a potential in acting as penetration enhancers include terpenes, azones, sulphoxides and pyrolidones [38–41].

The precise mechanisms through which penetration enhancers exert their actions are difficult to elucidate, and individual enhancers may have more than one effect. For example ethanol is thought to be able to extract lipids from the SC making it easier for drugs to diffuse across the membrane as well as improve drug partitioning into the SC [41]. Many of the small polar penetration enhancers are thought to work by increasing the solubility of a drug within the skin/membrane. [36]. Other enhancers are thought to be able to disrupt the packing of the intercellular lipids, allowing the drugs to be able to diffuse more easily through the SC increasing drug permeation. For example Harrison *et al.* have correlated the increase in the diffusion coefficient of a model compound across the stratum corneum with increased fluidity of the intercellular lipids [42]. Oleic acid has been suggested to be able to create pools in the intercellular lipids again improving drug diffusivity, and other fatty acid or fatty acid derivative enhancers such as IPM may act similarly, however they may also affect drug partitioning [39, 43–46]. Other possible mechanisms of action have been suggested for particular enhancers, for example DMSO in addition to other effects is believed to alter keratin protein structure in the stratum corneum which may affect drug permeation. Combining more than one penetration enhancer in a formulation may be able to produce synergistic improvements in drug permeation. Synergism has been observed with drugs such as nicardipine, clonazepam, tenoxicam, diclofenac sodium and nifedipine when using various combinations of penetration enhancing chemicals, in particular with propylene glycol since it is one of the most commonly studied penetration enhancers. Examples of enhancers that have shown synergestic effects on drug permeation with propylene glycol include transcutol, oleic acid, IPM and DMI [35, 36, 47–49]. Different penetration enhancers have shown different capabilities for improving drug permeation, and the performance of particular enhancers varies for different drugs, with considerable variation occurring even within a particular enhancer class. For example IPM can improve drug permeation significantly [34] but may be of limited usefulness in other cases where other esterified solvents such as propylene glycol laurate and glyceryl monocaprylate/caprate work well [50, 51]. Also whilst combining enhancers may synergistically improve penetration enhancement, this is not always the case [33, 34, 51, 52]. Thus currently, selection of these enhancer systems is often done on a trial-and-error basis. Care must also be taken with enhancer selection as issues may arise regarding particular enhancers or combinations of enhancers that can cause skin irritation limiting their use in topical products [53–56].

The dynamic nature of film-forming systems makes fully understanding drug delivery from these formulations challenging. Not only does the degree of drug saturation in the film change as the volatile components evaporate, but permeation of the residual solvent or penetration enhancer into the stratum corneum will also affect the degree of drug saturation. Moreover the presence of non-volatile components in the formulation may affect the evaporation rate of the volatile solvent, and the capabilities of any chemical penetration enhancers in the formulation along with their ability to work with the supersaturated system will all influence the overall drug permeation rate. Jones *et al.* found it difficult to determine the degree of drug saturation from these types of systems through measuring drug flux, instead opting for a mathematical equation that considered the amount of solvent that had evaporated from the system, whereas Leichtnam *et al.* compared a supersaturated film-forming system with that prepared by the co-solvent method to assess the degree of saturation in their system [15, 16]. Santos *et al.* have reported the effects of different residual solvents on the performance of a film-forming system for the delivery of oxybutynin [57]. Effective supersaturation was found to be produced when octyl salicylate was used as a non-volatile solvent, with a proportional relationship between the degree of drug saturation in the formulation and oxybutynin flux across the skin. In contrast, the same relationship was not observed when propylene glycol was the non-volatile solvent, seemingly as a result of different solvent permeation rates across the skin with propylene glycol diffusing much faster than octyl salicylate and the system therefore being unable to maintain the supersaturated state in the relative lack of residual solvent. The same authors found a similar effect with a fentanyl containing film-forming system, where IPM and octyl salicylate where found to be suitable enhancers for supersaturated systems, whereas with propylene glycol as the residual solvent, supersaturation did not occur which may again be possibly explained by the short residence time of the propylene glycol in the skin [58].

Reid *et al.* investigated the use of an aerosol spray film-forming system for delivering a topical corticosteroid that contained ethanol, PVP as a residual polymer as well as the propellant but without a non-volatile solvent [8]. They found that it was preferable to have a formulation that developed high levels of drug supersaturation at early stages after actuation rather than a system that slowly became supersaturated with ethanol evaporation over time and displayed less drug crystallisation. Addition of a residual solvent to this type of formulation was found to be beneficial, such that in dose matched systems sixfold more drug could be delivered in comparison to a commercial corticosteroid product when the aerosol spray included polyethylene glycol 400 (PEG 400) in the formulation as a residual solvent [59]. In this case the improvement in flux was not directly attributed to supersaturation or the penetration enhancement properties of the PEG 400 on the stratum corneum. Instead the PEG 400 seemed to delay ethanol evaporation, and the presence of this ethanol-rich PEG 400 residual phase seemed to be responsible for the increased drug permeation. In an earlier study the same authors have also shown non-volatile solvents like PEG 400 can be used to slow the rate of drug release but still deliver the same amount of drug, offering opportunities to adjust the drug release profile. These studies suggest PVP may be a suitable antinucleant polymer to stabilise supersaturated drugs in film-forming systems.

Other candidates that have been investigated include a cyclodextrin derivative (RAMEB) and vinylpyrollidone – vinylacetate copolymer for stabilising testosterone. These were selected on the basis of appearing to be beneficial when studied using differential scanning calorimetry and under *in vitro* conditions, supersaturation of between 1.4 and 2.6 DS were able to be obtained for up to 6h [3]. However this effect did not seem to be able to be maintained *in vivo* on a rat model. There are limited numbers of other investigations that have identified suitable anti-nucleants to provide drug delivery from supersaturated film-forming systems.

5.1.5   Advantages of Film-Forming Systems for Patient Use

As well as providing potential benefits with regards to drug delivery, film-forming systems offer a range of benefits over more conventional formulation types that [59] formulations are expected to provide a greater ease of use compared with traditional patch designs [60]. Moreover traditional patch formulations have previously shown skin irritation issues [61–64], and it is thought that these spray systems could provide less irritancy because of the high volatility of organic solvents [15] and the potentially non-occlusive nature of the formulation [60]. Occasionally there are problems with patches failing to adhere to the skin, such instances do not only represent an inconvenience to patients and a loss of therapeutic effect. Cases where children have died following ingestion of transdermal patches have been reported [65]. Related to this are other issues associated with the large quantities of drug that typically still remain in a transdermal after the patch has been used. The used patch offers something tangible from which drug can be extracted, with some internet forums specifically describing extraction procedures for used patches of drugs such as fentanyl and buprenorphine which are prone to abuse. Spray systems offer opportunities to circumvent these issues by being more efficient at delivering drugs and through not providing a physical patch to be discarded. Spray systems may also be more suitable for young children who usually have transdermal patches applied to their back to prevent them from removing it. In contrast to more common topical dosage forms for example creams, ointments and gels the film-forming systems again have advantages. Similarly to these formulations they are suitable also to achieve localised effects, but they can be applied more easily to large application areas. Additionally there is no need for the patient to contact the particular product with their hands, minimising the potential for transference of the product to other individuals and for prevent the spread of the condition if it is an infection. Moreover the rapidly drying/absorbing nature of the film-forming systems helps to minimise transference losses of product onto clothes or other people.

5.1.6   Therapeutic Applications

There are several examples of film-forming systems on the market with some showing considerable commercial success. It is not always possible to elucidate the reasons behind why specific commercial products are able to demonstrate better performance in comparison to other formulations as the relevant information is not available however the concepts discussed previously such as supersaturation or generation of a high drug thermodynamic activity are likely to be involved to some extent. Examples of the simplest design of film-forming system, cutaneous solutions include Lamisil Once and Actikerall. Lamisil Once is a cutaneous solution that contains terbinafine for the treatment of athlete's foot. The formulation consists of two different polymers, medium chain triglycerides and ethanol as the volatile solvent. On evaporation of the ethanol a drug containing polymeric film is formed on the skin surface. The main advantage of this product is that the formulation is effective following a single application, which contrasts with daily application for 1 week of other formulations such as cream and pump spray formulations of the same drug (both of which contain terbinafine at the same concentration) [66]. The ability to treat athlete's foot from a single application is beneficial for patients directly in terms of treatment speed but is also likely to lead to a reduction in treatment failure in response to poor patient compliance with treatment over a number of days or weeks. The film has been found to stay on the skin surface for 72 h with terbinafine concentrations above the minimum inhibitory concentration existing in the stratum corneum for up to 13 days [67, 68]. The formulation also benefits from the high drug thermodynamic activity generated by the evaporation of ethanol in the formulation as after 60 min 16–18% of the applied dose is present in the stratum corneum, which is considerably higher than what is typically expected from conventional topical products. In addition, under occluded conditions ethanol has been shown to act as a penetration enhancer increasing the concentration of terbinafine obtained in the stratum corneum. The ethanol in the formulation may also be able to have this effect to some extent under *in vivo* conditions. One aspect concerning Lamisil Once that limits its commercial potential is that it contains excipients that are not approved in the United States. A clinical trial of an aerosol spray-based system using approved excipients has demonstrated equivalent efficacy to Lamisil Once from a single application [69, 70]. Development of these highly effective formulations is not trivial; other pump spray terbinafine products exist which still require a week-long treatment duration. Another cutaneous solution, film-forming system is Actikerall which contains 5-fluorouracil and salicylic acid for the treatment of actinic keratosis. Although a direct comparator formulation does not exist for the Actikerall film-forming solution, the product was found to be superior to a commonly used diclofenac product and to have higher treatment clearance rates than those reported for other conventional 5-fluorouracil formulations [71]. As well as containing polymers to form a residual phase and ethanol and ethyl acetate as volatile solvents, the formulation contains the potent chemical penetration enhancer DMSO. The efficacy of the product has been ascribed to not only the constituents of the vehicle but also because the polymeric film formed following application is occlusive and therefore promotes penetration of the active ingredients.

Pump spray systems of topical corticosteroids have become available relatively recently and have been shown to be highly effective and superior to more conventional formulations. For example clobetasol spray which contains ethanol, is able to clinically outperform other formulations of the same drug and also improve patient's quality of life [72, 73]. Desoximetasone spray containing 23% isopropyl alcohol as the volatile ingredient is categorised as a Class 1 superpotent corticosteroid in the United States, whereas the cream and ointment formulations which contain the drug at the same concentration are in the lower Class 2 'potent' category [74]. Not only have both formulations shown clinical efficacy, but they are able to improve patient's quality of life and show low incidences of side effects, indicating that this penetration enhancement approach is suitable even for damaged skin [73, 75].

A diclofenac pump spray which forms a gel film on the surface of the skin has also been commercialised and has demonstrated clinical efficacy [76]. Using microdialysis it has been shown to be able to provide higher skeletal tissue concentrations than are achieved from a similar oral dose, with considerably lower plasma concentrations [77]. This suggests that the product is able to provide effective localised treatment whilst minimising systemic side effects. Direct comparison of the effectiveness of the spray with other topical diclofenac products is not available, however the spray based system is thought to be particularly rapidly absorbed. Within 30 min diclofenac plasma concentrations are a third of the maximum plasma concentrations obtained, which is likely to reflect the high thermodynamic activity of the drug in the formulation [78].

Transdermal film-forming systems have been found to have particular advantages to their use. Transdermal products containing estradiol and testosterone are popular; they provide relatively stable plasma concentrations and in the case of estradiol, the avoidance of first-pass metabolism is beneficial with regards to the proportions of estradiol to its metabolites that are obtained in the plasma [79–81]. Pump spray products of these hormones offer the opportunity to avoid the irritancy is sometimes associated with transdermal patch use and through using metred dosing avoids the difficulties encountered in accurately dosing with gels. These pump spray products contain ethanol and octyl salicylate and are absorbed rapidly into the skin, theoretically minimising the chance of transference of the product on to other individuals [82]. However even though the estradiol spray has been shown to have limited transference on to other individuals under study conditions, cases of transference to other individuals have been reported and care is needed to avoid inadvertent exposure to the drug [83].

5.2 Heated Systems

Thermophoresis, also referred to as the 'Soret effect', is defined as the motion of particles or molecules under a thermal gradient [84]. Although it has been known for a long time that heat can increase drug transport across the skin, the design of systems that utilise the effect to increase drug delivery is a relatively new strategy that has attracted a considerable amount of attention. This is at least partly associated with the development of a commercial product which has successfully used heat to achieve penetration enhancement. However the interest also stems from a need to better understand the effects of heat when applied to existing transdermal and topical systems as inadvertent heating of these systems in clinical practice can cause considerable patient morbidity. Heat is also applied to the skin in the absence of any drug for the treatment of conditions such as musculoskeletal pain and may have therapeutic benefit for some skin diseases, thus there may be opportunities provide synergy through using heat to improve drug delivery and to form part of treatment.

5.2.1 Mechanisms of Drug Penetration Enhancement

Perhaps the most expected effect that temperature would have on drug permeation across the skin is to increase the diffusion coefficient of the drug. Diffusion is the result of the random movement of molecules defined as Brownian motion, and the direct effect of temperature on this is shown in the Stokes Einstein equation:

$$D = \frac{k_B T}{6\pi\eta r} \tag{5.3}$$

where D is diffusivity, k_B is the Boltzman constant, T is the absolute temperature (Kelvin), η is the viscosity of the solvent and r is the radius of the diffusing species. Although the Stokes–Einstein equation suggests that D is directly proportional to T, this requires the other factors to remain constant, which is unlikely to be the case given the effects temperature can have on the viscosity of a medium. In fact Longsworth found that the increases in diffusivity of a series of compounds in aqueous solution, from 1 to 25°C were mainly due to a decrease in the viscosity of the solvent, with the kinetic energy changes, $k_B T$, accounting for less than 10% of the effect of temperature on D [85]. The Stokes–Einstein equation is designed to be used under isothermal conditions, and in examining the effect of heat on the transport of drugs across skin, experimental set-ups where *in vitro* experiments are performed at controlled temperatures are common. However in the clinical situation there is likely to be a thermal gradient as the core body temperature is relatively constant. When exposing particles to a thermal gradient as is likely to be the case in any heated topical drug delivery system, they show a 'drift mobility' (v_T) on top of the Brownian diffusion, which is linearly dependent on the temperature gradient (ΔT) with a proportionality constant D_T, known as thermal diffusion coefficient or thermophoretic mobility (Eq. 5.4) [86]:

$$v_T = -D_T \Delta T \tag{5.4}$$

Though the physical mechanisms behind the thermophoretic effect have not been fully elucidated yet, the 'thermophoretic driving force' is thought to mainly originate from the changes at the interfacial layer between particles and solvent brought by the thermal gradient in the system [86]. This means that this 'thermal force' is system-specific, and therefore its magnitude is expected to vary depending on the vehicle response to ΔT and the dependence of the molecule-vehicle interaction on temperature.

5.2.2   Partitioning

Partitioning behaviour is thermodynamically governed and often affected by temperature. The partition coefficient used in Fick's first law is not a true equilibrium partition coefficient; however the use of Fick's law requires the concentration, or activity of a drug inside the superficial layer of the stratum corneum. As this is difficult to measure the concentration of the drug in the donor phase is used alongside the partition coefficient (K) which should be a constant under steady-state conditions. However the relationship between the concentration in the donor phase and inside the membrane may be expected to change with temperature.

One way of exploring how temperature influences K is to a use Van't Hoff plot ($\ln K$ vs. $1/T$), the slope of which yields $-\Delta H/R$:

$$\ln K = -\frac{\Delta H}{RT} + \frac{\Delta S}{R} \tag{5.5}$$

where R is the gas constant and ΔH the enthalpy change associated with the partitioning process. The effect of temperature on the partitioning of the drug from the formulation into the skin will depend on the net change in interactions between the drug and the 'solvent medium' in the different environments. As an example the work of Baena *et al.* can be used to illustrate this [87]. They performed a Van't Hoff analysis of the acetanilide partitioning from aqueous solution into organic solvents with different hydrogen bond forming capabilities. They found that drug transfer in into cyclohexane, which does not form hydrogen bonds was positive ($\Delta H > 0$), with K increasing as function of temperature whereas acetanilide uptake in to IPM (which is a hydrogen bond acceptor), which is sometimes used to predict SC/water partition coefficient [88, 89], was found to be enthalpically driven ($\Delta H < 0$) and therefore favoured at lower temperature. The exothermic nature of the drug transfer in IPM ($\Delta H < 0$) is likely to reflect the formation H-bonds between acetanilide and IPM [87], which does not occur in cyclohexane. This means that if drugs interact strongly with an immiscible solvent (or, e.g. components of the stratum corneum), then its uptake would be enthalpically driven and facilitated with decreasing temperature. As temperature improves diffusion across the skin but potentially may reduce partitioning it has been suggested that temperature cycling (i.e. heating and cooling) may improve drug transport across the skin by promoting K at lower temperatures and D at higher temperatures [90]. Conversely, if drug partitioning is endothermic ($\Delta H > 0$), then the application of heat would result in having a beneficial effect on K and D simultaneously. As for a spontaneous process to occur, the Gibbs' free energy (ΔG) must be negative, it is expected that endothermic drug uptake processes would be entropically driven ($\Delta S > 0$):

$$\Delta G = \Delta H - T\Delta S \tag{5.6}$$

Lin *et al.* when conducting a thermodynamic study on the uptake of a series of amino acids and dipeptides from acetate buffer solution into SC lipids vesicles found that all the drug transfer processes were endothermic ($\Delta H > 0$) and associated with a positive change in entropy ($\Delta S > 0$) [91]. A recent investigation into the effect of temperature on drug partitioning into the skin has been performed by Wood *et al.* who found that heat causes lidocaine uptake from an aqueous buffer solution to increase significantly by 1.2-fold when moving from 32 to 45°C, indicating the endothermic nature of lidocaine transfer into the skin [92].

In addition to influencing drug transfer in the skin, heat can modify the uptake of formulation ingredients such as penetration enhancers into the barrier. If their uptake into the SC is increased, it would be expected to further improve drug permeation [93]. Examples of increased solvent/enhancer uptake into model membranes and improvements in drug transport rates have been observed [94, 95].

One other effect of heat on the delivery of a drug from a formulation is the effects it may have on the thermodynamic activity of the drug in the formulation. With *in vitro* experiments it is common to negate this effect by using a suspension of a drug at the particular temperature the experiment is performed at thus the thermodynamic activity is kept constant. However in a more realistic scenario when the temperature of a formulation is increased on the skin *in vivo*, the solubility of the drug would be expected to change with a consequent change in thermodynamic activity. In general a drug's solubility in a vehicle increases with increased temperature, and so drug thermodynamic activity and flux across the membrane would be expected to decrease, though inclusion of volatile solvents in the formulation may mitigate this effect. Although not common, some drugs possess an inverse relationship between temperature and solubility (e.g. lidocaine and erythromycin). In these cases heat can potentially be used to lower the solubility of the drug in the vehicle, increasing drug thermodynamic activity, potentially producing supersaturation and increasing drug flux across skin.

5.2.3   Effects of Heat on Skin

Heat also has effects on the skin which affects drug absorption across it. One effect is on the intercellular lipids of the stratum corneum. The bricks and mortar model of the stratum corneum describes the keratinocytes as flattened bricks imbedded in an intercellular lipid matrix which consists of a complex mixture of ceramides, free fatty acids and cholesterol [96–100]. These lipids form two lamellar phases between corneocytes with repeat distances of approximately 6 (short periodicity phase, SPP) and 13 nm (long periodicity phase, LPP) [101, 102] (Figure 5.1). Within these lamellae, the SC lipids are highly organised, mostly in an orthorhombic packing arrangement, which is thought to have an important role in regulating the SC barrier; for example in skin conditions such as atopic eczema and lamellar ichthyosis, impaired skin barrier properties have been associated with altered lateral lipid packing, with an increased proportion the lipids being organised in the less dense hexagonal form in comparison to the more usual orthorhombic form [103, 104]. Indeed for both healthy and diseased human skin, trans-epidermal water loss (TEWL) has been found to be significantly increased when the proportion of the orthorhombic lattice packing is reduced [104, 105].

This implies therefore that a shift from an orthorhombic to a more penetrable hexagonal lipid organisation may also offer the possibility of increasing drug transport across the skin, given that the intercellular lipid pathway is thought to be the most important pathway for the permeation of drug molecules across the stratum corneum. This can be achieved by

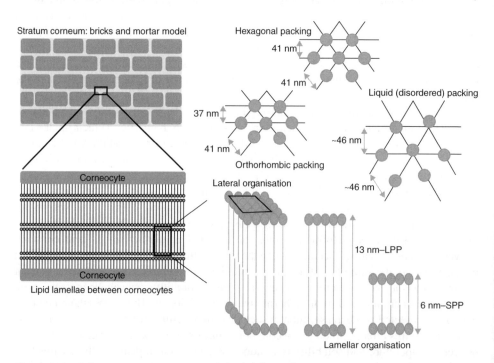

Figure 5.1 *Schematic representation of the bricks and mortar structure of the SC, showing the lamellar and possible lateral organisation of the intercellular lipids. Redrawn from [103] and [104].*

heating the stratum corneum to approximately 40°C, which has been shown to reversibly alter the packing of the lipids from the orthorhombic to the hexagonal form [106]. Drug diffusion across a complex structure such as, skin, is highly dependent on the permeant molecular volume [107], and it is therefore likely that the increase in free volume between the lipid alkyl chains would facilitate diffusion of drug molecules across the SC.

Wood *et al.* have found that the diffusion rate of lidocaine increased significantly, 1.4- and 6.2-fold in the vehicle and across skin respectively when the temperature was increased from 32 to 45°C [92]. They attributed this difference to the orthorhombic-hexagonal transition in SC lipids, decreasing the diffusional resistance of the stratum corneum at the higher temperature. However not all data have clearly pointed to the significance of the orthorhombic to hexagonal transition having a large effect on transdermal penetration. For example, Groen *et al.* found that, even though the flux of benzoic acid through human SC increases 4.5-fold as result of the application of heat (when increased from 32 to 46°C), an Arrhenius plot of steady-state flux yielded a straight line suggesting that no significant change in the properties of the membrane with regards to permeant diffusion occurred over this temperature range [108]. Akomeah *et al.* have reported that the influence of heat on the skin's barrier properties accounts for only ≤34% of the total enhancement effect of temperature on the epidermal diffusivity of caffeine, methyl paraben and butyl paraben [109]. The remaining effect (up to 66%) appeared to be related to the increased molecular motion of the permeants. Further structural and conformational changes in SC lipids are known to occur at temperatures higher than 40°C, which potentially could result in greater increases in drug permeability, including transitions at 65–75°C and 78–86°C that are thought to be reversible [110–116]. However it is unlikely that these temperatures could be used to enhance drug delivery across the skin without causing discomfort or damage to the skin.

5.2.4 Dermal Clearance

As well as altering the lipid packing of the stratum corneum, the application of heat is able to produce changes deeper in the skin, notably causing vasodilation in the dermis, which increases blood flow through the skin. Vasodilation represents a cardiovascular response to a heat stress [117] and potentially could cause drug delivery into the systemic circulation to increase as result of an increased permeant clearance from the dermis [118, 119]. This effect may be beneficial when attempting to deliver drugs transdermally but is likely to have negative effects for topical delivery where the intention is to try to achieve high localised concentrations of drug. An early study on the effect of temperature on blood flow in the forearm was conducted by Barcroft and Edholm where the temperature of the arm was regulated by submersion in water. Temperatures higher than 35°C resulted in vasodilation and an increase in blood flow [120]. Both the temperature that the skin is exposed to and the duration of the elevated temperature contributed to the extent of vasodilatation with blood flow increasing over sevenfold between 32 and 45°C [120]. More recently, Petersen *et al.* have observed similar increases in blood flow (~ninefold) when using an intermittent heat exposure (5 min at 43°C and 5 min with no heat) over a 30 min period. They attempted to correlate the blood flow changes with the transdermal delivery of nicotine; the nicotine content remaining in a transdermal patch was 13-fold lower when the same heating pattern was used [121]. Similarly other studies have looked at changes in blood flow in response to heat, for example Klemsdal *et al.* exposed volunteers to either 15 min of heating using a

250 W infrared bulb or cooling using ice and used these changes to explain a twofold increase and moderate decrease in nitro-glycerine plasma levels on heating and cooling the skin respectively [119]. However it is difficult to clearly distinguish to what extent the increased drug delivery in response to heat is related to increased blood flow in comparison to the other effects heat can have on drug transport across skin. Attempts have been made to minimise dermal clearance through causing vasoconstriction, particularly for NSAID drugs whose site of action is beneath the vasculature of the dermis. For example Sammeta and Murthy used skin cooling to reduce dermal clearance coupled with iontophoresis to improve diclofenac permeation across skin [122]. Higaki *et al.* used a vasoconstrictor (phenylephrine) to achieve a similar effect to reduce dermal clearance [123]. Both strategies appear to have had benefit increasing the amount of drug delivered to the joints, suggesting that the effects of dermal clearance can be significant for topical drug delivery. Increased dermal clearance in response to heat may be of benefit for transdermal products and whether it is a limiting issue for topical drug delivery is likely to depend on the magnitude of the increased clearance effect relative to the heat induced improvement in drug transport across the stratum corneum. Selection of optimal levels of the intensity and duration of heat may be able to influence this balance and help to mitigate any effect of increased dermal clearance on topical drug delivery.

5.2.5   The Effects of Heat on the Permeation of Drugs Across Skin

Early published work showing the effect of temperature on solute permeation across the skin dates back to 1930s when it was found that increasing the skin temperature from 27 to 44°C increased the *in vivo* permeation of methyl salicylate by a factor of 2.4 [124]. Further work in the 1950s and 1960s confirmed these findings; for example, Piotrowski observed a fourfold increase the *in vivo* aniline permeation across the skin between with an increase in skin temperature from 29.8 to 35.0°C, whilst Fritsch and Stoughton reported that the radiolabelled acetylsalicylic acid transport *in vitro* was enhanced up to 15-fold for a 30°C rise in temperature (from 10 to 40°C) [125, 126]. A number of studies have suggested that increasing temperature is more effective for promoting skin permeation of relatively lipophilic substances. For example Blank *et al.* found the permeation of lipid-soluble alcohols (with carbon chain lengths between six and eight carbons) across the skin were increased to a greater extent than those with shorter chain lengths and Knutson *et al.* observed enhanced skin drug flux of lipophilic permeants such as, butanol and hydrocortisone, in response to heat [110, 127]. In both cases decreased intercellular lipid viscosity in response to increased temperature was suggested to be a significant causative factor. These explanations have been supported by the findings of Ogiso *et al.* [128] who found a strong correlation between the transport of terolidine, a lipophilic drug (log P 4.46) with heat induced SC lipid packing alterations . Akomeah *et al.* investigated the temperature dependence of skin transport using three permeants of different lipophilicities but with similar molecular weights [109]. Caffeine, methyl paraben and butyl paraben were used, which have log P values of −0.07, 1.96 and 3.57, respectively. The k_p of these compounds increased 2.7, 3.4 and 4.9-fold, respectively, over the same temperature range, again suggesting that heat has a greater effect on the permeation of lipophilic drugs through the stratum corneum. Although it is clear the permeation of more hydrophilic substances is still enhanced, just to a lesser extent.

It has also been found that the heating profile can have a significant effect on drug transport. For example Wood *et al.* found that heat produced significant increases in the drug release rate from a hydrophilic gel with relationships between the time to maximum temperature as well as the maximum temperature on the drug transport rate being apparent [129]. In fact, the application of a slow and sustained level of heat may actually reduce drug transport compared to when no heat was applied, an effect which was attributed to the influence of heat on drug release from the formulation [129].

Similarly to film-forming systems, heat can be used in conjunction with other enhancement approaches, with chemical penetration enhancers again being an obvious choice and their combination has been found to be able to synergistically improve drug permeation across the skin. Wood *et al.* found that the combined approach improved lidocaine permeation across the skin 75-fold after 10 min, whereas individually heat and the enhancers improved drug permeation only 2.7 and 7.5-fold, respectively [130]. They also found that this approach is also suitable for relatively hydrophilic drugs. Aciclovir permeation could be improved 30-fold by using both heat and chemical penetration enhancers whilst individually they were found to improve permeation 3.2 and 14.7-fold, respectively. These data provide strong evidence that the use of a thermophoretic system with specific chemical penetration enhancers would be able to considerably improve skin permeation of a variety of different drug molecules, potentially opening up new avenues of treatment.

5.2.6 Strategies for Generating Heat

Developing a heating system for application to skin presents challenges in that the device should not be bulky and needs to be able to deliver a controlled amount of heat, potentially for several hours. Currently the only device that is available for generating low, physiologically tolerable temperatures (<45°C), when attempting to optimise percutaneous absorption, is Controlled Heat-Assisted Drug Delivery (CHADD®), which is used in the only commercial product which uses heat to improve drug delivery across skin. However there are limitations with this system, and work has been performed investigating other methods to generate heat that would be compatible with a topical/transdermal drug delivery system.

CHADD was developed by Zars Pharma Ltd and is capable of generating a moderate, physiologically tolerable heat level (<45°C) at the application site, which can be maintained for several hours [129]. The CHADD patch uses the oxidation of iron to generate heat on the surface of the skin. It consists of three main layers: the first layer closest to the skin is a pressure-sensitive adhesive, the second layer is a pouch containing the components for the heating reaction (iron, carbon, sodium chloride, vermiculite and water) and a third outer layer is a semi-permeable membrane that indirectly regulates the temperature by controlling oxygen ingress. On removal from its packaging and contact with oxygen, heat generation is initiated and typically a constant temperature of approximately 42°C is produced and can be maintained for several hours at the site of application [131]. The reaction is:

$$4\,Fe + 3\,O_2 \rightarrow 2\,Fe_2O_3 \tag{5.7}$$

Depending on the ratio of the different constituents that make up the heating system, different heating profiles can be generated [129]. Although CHADD is the only system that uses heat to improve drug delivery across skin, there are a number of similar systems that use the same mechanism to produce heat for application to the skin that don't contain drugs. Examples include Nurofen Express Heat Patches and Thermacare.

There are issues with the use of this mechanism of heating which impacts on its suitability for topical drug delivery, for example there is a risk of dose variability and burns as a result of the inherent variability of the exothermic oxidation of iron with changes in air oxygen and moisture content and pressure [129, 132]. For example, a case of a second degree burn has been documented, after a patient used Thermacare on their shoulder for 8 h as is directed [133].

As a result of this, effort has been invested in identifying other systems that could potentially improve drug delivery through generating heat on the skin surface, with phase change materials (PCMs) in particular being investigated [134]. An efficient class of PCMs are supercooled salt hydrates and supercooled salt solutions, which have an excellent ability to store heat, high thermal conductivity and are relatively inexpensive [135]. Supercooled systems are metastable solutions, whereby on the addition of a nucleating agent, a liquid to solid phase change (i.e. crystallisation) is initiated, releasing the latent energy of crystallisation. Supercooled salts can be produced by dissolving the material in an (aqueous) liquid, with the aid of heat. Upon cooling, the salt passes its melting point but does not crystallise, offering the chance of storing the absorbed energy (heat of fusion) in the system [136]. The use of supercooled salts to generate heat is not new and has been employed in portable heating devices since the late 1970s [137], though their use for the purpose of enhancing drug delivery is relatively novel [134]. Sodium acetate tri-hydrate (SAT) is the most commonly used supercooled salt in portable heating devices, which are generally constructed from a flexible plastic bag and contain a trigger such as a flexible metal disc along with the supercooled salt solution [137]. On flexing the trigger to create a nucleation point the sodium (Na^+) and acetate (CH_3COO^-) ions that are solubilised within the water undergo thermodynamic stabilisation to SAT and as a result heat is released (Eq. 5.8).

$$Na^+_{(aq)} + CH_3COO^-_{(aq)} \rightarrow NaCH_3COO.3H_2O(s) \tag{5.8}$$

This heat release results in the increased temperature of the medium; in the case of SAT the maximum temperature that can be obtained is approximately 57°C [138]. This is because energy can only be recovered up to the melting point of the phase change material. The temperatures that can be attained from supercooled salts are comparable to the temperatures seen with devices based on iron oxidation and their heat release profiles can be modified to provide optimal heating profiles for therapeutic applications by controlling the mass of salt which crystallises, the rate of crystallisation and the relative heat capacities of the supercooled solution and the crystalline material [134]. Portable heating devices are currently available, for example ThermaClick and are used for pain relief, muscle relaxation, cosmetic treatments and as hand warmers. Their current use in heating pads suggests that they may be a suitable and safe strategy to provide heat enhanced drug delivery across skin.

Currently there are no regulatory limits on the extent or duration of heat generated by thermophoretic drug delivery systems, though given their potential to cause, discomfort, pain and cutaneous damage, care in their design is necessary. Cutaneous damage has been shown to be controlled by both the intensity of the temperature to which the skin is exposed as well as the exposure time [139, 140]. For example Moritz and Henriques found that epidermal damage could be caused by a temperature of 44°C over a period of 6 h but that time required to cause this decreased by approximately one-half, for each degree rise, between 44 and 51°C [139].

As well as potentially causing damage to the skin, heat can also cause patient discomfort which would limit the usefulness of any potential product. This issue is complicated through variation in individual's pain thresholds and several studies have investigated factors which influence these. In general there does not seem to be differences in the heat pain threshold of individuals of different ages, however it has been suggested that women have lower heat pain thresholds than men and differences in the heat pain threshold have been observed at different body sites, for example with the hands having a lower threshold value than the feet [141–143]. In addition to heat intensity, the sensation of pain on the application of heat is also known to be dependent on the rate at which the thermal stimulus increases, with the magnitude of the pain sensation increasing with a higher heating rate [144]. However although differences in these factors have been found to be significant, the differences are often small in magnitude, with the pain threshold commonly being reported to be within the range of 44–47°C.

5.2.7 Therapeutic Applications

Currently, there is only one licensed topical medicine that uses heat (with CHADD technology) to aid the delivery of drugs. It is used for dermal anaesthesia and is known as Synera in the United States and Rapydan in Europe. The patch contains a eutectic drug reservoir of lidocaine (70 mg) and tetracaine (70 mg) with the main advantage of the product being the quicker onset of action in comparison to other dermal anaesthetic products. It is recommended to be applied for 30 min prior to any procedure to have a suitable effect, in comparison to the 45–60 min recommended for other products. When compared to the leading topical anaesthetic cream EMLA (a eutectic mixture of lidocaine and prilocaine), the pain relief provided by the Synera patch was found to be sufficient after just 10 min of application and was more effective at pain relief than EMLA cream over the first hour after application [145]. EMLA is recommended to be applied for 1 h prior to any procedure being commenced. The heated patch has also been compared to another topical anaesthetic Ametop gel which contains tetracaine and is usually recommended for application for 45 min prior to any procedure. Again the heated patch was found to have a more rapid onset of action with the anaesthetic effect provided being similar [146]. The effectiveness of this heated patch in having a localised effect suggests that increased dermal clearance in response to heat may not necessarily limit its use in improving drug delivery from topical treatments. The heated patch appears to have similar efficacy to the products already available, with a faster onset of action. However the cost of the product has limited its use, with EMLA and Ametop being significantly cheaper [146]. However the shorter onset time has the potential to reduce overall costs allowing a faster turnaround of patients depending on the nature of the procedure they are to undergo.

The CHADD heating system has been investigated to examine its effect on transdermal delivery by placing the heating system above existing patches. Fentanyl plasma concentrations could be increased from three- to fourfold using this approach [147, 148]. Negative consequences have been observed with inadvertent exposure of transdermal systems to heat, notably with fentanyl. For example behaviour such as sunbathing when wearing a transdermal fentanyl patch has increased fentanyl concentrations to such an extent that they have caused coma and respiratory depression [149]. Increases in the drug plasma concentrations of transdermal patch wearers have been observed with other drugs, for example nitro-glycerine plasma concentrations increased from 2.3 to 7.3 nmol/l when 12 healthy subjects were exposed to a 20 min sauna, which caused their skin temperature to increase up to 39°C [150]. These examples highlight the importance of being aware of the clinical manifestations that unintentional exposure to heat can cause but also give a clear indication of the potential of heat to increase drug delivery across the skin even from systems that have not been specifically designed to utilise it.

As well as being able to improve drug delivery across the skin, potentially heat may have a role in the treatment of particular conditions, suggesting that products could be designed to use both the drug delivery enhancement of heat and its therapeutic benefits. For example, it is well established that applying heat to a painful area aids in reducing the sensation of pain [151]. As such, a number of heat therapy devices, which contain no active pharmaceutical ingredients but effectively alleviate pain, have been developed and are available in the market [152, 153]. Although not fully understood the ability of topical heat to increase the activity of the small unmyelinated C-fibres, which inhibits the transmission of nociceptive signals, is thought to be the mechanism behind this effect [154]. Another potential therapeutic benefit of the application of heat relates to its ability to promote the synthesis of heat shock proteins (HSPs) [155]. HSPs are generally expressed under stress conditions as, hypoxia, heat, UV-radiation, toxic chemicals or inflammation and have a key role in protecting cells against potentially lethal stimuli. As such, use of this response as a treatment mechanism has been proposed as a means of treating disease [156]. For example, it has been suggested the use of heat might find application in dermatology for the prevention of skin photo-damage or wound healing [156, 157]. It is likely that immunomodulatory properties of the HSPs, for example induction of the production of anti-inflammatory cytokines, may have a key role in reducing inflammatory damage [158, 159]. A strategy that combines the use of heat as a therapeutic agent along with its ability to improve drug delivery across the skin would seem to be promising approach to improving patient treatment for a variety of conditions.

5.3   Conclusions

Both film-forming and heated system approaches have shown great potential for improving drug delivery across the skin and have been able to show good patient acceptability, key attributes required for the development of products that improve patient treatment. Both approaches can be used for a range of different types of drug molecules and whilst they are unlikely to be useful for delivering large molecular weight molecules such as proteins, for many others they are likely to be able to either offer improved efficacy over current treatments or indeed be able to successfully deliver molecules that are ineffective in conventional dosage forms. Despite the usefulness of both of these approaches there still remains considerable scope to improve and develop their use. For film-forming systems a better understanding of how to stabilise the supersaturated state coupled with optimal delivery of both penetration enhancers and the drug is required to fully realise the usefulness of these systems. For thermophoresis, a better understanding of the effect of heat on drug absorption, particularly with regards to obtaining synergy with use of chemical penetration enhancers is needed. Investigations of the effect of heating intensity and duration on drug delivery and their effects on dermal clearance will also be of use to optimise this approach. Heated systems also offer an opportunity to develop treatment options that utilise heat therapeutically as well as to enhance the delivery of drugs across the skin, thereby synergistically improving patient treatment.

References

[1] Coldman, M.F., Poulsen, B.J. and Higuchi, T. (1969) Enhancement of percutaneous absorption by the use of volatile : nonvolatile systems as vehicles. *J. Pharm. Sci.*, **58** (9), 1098–1102.
[2] Feldman, R.J. and Maibach, H.I. (1966) Percutaneous penetration of 14C hydrocortisone in man. II. Effect of certain bases and pretreatments. *Arch. Dermatol.*, **94**, 649–651.
[3] Leichtnam, M.L., Rolland, H., Wüthrich, P. and Guy, R.H. (2007) Impact of antinucleants on transdermal delivery of testosterone from a spray. *J. Pharm. Sci.*, **96** (1), 84–92.
[4] Reid, M., Brown, M. and Jones, S. (2008) Manipulation of corticosteroid release from a transiently supersaturated topical metered dose aerosol using a residual miscible co-solvent. *Pharm. Res.*, **25** (11), 2573–2580.
[5] Higuchi, T. (1960) Physical chemical analysis of percutaneous absorption process from creams and ointments. *J. Soc. Cosmet. Chem.*, **11** (2), 85–97.
[6] Twist, J.N. and Zatz, J.L. (1986) Influence of solvents on paraben permeation through idealized skin model membranes. *J. Soc. Cosmet. Chem.*, **37** (6), 429–444.
[7] Davis, A. (1990) Topical drug release system. US Patent US 4,767,751 A.
[8] Reid, M.L., Jones, S.A. and Brown, M.B. (2009) Transient drug supersaturation kinetics of beclomethasone dipropionate in rapidly drying films. *Int. J. Pharm.*, **371** (1–2), 114–119.
[9] Wiedersberg, S., Leopold, C.S. and Guy, R.H. (2008) Bioavailability and bioequivalence of topical glucocorticoids. *Eur. J. Pharm. Biopharm.*, **68** (3), 453–466.
[10] Davis, A.F. and Hadgraft, J. (1991) Effect of supersaturation on membrane-transport 1. Hydrocortisone acetate. *Int. J. Pharm.*, **76** (1–2), 1–8.
[11] Fang, J.Y., Kuo, C.T., Huang, Y.B. *et al.* (1999) Transdermal delivery of sodium nonivamide acetate from volatile vehicles: effects of polymers. *Int. J. Pharm.*, **176** (2), 157–167.
[12] Hou, H. and Siegel, R.A. (2006) Enhanced permeation of diazepam through artificial membranes from supersaturated solutions. *J. Pharm. Sci.*, **95** (4), 896–905.
[13] Iervolino, M. (2001) Penetration enhancement of ibuprofen from supersaturated solutions through human skin. *Int. J. Pharm.*, **212** (1), 131–141.
[14] Iervolino, M., Raghavan, S.L. and Hadgraft, J. (2000) Membrane penetration enhancement of ibuprofen using supersaturation. *Int. J. Pharm.*, **198** (2), 229–238.
[15] Jones, S.A., Reid, M.L. and Brown, M.B. (2009) Determining degree of saturation after application of transiently supersaturated metered dose aerosols for topical delivery of corticosteroids. *J. Pharm. Sci.*, **98** (2), 543–554.
[16] Leichtnam, M.L., Rolland, H., Wüthrich, P. and Guy, R.H. (2006) Enhancement of transdermal testosterone delivery by supersaturation. *J. Pharm. Sci.*, **95** (11), 2373–2379.
[17] Moser, K., Kriwet, K., Froehlich, C. *et al.* (2001) Supersaturation: enhancement of skin penetration and permeation of a lipophilic drug. *Pharm. Res.*, **18** (7), 1006–1011.
[18] Pellett, M.A., Castellano, S., Hadgraft, J. and Davis, A.F. (1997) The penetration of supersaturated solutions of piroxicam across silicone membranes and human skin in vitro. *J. Control. Release*, **46** (3), 205–214.
[19] Raghavan, S.L., Kiepfer, B., Davis, A.F. *et al.* (2001) Membrane transport of hydrocortisone acetate from supersaturated solutions; the role of polymers. *Int. J. Pharm.*, **221** (1–2), 95–105.
[20] Cui, Y. and Frank, S.G. (2006) Characterization of supersaturated lidocaine/polyacrylate pressure sensitive adhesive systems: thermal analysis and FT-IR. *J. Pharm. Sci.*, **95** (3), 701–713.

[21] Megrab, N.A., Williams, A.C. and Barry, B.W. (1995) Estradiol permeation through human skin and silastic membrane – effects of propylene-glycol and supersaturation. *J. Control. Release*, **36** (3), 277–294.

[22] Moser, K., Kriwet, K., Kalia, Y.N. and Guy, R.H. (2001) Enhanced skin permeation of a lipophilic drug using supersaturated formulations. *J. Control. Release*, **73** (2–3), 245–253.

[23] Kondo, S., Yamasaki-Konishi, H. and Sugimoto, I. (1987) Enhancement of transdermal delivery by superfluous thermodynamic potential. II. In vitro-in vivo correlation of percutaneous nifedipine transport. *J. Pharmacobiodyn.*, **10** (11), 662.

[24] Kemken, J., Ziegler, A. and Müller, B.W. (1992) Influence of supersaturation on the pharmacodynamic effect of bupranolol after dermal administration using microemulsions as vehicle. *Pharm. Res.*, **9** (4), 554–558.

[25] Schwarb, F.P., Imanidis, G., Smith, E.W. *et al.* (1999) Effect of concentration and degree of saturation of topical fluocinonide formulations on in vitro membrane transport and in vivo availability on human skin. *Pharm. Res.*, **16** (6), 909–915.

[26] Moser, K., Kriwet, K., Kalia, Y.N. and Guy, R.H. (2001) Stabilization of supersaturated solutions of a lipophilic drug for dermal delivery. *Int. J. Pharm.*, **224** (1–2), 169–176.

[27] Raghavan, S.L., Trividic, A., Davis, A.F. and Hadgraft, J. (2001) Crystallization of hydrocortisone acetate: influence of polymers. *Int. J. Pharm.*, **212** (2), 213–221.

[28] Lindfors, L., Forssén, S., Westergren, J. and Olsson, U. (2008) Nucleation and crystal growth in supersaturated solutions of a model drug. *J. Colloid Interface Sci.*, **325** (2), 404–413.

[29] Usui, F., Maeda, K., Kusai, A. *et al.* (1997) Inhibitory effects of water-soluble polymers on precipitation of RS-8359. *Int. J. Pharm.*, **154** (1), 59–66.

[30] Simonelli, A.P., Mehta, S.C. and Higuichi, W.I. (1970) Inhibition of sulfathiazole crystal growth by polyvinylpyrollidone. *J. Pharm. Sci.*, **59**, 633–638.

[31] Ziller, K.H. and Rupprecht, H. (1988) Control of crystal-growth in drug suspensions 1. design of a control unit and application to acetaminophen suspensions. *Drug Dev. Ind. Pharm.*, **14** (15–17), 2341–2370.

[32] Ogiso, T., Iwaki, M. and Paku, T. (1995) Effect of various enhancers on transdermal penetration of indomethacin and urea, and relationship between penetration parameters and enhancement factors. *J. Pharm. Sci.*, **84** (4), 482–488.

[33] Liu, P., Cettina, M. and Wong, J. (2009) Effects of isopropanol-isopropyl myristate binary enhancers on in vitro transport of estradiol in human epidermis: a mechanistic evaluation. *J. Pharm. Sci.*, **98** (2), 565–572.

[34] Gorukanti, S.R., Li, L.L. and Kim, K.H. (1999) Transdermal delivery of antiparkinsonian agent, benztropine I. Effect of vehicles on skin permeation. *Int. J. Pharm.*, **192** (2), 159–172.

[35] Mura, P., Faucci, M.T., Bramanti, G. and Corti, P. (2000) Evaluation of transcutol as a clonazepam transdermal permeation enhancer from hydrophilic gel formulations. *Eur. J. Pharm. Sci.*, **9** (4), 365–372.

[36] Aboofazeli, R., Zia, H. and Needham, T.E. (2002) Transdermal delivery of nicardipine: an approach to in vitro permeation enhancement. *Drug Deliv.*, **9** (4), 239–247.

[37] Leichtnam, M.L., Rolland, H., Wüthrich, P. and Guy, R.H. (2006) Identification of penetration enhancers for testosterone transdermal delivery from spray formulations. *J. Control. Release*, **113** (1), 57–62.

[38] Goodman, M. and Barry, B.W. (1989) Lipid-protein-partitioning (LPP) theory of skin enhancer activity: finite dose technique. *Int. J. Pharm.*, **57** (1), 29–40.

[39] Mittal, A., Sara, U.V., Ali, A. and Aqil, M. (2008) The effect of penetration enhancers on permeation kinetics of nitrendipine in two different skin models. *Biol. Pharm. Bull.*, **31** (9), 1766–1772.

[40] Lee, P.J., Langer, R. and Shastri, V.P. (2005) Role of n-methyl pyrrolidone in the enhancement of aqueous phase transdermal transport. *J. Pharm. Sci.*, **94** (4), 912–917.

[41] Williams, A.C. and Barry, B.W. (2004) Penetration enhancers. *Adv. Drug Deliv. Rev.*, **56** (5), 603–618.

[42] Harrison, J.E., Watkinson, A.C., Green, D.M. *et al.* (1996) The relative effect of Azone(R) and Transcutol(R) on permeant diffusivity and solubility in human stratum corneum. *Pharm. Res.*, **13** (4), 542–546.

[43] Touitou, E. (1988) Skin permeation enhancement by normal-decyl methyl sulfoxide – effect of solvent systems and insights on mechanism of action. *Int. J. Pharm.*, **43** (1–2), 1–7.

[44] Barry, B. (1987) Mode of action of penetration enhancers in human skin. *J. Control. Release*, **6** (1), 85–97.

[45] Ongpipattanakul, B., Burnette, R.R., Potts, R.O. and Francoeur, M.L. (1991) Evidence that oleic acid exists in a separate phase within stratum corneum lipids. *Pharm. Res.*, **8** (3), 350–354.

[46] Guo, H., Liu, Z., Li, J. *et al.* (2006) Effects of isopropyl palmitate on the skin permeation of drugs. *Biol. Pharm. Bull.*, **29** (11), 2324–2326.

[47] Larrucea, E., Arellano, A., Santoyo, S. and Ygartua, P. (2001) Combined effect of oleic acid and propylene glycol on the percutaneous penetration of tenoxicam and its retention in the skin. *Eur. J. Pharm. Biopharm.*, **52** (2), 113–119.

[48] Arellano, A., Santoyo, S., Martín, C. and Ygartua, P. (1999) Influence of propylene glycol and isopropyl myristate on the in vitro percutaneous penetration of diclofenac sodium from carbopol gels. *Eur. J. Pharm. Sci.*, **7** (2), 129–135.

[49] Squillante, E., Maniar, A., Needham, T. and Zia, H. (1998) Optimization of in vitro nifedipine penetration enhancement through hairless mouse skin. *Int. J. Pharm.*, **169** (2), 143–154.

[50] Cornwell, P.A., Tubek, J., van Gompel, H.A.H.P. *et al.* (1998) Glyceryl monocaprylate/caprate as a moderate skin penetration enhancer. *Int. J. Pharm.*, **171** (2), 243–255.

[51] Gwak, H.S. and Chun, I.K. (2002) Effect of vehicles and penetration enhancers on the in vitro percutaneous absorption of tenoxicam through hairless mouse skin. *Int. J. Pharm.*, **236** (1–2), 57–64.

[52] Alberti, I., Kalia, Y.N., Naik, A. *et al.* (2001) Effect of ethanol and isopropyl myristate on the availability of topical terbinafine in human stratum corneum, in vivo. *Int. J. Pharm.*, **219** (1–2), 11–19.

[53] Touitou, E., Godin, B., Karl, Y. *et al.* (2002) Oleic acid, a skin penetration enhancer, affects langerhans cells and corneocytes. *J. Control. Release*, **80** (1), 1–7.

[54] Montes, L.F., Day, J.L., Wand, C.J. and Kennedy, L. (1967) Ultrastructural changes in the horny layer following local application of dimethyl sulfoxide. *J. Invest. Dermatol.*, **48** (2), 184–196.

[55] Phillips, C.A. and Michniak, B.B. (1995) Topical application of Azone analogs to hairless mouse skin: a histopathological study. *Int. J. Pharm.*, **125** (1), 63–71.

[56] Boelsma, E., Tanojo, H., Boddé, H.E. and Ponec, M. (1996) Assessment of the potential irritancy of oleic acid on human skin: evaluation in vitro and in vivo. *Toxicol. In Vitro*, **10** (6), 729–742.

[57] Santos, P., Watkinson, A.C., Hadgraft, J. and Lane, M.E. (2010) Oxybutynin permeation in skin: the influence of drug and solvent activity. *Int. J. Pharm.*, **384** (1–2), 67–72.

[58] Santos, P., Watkinson, A.C., Hadgraft, J. and Lane, M.E. (2012) Influence of penetration enhancer on drug permeation from volatile formulations. *Int. J. Pharm.*, **439** (1–2), 260–268.

[59] Reid, M.L., Benaouda, F., Khengar, R. *et al.* (2013) Topical corticosteroid delivery into human skin using hydrofluoroalkane metered dose aerosol sprays. *Int. J. Pharm.*, **452** (1–2), 157–165.

[60] Morgan, T.M., Reed, B.L. and Finnin, B.C. (1998) Enhanced skin permeation of sex hormones with novel topical spray vehicles. *J. Pharm. Sci.*, **87** (10), 1213–1218.

[61] Horning, J.R., Zawada, E.T., Jr, Simmons, J.L. *et al.* (1988) Efficacy and safety of two-year therapy with transdermal clonidine for essential hypertension. *Chest J.*, **93** (5), 941–945.

[62] Galer, B.S., Rowbotham, M.C., Perander, J. and Friedman, E. (1999) Topical lidocaine patch relieves postherpetic neuralgia more effectively than a vehicle topical patch: results of an enriched enrollment study. *Pain*, **80** (3), 533–538.

[63] Kakubari, I., Sasaki, H., Takayasu, T. *et al.* (2006) Effects of ethylcellulose and 2-octyldodecanol additives on skin permeation and irritation with ethylene-vinyl acetate copolymer matrix patches containing formoterol fumarate. *Biol. Pharm. Bull.*, **29** (8), 1717.

[64] Segers, K., Cytryn, E. and Surquin, M. (2012) Do local meteorological conditions influence skin irritation caused by transdermal rivastigmine?: a retroprospective, pilot study. *J. Clin. Psychopharmacol.*, **32** (3), 412.

[65] Teske, J., Weller, J.P., Larsch, K. *et al.* (2007) Fatal outcome in a child after ingestion of a transdermal fentanyl patch. *Int J Legal Med*, **121** (2), 147–151.

[66] Ortonne, J.P., Korting, H.C., Viguié-Vallanet, C. *et al.* (2006) Efficacy and safety of a new single-dose terbinafine 1% formulation in patients with tinea pedis (athlete's foot): a randomized, double-blind, placebo-controlled study. *J. Eur. Acad. Dermatol. Venereol.*, **20** (10), 1307–1313.

[67] Schafer-Korting, M., Schoellmann, C. and Korting, H.C. (2008) Fungicidal activity plus reservoir effect allow short treatment courses with terbinafine in tinea pedis. *Skin Pharmacol. Physiol.*, **21** (4), 203–210.

[68] Kienzler, J.L., Queille-Roussel, C., Mugglestone, C. *et al.* (2007) Stratum corneum pharmacokinetics of the anti-fungal drug, terbinafine, in a novel topical formulation, for single-dose application in dermatophytoses. *Curr. Med. Res. Opin.*, **23** (6), 1293–1302.

[69] Evans, C., Brown, M.B., Traynor, M.J. *et al.* (2010) A phase II study of the efficacy, tolerability and consumer acceptability of a 1% w/w terbinafine topical spray versus a currently marketed terbinafine product in the once only treatment of tinea pedis. *J. Pharm. Pharmacol.*, **62** (6), 805.

[70] MedPharm Ltd (2009) *MedSpray® 'Patch in a Can®' Technology Successfully Completes Phase II Clinical Evaluation in Athlete's Foot*, http://www.medpharm.co.uk/uploads/media/Press_Release_Medspray_AF_Phase_II_completion_l_04.pdf (accessed 5 April 2013).

[71] Stockfleth, E., Kerl, H., Zwingers, T. and Willers, C. (2011) Low-dose 5-fluorouracil in combination with salicylic acid as a new lesion-directed option to treat topically actinic keratoses: histological and clinical study results. *Br. J. Dermatol.*, **165** (5), 1101–1108.

[72] Bhutani, T., Koo, J. and Maibach, H.I. (2012) Efficacy of clobetasol spray: factors beyond patient compliance. *J. Dermatolog. Treat.*, **23** (1), 11–15.

[73] Menter, M.A., Caveney, S.W. and Gottschalk, R.W. (2012) Impact of clobetasol propionate 0.05% spray on health-related quality of life in patients with plaque psoriasis. *J. Drugs Dermatol.*, **11** (11), 1348–1354.

[74] National Psoriasis Foundation (2009) *Topical Treatments for Psoriasis, Including Steroids*, http://www.psoriasis.org/document.doc?id=164 (accessed 18 January 2012).

[75] Kircik, L., Lebwohl, M.G., Del Rosso, J.Q. *et al.* (2013) Clinical study results of desoximetasone spray, 0.25% in moderate to severe plaque psoriasis. *J. Drugs Dermatol.*, **12** (12), 1404–1410.

[76] Predel, H.G., Giannetti, B., Seigfried, B. *et al.* (2013) A randomized, double-blind, placebo-controlled multicentre study to evaluate the efficacy and safety of diclofenac 4% spray gel in the treatment of acute uncomplicated ankle sprain. *J. Int. Med. Res.*, **41** (4), 1187–1202.

[77] Brunner, M., Dehghanyar, P., Seigfried, B. *et al.* (2005) Favourable dermal penetration of diclofenac after administration to the skin using a novel spray gel formulation. *Br. J. Clin. Pharmacol.*, **60** (5), 573–577.

[78] Electronic Medicines Compendium (2012) *Summary of Product Characteristics: Voltarol 4% Cutaneous Spray*, http://www.medicines.org.uk/emc/medicine/25679/SPC/Voltarol+Active+4++cutaneous+Spray (accessed 10 June 2014).

[79] Kopper, N.W., Gudeman, J. and Thompson, D.J. (2009) Transdermal hormone therapy in postmenopausal women: a review of metabolic effects and drug delivery technologies. *Drug Des. Devel. Ther.*, **2**, 193–202.

[80] Ullah, M.I., Riche, D.M. and Koch, C.A. (2014) Transdermal testosterone replacement therapy in men. *Drug Des. Devel. Ther.*, **8**, 101–112.

[81] Morton, T.L., Gattermeir, D.J., Petersen, C.A. *et al.* (2009) Steady-state pharmacokinetics following application of a novel transdermal estradiol spray in healthy postmenopausal women. *J. Clin. Pharmacol.*, **49** (9), 1037–1046.

[82] Schumacher, R.J., Gattermeir, D.J., Peterson, C.A. *et al.* (2009) The effects of skin-to-skin contact, application site washing, and sunscreen use on the pharmacokinetics of estradiol from a metered-dose transdermal spray. *Menopause*, **16** (1), 177–183.

[83] Gupta, R., Patra, K., Dariya, V. *et al.* (2010) Spray it with caution: precocious puberty from maternal transdermal estradiol spray. *J. Invest. Med.*, **58** (2), 411.

[84] Duhr, S. and Braun, D. (2006) Why molecules move along a temperature gradient. *Proc. Natl. Acad. Sci. U. S. A.*, **103** (52), 19678–19682.

[85] Longsworth, L. (1954) Temperature dependence of diffusion in aqueous solutions. *J. Phys. Chem.*, **58** (9), 770–773.

[86] Piazza, R. (2008) Thermophoresis: moving particles with thermal gradients. *Soft Matter*, **4** (9), 1740–1744.

[87] Baena, Y., Pinzón, J.A., Barbosa, H.J. and Martínez, F. (2005) Thermodynamic study of the transfer of acetanilide and phenacetin from water to different organic solvents. *Acta Pharm.*, **55** (2), 195–205.

[88] Jaiswal, J., Poduri, R. and Panchagnula, R. (1999) Transdermal delivery of naloxone: ex vivo permeation studies. *Int. J. Pharm.*, **179** (1), 129–134.

[89] Barry, B.W. (1983) Dermatological Formulations: Percutaneous Absorption, Marcel Dekker Inc, New York.

[90] Burgess, S., O'Neill, M.A.A., Beezer, A.E. *et al.* (2005) Thermodynamics of membrane transport and implications for dermal delivery. *J. Drug Deliv. Sci. Technol.*, **15** (4), 325.

[91] Lin, R.-Y., Chen, W.-Y. and Liao, C.-W. (1998) Entropy-driven binding/partition of amino acids/dipeptides to stratum corneum lipid vesicles. *J. Control. Release*, **50** (1–3), 51–59.

[92] Wood, D.G., Brown, M.B. and Jones, S.A. (2012) Understanding heat facilitated drug transport across human epidermis. *Eur. J. Pharm. Biopharm.*, **81** (3), 642–649.

[93] Twist, J.N. and Zatz, J.L. (1988) Membrane–solvent–solute interaction in a model permeation system. *J. Pharm. Sci.*, **77** (6), 536–540.

[94] McAuley, W.J., Oliveira, G., Mohammed, D. *et al.* (2010) Thermodynamic considerations of solvent/enhancer uptake into a model membrane. *Int. J. Pharm.*, **396** (1–2), 134–139.

[95] Oliveira, G., Beezer, A.E., Hadgraft, J. and Lane, M.E. (2010) Alcohol enhanced permeation in model membranes. Part I. Thermodynamic and kinetic analyses of membrane permeation. *Int. J. Pharm.*, **393** (1–2), 61–67.

[96] Gray, G.M. and Yardley, H.J. (1975) Different populations of pig epidermal cells: isolation and lipid composition. *J. Lipid Res.*, **16** (6), 441–447.

[97] Gray, G.M. and White, R.J. (1978) Glycosphingolipids and ceramides in human and pig epidermis. *J. Invest. Dermatol.*, **70** (6), 336–341.

[98] Lampe, M.A., Burlingame, A.L., Whitney, J. *et al.* (1983) Human stratum-corneum lipids – characterization and regional variations. *J. Lipid Res.*, **24** (2), 120–130.

[99] van Smeden, J., Hoppel, L., van der Heijden, R. *et al.* (2011) LC/MS analysis of stratum corneum lipids: ceramide profiling and discovery. *J. Lipid Res.*, **52** (6), 1211–1221.

[100] van Smeden, J., Hankemeije, R.T., Vreeken, R.J. and Bouwstra, J. (2010) The detailed lipid composition in human stratum corneum. *J. Pharm. Pharmacol.*, **62** (6), 807–807.

[101] Madison, K.C., Swartzendruber, D.C., Wertz, P.W. and Downing, D.T. (1987) Presence of intact intercellular lipid lamellae in the upper layers of the stratum corneum. *J. Invest. Dermatol.*, **88** (6), 714–718.

[102] Bouwstra, J.A., Gooris, G.S., van der Spek, J.A. and Bras, W. (1991) Structural investigations of human stratum corneum by small-angle x-ray scattering. *J. Invest. Dermatol.*, **97** (6), 1005–1012.

[103] Pilgram, G.S.K., Vissers, D.C., van der Meulen, H. *et al.* (2001) Aberrant lipid organization in stratum corneum of patients with atopic dermatitis and lamellar ichthyosis. *J. Invest. Dermatol.*, **117** (3), 710–717.

[104] Janssens, M., van Smeden, J., Gooris, G.S. *et al.* (2012) Increase in short-chain ceramides correlates with an altered lipid organization and decreased barrier function in atopic eczema patients. *J. Lipid Res.*, **53** (12), 2755–2766.

[105] Damien, F. and Boncheva, M. (2009) The extent of orthorhombic lipid phases in the stratum corneum determines the barrier efficiency of human skin in vivo. *J. Invest. Dermatol.*, **130** (2), 611–614.

[106] Boncheva, M., Damien, F. and Normand, V. (2008) Molecular organization of the lipid matrix in intact Stratum corneum using ATR-FTIR spectroscopy. *Biochim. Biophys. Acta Biomembr.*, **1778** (5), 1344–1355.

[107] Mitragotri, S., Anissimov, Y.G., Bunge, A.L. *et al.* (2011) Mathematical models of skin permeability: an overview. *Int. J. Pharm.*, **418** (1), 115–129.

[108] Groen, D., Poole, D.S., Gooris, G.S. and Bouwstra, J.A. (2011) Is an orthorhombic lateral packing and a proper lamellar organization important for the skin barrier function? *Biochim. Biophys. Acta Biomembr.*, **1808** (6), 1529–1537.

[109] Akomeah, F., Nazir, T., Martin, G.P. and Brown, M.B. (2004) Effect of heat on the percutaneous absorption and skin retention of three model penetrants. *Eur. J. Pharm. Sci.*, **21** (2–3), 337–345.

[110] Knutson, K., Potts, R.O., Guzek, D.B. *et al.* (1985) Macro-and molecular physical-chemical considerations in understanding drug transport in the stratum corneum. *J. Control. Release*, **2**, 67–87.

[111] Van Duzee, B.F. (1975) Thermal analysis of human stratum corneum. *J. Invest. Dermatol.*, **65** (4), 404–408.

[112] Bouwstra, J.A., de Vries, M.A., Gooris, G.S. *et al.* (1991) Thermodynamic and structural aspects of the skin barrier. *J. Control. Release*, **15** (3), 209–219.

[113] Cornwell, P.A., Barry, B.W., Bouwstra, J.A. and Gooris, G.S. (1996) Modes of action of terpene penetration enhancers in human skin; Differential scanning calorimetry, small-angle X-ray diffraction and enhancer uptake studies. *Int. J. Pharm.*, **127** (1), 9–26.

[114] Gay, C.L., Guy, R.H., Golden, G.M. *et al.* (1994) Characterization of low-temperature (i.e., < 65 degrees C) lipid transitions in human stratum corneum. *J. Invest. Dermatol.*, **103** (2), 233–239.

[115] Silva, C.L., Nunes, S.C.C., Eusébio, M.E.S. *et al.* (2006) Thermal behaviour of human stratum corneum. *Skin Pharmacol. Physiol.*, **19** (3), 132–139.

[116] Silva, C.L., Nunes, S.C.C., Eusébio, M.E.S. *et al.* (2006) Study of human stratum corneum and extracted lipids by thermomicroscopy and DSC. *Chem. Phys. Lipids*, **140** (1–2), 36–47.

[117] Wyss, C.R. and Rowell, L.B. (1976) Lack of humanlike active vasodilation in skin of heat-stressed baboons. *J. Appl. Physiol.*, **41** (4), 528–531.

[118] Rowland, M. (1984) Protein binding and drug clearance. *Clin. Pharmacokinet.*, **9** (1), 10–17.

[119] Klemsdal, T.O., Gjesdal, K. and Bredesen, J.E. (1992) Heating and cooling of the nitroglycerin patch application area modify the plasma level of nitroglycerin. *Eur. J. Clin. Pharmacol.*, **43** (6), 625–628.

[120] Barcroft, H. and Edholm, O. (1943) The effect of temperature on blood flow and deep temperature in the human forearm. *J. Physiol.*, **102** (1), 5–20.

[121] Petersen, K.K., Rousing, M.L., Jensen, C. *et al.* (2011) Effect of local controlled heat on transdermal delivery of nicotine. *Int. J. Physiol. Pathophysiol. Pharmacol.*, **3** (3), 236.

[122] Sammeta, S. and Murthy, S. (2009) "ChilDrive": a technique of combining regional cutaneous hypothermia with iontophoresis for the delivery of drugs to synovial fluid. *Pharm. Res.*, **26** (11), 2535–2540.

[123] Higaki, K., Nakayama, K., Suyama, T. *et al.* (2005) Enhancement of topical delivery of drugs via direct penetration by reducing blood flow rate in skin. *Int. J. Pharm.*, **288** (2), 227–233.

[124] Brown, E.W. and Scott, W.O. (1934) The absorption of methyl salicylate by the human skin. *J. Pharmacol. Exp. Ther.*, **50** (1), 32–50.

[125] Fritsch, W.C. and Stoughton, R.B. (1963) The effect of temperature and humidity on the penetration of C14 acetylsalicylic acid in excised human skin. *J. Invest. Dermatol.*, **41**, 307–11.

[126] Piotrowski, J. (1957) Quantitative estimation of aniline absorption through the skin in man. *J. Hyg. Epidemiol. Microbiol. Immunol.*, **1** (1), 23.

[127] Blank, I.H., Scheuplein, R.J. and MacFarlane, D.J. (1967) Mechanism of percutaneous absorption. 3. The effect of temperature on the transport of non-electrolytes across the skin. *J. Invest. Dermatol.*, **49** (6), 582–589.

[128] Ogiso, T., Hirota, T., Iwaki, M. *et al.* (1998) Effect of temperature on percutaneous absorption of terodiline, and relationship between penetration and fluidity of the stratum corneum lipids. *Int. J. Pharm.*, **176** (1), 63–72.

[129] Wood, D.G., Brown, M.B. and Jones, S.A. (2011) Controlling barrier penetration via exothermic iron oxidation. *Int. J. Pharm.*, **404** (1), 42–48.

[130] Wood, D.G., Jones, S.A. and Brown, M.B. (2012) Drug delivery formulations. WO 2012175996 A2.

[131] Stanley, T., Hull, W., and Rigby, L. (2002) Transdermal drug patch with attached pocket for controlled heating device. US Patent 20,020,004,066 A1.

[132] Raleigh, G., Rivard, R. and Fabus, S. (2005) Air-activated chemical warming devices: effects of oxygen and pressure. *Undersea Hyperb. Med.*, **32** (6), 445–449.

[133] FDA (2010) *Maude Adverse Event Report: Thermacare/Pfizer Inc. Thermacare Heatwrapneck, Shoulder and Wrist Heat Wraps*, http://www.accessdata.fda.gov/scripts/cdrh/cfdocs/cfmaude/detail.cfm?mdrfoi__id=1620202 (accessed 2 February 2015).

[134] Wood, D.G., Jones, S.A., Murnane, D. and Brown, M.B. (2011) Characterization of latent heat-releasing phase change materials for dermal therapies. *J. Phys. Chem. C*, **115** (16), 8369–8375.

[135] Farid, M.M., Khudhair, A.M., Razack, S.A.K. and Al-Hallaj, S. (2004) A review on phase change energy storage: materials and applications. *Energy Conv. Manag.*, **45** (9), 1597–1615.

[136] Young, S. and Mitchell, J. (1904) A study of the supercooled fusions and solutions of sodium thiosulphate. *J. Am. Chem. Soc.*, **26** (11), 1389–1413.

[137] Stanley, J. and Hoerner, G.L. (1978) Reusable heat pack containing supercooled solution and means for activating same. US Patent 4,077,390 A.

[138] Sandnes, B. (2008) The physics and the chemistry of the heat pad. *Am. J. Phys.*, **76**, 546.

[139] Moritz, A.R. and Henriques, F., Jr (1947) Studies of thermal injury: II. The relative importance of time and surface temperature in the causation of cutaneous burns. *Am. J. Pathol.*, **23** (5), 695.

[140] Page, E.H. and Shear, N.H. (1988) Temperature-dependent skin disorders. *J. Am. Acad. Dermatol.*, **18** (5, Part 1), 1003–1019.

[141] Kenshalo, D.R. (1986) Somesthetic sensitivity in young and elderly humans. *J. Gerontol.*, **41** (6), 732–742.

[142] Heft, M., Cooper, B.Y., O'Brien, K.K. *et al.* (1996) Aging effects on the perception of noxious and non-noxious thermal stimuli applied to the face. *Aging*, **8** (1), 35.

[143] Kuhtz-Buschbeck, J.P., Andresen, W., Göbel, S. *et al.* (2010) Thermoreception and nociception of the skin: a classic paper of Bessou and Perl and analyses of thermal sensitivity during a student laboratory exercise. *Adv. Physiol. Educ.*, **34** (2), 25–34.

[144] Yarnitsky, D., Simone, D.A., Dotson, R.M. *et al.* (1992) Single C nociceptor responses and psychophysical parameters of evoked pain: effect of rate of rise of heat stimuli in humans. *J. Physiol.*, **450** (1), 581–592.

[145] Sawyer, J., Febbraro, S., Masud, S. *et al.* (2009) Heated lidocaine/tetracaine patch (Synera™, Rapydan™) compared with lidocaine/prilocaine cream (EMLA®) for topical anaesthesia before vascular access. *Br. J. Anaesth.*, **102** (2), 210–215.

[146] Ravishankar, N., Elliot, S.C., Beardow, Z. and Mallick, A. (2012) A comparison of Rapydan® patch and Ametop® gel for venous cannulation*. *Anaesthesia*, **67** (4), 367–370.

[147] Shomaker, T.S., Zhang, J. and Ashburn, M.A. (2000) Assessing the impact of heat on the systemic delivery of fentanyl through the transdermal fentanyl delivery system. *Pain Med.*, **1** (3), 225–230.

[148] Ashburn, M.A., Ogden, L.L., Zhang, J. *et al.* (2003) The pharmacokinetics of transdermal fentanyl delivered with and without controlled heat. *J. Pain*, **4** (6), 291–297.

[149] Sindali, K., Sherry, K., Sen, S. and Dheansa, B. (2012) Life-threatening coma and full-thickness sunburn in a patient treated with transdermal fentanyl patches: a case report. *J. Med. Case. Rep.*, **6** (1), 220.

[150] Barkve, T.F., Langseth-Manrique, K., Bredesen, J.E. and Gjesdal, K. (1986) Increased uptake of transdermal glyceryl trinitrate during physical exercise and during high ambient temperature. *Am. Heart J.*, **112** (3), 537–541.

[151] Garra, G., Singer, A.J., Leno, R. *et al.* (2010) Heat or cold packs for neck and back strain: a randomized controlled trial of efficacy. *Acad. Emerg. Med.*, **17** (5), 484–489.

[152] Nadler, S.F., Steiner, D.J., Erasala, G.N. *et al.* (2003) Continuous low-level heatwrap therapy for treating acute nonspecific low back pain. *Arch. Phys. Med. Rehabil.*, **84** (3), 329–334.

[153] Mayer, J.M., Ralph, L., Look, M. *et al.* (2005) Treating acute low back pain with continuous low-level heat wrap therapy and/or exercise: a randomized controlled trial. *Spine J.*, **5** (4), 395–403.

[154] Kanui, T.I. (1987) Thermal inhibition of nociceptor-driven spinal cord neurones in the cat: a possible neuronal basis for thermal analgesia. *Brain Res.*, **402** (1), 160–163.

[155] Van der Zee, J. (2002) Heating the patient: a promising approach? *Ann. Oncol.*, **13** (8), 1173–1184.

[156] Jonak, C., Klosner, G. and Trautinger, F. (2006) Heat shock proteins in the skin. *Int. J. Cosmet. Sci.*, **28** (4), 233–241.

[157] Polla, B. (1990) Heat (shock) and the skin. *Dermatology*, **180** (3), 113–117.

[158] Van Eden, W., Van der Zee, R. and Prakken, B. (2005) Heat-shock proteins induce T-cell regulation of chronic inflammation. *Nat. Rev. Immunol.*, **5** (4), 318–330.

[159] Colaco, C.A., Bailey, C.R., Walker, K.B. and Keeble, J. (2013) Heat shock proteins: stimulators of innate and acquired immunity. *BioMed. Res. Int.*, **2013**, 11.

6

Nanotechnology-Based Applications for Transdermal Delivery of Therapeutics

Venkata K. Yellepeddi

College of Pharmacy, Roseman University of Health Sciences, South Jordan, UT, USA
College of Pharmacy, University of Utah, Salt Lake City, UT, USA

6.1   Introduction

Nanotechnology as a scientific discipline involves the design, synthesis and application of materials with at least one dimension in the size range of 1–100 nm. National Nanotechnology Initiative, USA, defines nanoparticles as a particle with all dimensions between 1 and 100 nm [1]. Major applications of nanoparticles include encapsulation of poorly soluble drugs, protection of therapeutic molecules, modification of blood circulation and tissue distribution and targeting of therapeutics and diagnostic agents to cancer [2, 3]. Because of huge success in biomedical applications, nanotechnology was also investigated for its use in the development of formulations that can successfully overcome skin barriers. Furthermore, increased use of skin for cutaneous and percutaneous delivery of therapeutics led to discovery of new transdermal delivery strategies involving nanotechnology. The discussion of nanotechnology in dermatology also gained importance due to toxicological concerns related to penetration of environmental pollutants and other toxic nanomaterials through skin [4]. This chapter intends to provide concise and current information about applications of nanotechnology for topical and transdermal drug delivery. For a comprehensive review of this topic readers can refer to reviews published elsewhere [5–9].

Novel Delivery Systems for Transdermal and Intradermal Drug Delivery, First Edition.
Ryan F. Donnelly and Thakur Raghu Raj Singh.
© 2015 John Wiley & Sons, Ltd. Published 2015 by John Wiley & Sons, Ltd.

6.1.1   Skin Structure

Skin is the largest and most complex organ of the human body, and its principal physiological function is to defend the body from the external environment [10]. Skin comprises three layers: epidermis, dermis and hypodermis. Only epidermis and dermis are layers relevant to penetration of nanoparticles through skin. Epidermis is involved in imparting defensive features to the skin, and dermis is involved in bringing nutrients and oxygen to skin via capillary anastomoses [11]. Epidermis is classified histologically as a stratified squamous epithelial layer in which keratinocytes are interspersed in five different strata: stratum basale (attached to dermis), stratum spinosum, stratus granulosum, stratum lucidium and stratum corneum (SC) which is in contact with the external environment.

SC (horny layer) is the outermost layer of the epidermis, which is responsible for its defensive functions. In SC keratinocytes are completely differentiated into enucleated and keratin-filaggrin-filled cells, called corneocytes. SC macrostructure in humans generally refers to cross-sectional organisation of corneocyte described by the brick-and-mortar model. Where, 'bricks' refer to alternate corneocytes (15–30%) and 'mortar' refers to a lipidic matrix surrounding the corneocytes [12]. SC microstructure in humans is a supramolecular organisation of intercorneocyte lipids that are assembled in parallel head–head, tail–tail repeating bilayers. These repeating bilayers consist of both hydrophilic and lipophilic regions. The barrier action of SC is a result of collective cooperation and interactions between SC macro- and microstructure. The intercellular lipids of SC comprise of cholesterol and its derivatives, ceramides, free fatty acids and triglycerides [13]. Approximately, 80% of these lipids are nonpolar, and the presence of hydroxyl groups on the heads of lipids imparts hydrophillicity by engaging hydrogen bond interactions with adjacent lipids or water. SC also has aqueous pores (hydrophilic regions) approximately 36 nm in diameter and identifies the trans-epidermal intercellular polar (hydrophilic) route of skin absorption. The lipophilic central area which is approximately ≤6.94 nm is the tail–tail region known as the trans-epidermal intercellular apolar (lipophilic) route of skin absorption [14]. Therefore, depending on the physicochemical characteristics of the nanoparticle the penetration may involve either polar or lipidic routes. However, it was shown that the penetration of water and polar molecules through SC is poor, suggesting that penetration of nanoparticles by polar route is highly unlikely. SC, as a whole, is generally referred to as a lipophilic stratum and viable epidermis as hydrophilic because of higher amount of water. Because of this, a nonhomogenously distributed change in hydrophobicity is observed continuing from the SC to the stratum granulosum. This is an additional defensive strategy of epidermis which limits the penetration of lipophilic agents in the viable epidermis. A pH gradient ranging from 4.5 and 5.5 to neutral exists from the surface of SC to stratum granulosum. This SC pH which is referred to as 'acid mantle' of skin is involved in antimicrobial defense, permeability barrier homeostasis, SC integrity and cohesiveness, and desquamation processes [5].

The other skin appendages such as sweat glands and pilosebaceous units which open on skin surfaces can allow the penetration of nanoparticles. Sweat glands are coiled tubular glands extending from SC to dermis which are 2–5 mm in length, and are involved in thermoregulation and excretion of body waste [11]. Pilosebaceous units comprise of hair infundibulum which is hair follicle with one hair which serves as a route to expel the sebum of associated sebaceous gland. Eventhough, sweat glands and hair follicles provide large openings on skin surface, they have low density (sweat glands: 0.01%, hair follicles: 0.1%) [15], which may be a limiting factor for the utilisation of trans-follicular route for delivery of nanoparticles.

6.1.2 Skin Sites for Nanoparticle Delivery

To understand further the importance of nanotechnology in topical and transdermal delivery, it is necessary to focus on potential sites on the skin responsible for nanoparticle absorption. The major sites of skin responsible for nanoparticle absorption are through the lipid layers of stratum corneum, furrows or dermatoglyphs and opening of hair follicles [6, 8]. Figure 6.1 depicts the skin sites through which nanoparticles can penetrate.

A review of nanoparticle penetration through skin sites was provided by the Scientific Committee on Consumer Products (SCCP) [16] in their report. The report summarised the following conclusions:

1. There is evidence of some skin penetration into viable tissues (mainly into the stratum spinosum in the epidermal layer, but eventually also into the dermis) for very small particles (<10 nm), such as functionalised fullerenes and quantum dots.
2. When using accepted skin penetration protocols (intact skin), there is no conclusive evidence for skin penetration into viable tissue for particles of about 20 nm and larger primary particle size as used in sunscreens with physical UV filters.

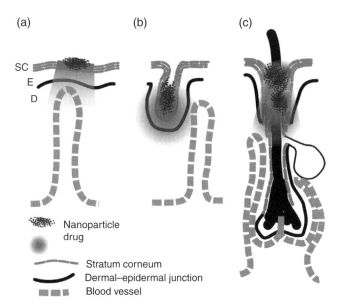

Figure 6.1 *Three pathways for nanoparticle penetration through skin. The three major sites involved in nanoparticle delivery through skin are stratum corneum (SC) surface (a), furrows (dermatoglyphs) (b) and openings of hair follicles (infundibulum) (c). D and E are dermis and epidermis. Reproduced from Ref. [6] with permission from Elsevier.*

3. The aforementioned statements on skin penetration apply to healthy skin (human and porcine). There is the absence of appropriate information for skin with impaired barrier function, for example atopic skin or sunburned skin. A few data are available on psoriatic skin.
4. There is evidence that some mechanical effects (e.g. flexing) on skin may have an effect on nanoparticle penetration.
5. There is no information on the transadnexal penetration for particles under 20 nm. Nanoparticles of 20 nm and above penetrate deeply into hair follicles, but no penetration into viable tissue has been observed.

However, the published scientific data by various research groups across the world is very trivial and incongruent with the SCCP conclusions. For example, Baroli *et al.* and Ryman-Rasmussen *et al.* have reported that iron nanoparticles and quantum dot nanoparticles whose size is greater than 20 nm were able to penetrate into viable epidermis [17, 18]. In contrast, Prow *et al.* reported that gold, silver and quantum dots do consistently penetrate through SC, but not into the viable epidermis [6]. Therefore, more mechanistic studies are required to understand the penetration and localisation of nanoparticles into various sites of skin. Furthermore, there is clear lack of understanding on penetration of nanoparticle through skin whose structure and functionality is compromised due to pathological conditions.

6.1.3   Skin as a Barrier for Nanoparticle Penetration

The major physiological function of the mammalian skin is to prevent the invasion of foreign particles or microorganisms into human body, thus providing a first line of defence against threats from external environment. Therefore, in order to successfully penetrate nanoparticles through skin, it is important to overcome the natural defense barrier of the skin.

The SC is the major physical barrier of the skin, for penetration of nanoparticle through skin, diffusion through SC is the rate-limiting step [6]. Transport of substances through SC mainly occurs by passive diffusion by three possible routes. These are transcellular, the intercellular and the appendageal routes. For nanoparticles, the intercellular lipid channels present an additional barrier when compared with small molecules as their size was reported to be 19 nm by van der Merwe *et al.* [19] and 75 nm by Baroli *et al.* [17]. The continuous turnover of SC by upward migration and sloughing of corneocytes results in a constantly renewable barrier, providing an inherent defensive mechanism for skin [20, 21]. This barrier is another major impediment for penetration of nanoparticle through skin.

The pH of the surface of the skin is around 4.2–5.6; and due to its acidic nature, it is commonly referred to as 'acid mantle'. The major functions of acid mantle include antimicrobial defense, maintenance of permeability barrier, preservation of optimal corneocyte integrity and cohesion and restriction of inflammation. Thus, acid mantle of skin may pose a barrier for nanoparticle penetration as this acidic pH can alter the physico-chemical properties of nanoparticles which may confound the penetration of nanoparticle into the skin. For example, for zinc oxide nanoparticles the acidic pH greatly affects the nanoparticle aggregation and dissolution kinetics which directly influences the penetrability of these nanoparticles to skin [22].

The viable epidermis is another important barrier for NP penetration because of presence of tight junctions and metabolic deactivation of foreign substances by enzymes [23, 24]. The functional tight junctions and tight junction proteins were reported to be present in mammalian stratum granulosum, epithelial layers and follicles [25–27]. Tight junctions were considered to be important components of the epidermal barrier system impeding the penetration of nanoparticles. Many metabolic enzymes are located in the basal layer of the viable epidermis and in the extracellular spaces of the SC and skin appendages [24]. These enzymes by virtue of their proteolytic properties may degrade the NPs and inhibit their further penetration and activity in skin.

6.1.4   Physicochemical Characteristics of NPs for Penetration through Skin

Since four decades, the research in the field of transdermal drug delivery has shown that the most important physicochemical properties of the drug molecules that influence their skin permeation are its solubility, pK_a, O/W partition coefficient, molecular weight, metabolism and binding to skin components and diffusion coefficient (D) [5]. However, for nanoparticle penetration there exists a separate set of physicochemical properties responsible for NP interaction with skin components. Since NPs are formulated in vehicles with various ingredients it is also important to understand the influence of physicochemical properties of the dispersing vehicle used in nanoparticle formulations.

The most important physicochemical property of the NP influencing its penetration is the dimension of NP. The dimensions of the NP influence the ability of the NP to enter the skin, the selection of the route for penetration and the coefficient of diffusion in the dispersing vehicle and in skin [5]. Shape of the NP and superficial properties are other physicochemical properties that further influence interactions with skin components. Specifically, the 'shape deformability' and 'dimension versus orientation' must be considered for NP penetration through skin [5]. Stability of NPs is another important physicochemical parameter that influences polymeric, lipidic, polysaccharidic and metallic nanoparticles. For example, in the case of metallic NPs chemical reactions such as oxidation, hydrolysis and deamination will affect their stability and can alter their superficial properties, thus altering its interactions with skin components. Solubility is another parameter that governs the NP skin absorption. In many cases, NPs are dispersed in the vehicle rather than dissolved in the vehicle that contains them. Therefore, it is assumed that NPs sized 1 nm or below will act like a small molecule rather than a large particle [5, 28].

The physicochemical properties of the dissolving or dispersing vehicles are of great importance for formulations that contain NPs. Because of the potential enhancing effects, NPs are formulated in vehicles with different physicochemical properties. For example, liposomes and lipidic NPs are dispersed in aqueous solutions [29], metallic and magnetic NPs are dispersed in organic solvent-based vehicles [17, 30] and sunscreen or particle-containing NPs are formulated in cream- or gel-based vehicles [31]. Nonetheless, it is worthwhile to mention that very limited data exists in the literature that describes the influence of dispersing vehicle on NP penetration through skin.

6.2   Nanocarriers for Topical and Transdermal Delivery

Because of the advancements in the field of biomaterials and their characterisation, the field of nanotechnology has seen a sharp increase in the number of nanocarriers used for biomedical applications. Many of these biomaterials were designed specifically for the required medical application with a desirable toxicity profile. One such application involved

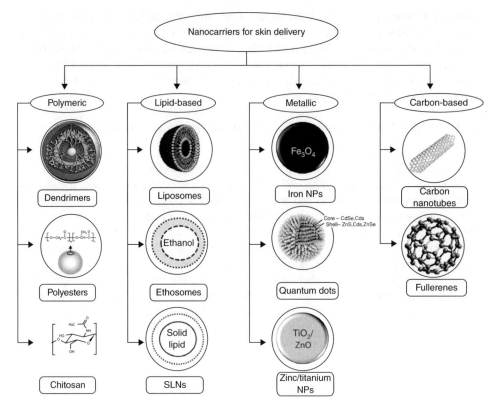

Figure 6.2   *Schematic representation of various nanocarriers used for transdermal delivery of therapeutics.*

the use of these materials for dermatological purposes. Based on the desirable properties described in Section 6.1.4, many nanocarriers were investigated for their ability to deliver therapeutics to skin. However, a complete listing of nanomaterials used for dermatological and transdermal purposes is out of the scope of the review. Only, major classes of nanomaterials currently used in skin delivery will be discussed in this section. Figure 6.2 is schematic representation of various nanocarriers used for transdermal applications.

6.2.1   Polymeric Nanoparticles

6.2.1.1   *Dendrimers*

Dendrimers are 'tree-like' macromolecules originating from a core initiator molecule. Dendrimers were initially defined as 'arborols' which means monocascade spheres. Poly(amido)amine (PAMAM) dendrimers belong to the first class of Starburst® dendrimers which are widely used in biomedical applications [32]. PAMAM dendrimers because of their easily modifiable surface characteristics have shown promise in targeted delivery of various chemotherapeutic agents to cancer [33, 34]. Dendrimers have shown to interact with biological membranes and the membrane permeability, biodistribution and toxicity of dendrimers are dependent on their surface-functional groups, charge, size and generation [35, 36]. Because of their success in permeation through various biological barriers dendrimers were investigated for their applications in delivery of therapeutics through skin.

The reviews by Mignani *et al.* [37] and Sun *et al.* [38] provided comprehensive discussion on this topic. According to those reviews, dendrimers used in skin delivery mainly act by three possible mechanisms. The first mechanism involved the use of dendrimers as drug-release modifiers, second is vehicle-dependent penetration enhancement approach and the third involves penetration through hair follicles on skin surface [38]. Various research groups have demonstrated that the skin penetration of PAMAM dendrimers is directly related to their size, molecular weight, surface charge and hydrophobicity [39, 40]. Their reports suggested that smaller generation PAMAM dendrimers penetrate skin layers more effectively than larger ones. It was also reported that surface modification of cationic PAMAM dendrimers with either acetyl group or carboxyl group enhanced skin permeation. Figure 6.3 depicts the internalisation pattern of PAMAM dendrimers with various surface groups. In addition, conjugation of hydrophobic oleic acid to cationic dendrimers increased their 1-octanol/PBS partition coefficient, resulting in enhanced skin permeation and retention [40]. Table 6.1 provides a list of dendrimers currently used for transdermal applications.

Eventhough, PAMAM dendrimers dominate the field of dendrimers for skin delivery, new dendrimers such as Janus-type dendrimers are recently investigated for their

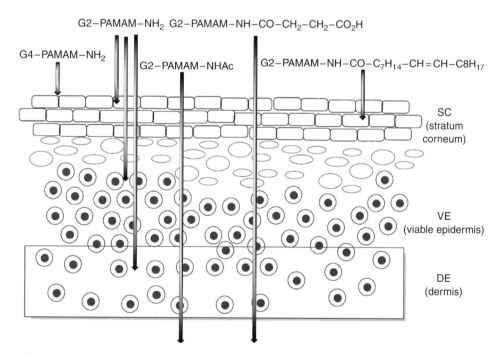

Figure 6.3 *Schematic representation of internalisation pattern of PAMAM dendrimers with various surface groups. Reproduced with permission from Ref. [40], © 2012, American Chemical Society and from Ref. [37] with permission from Elsevier.*

Table 6.1 *List of dendrimers used in transdermal delivery of therapeutics*

Dendrimer type	Drug/therapeutic	Experimental model	Reference
Polypropylene imine G 5.0	Dithranol	*In vitro* skin permeation	[41]
PAMAM G 2.5 and 3.5	8-Methoxypsoralen	*In vitro* skin permeation	[42]
PAMAM G1–G5 NH$_2$	Ketoprofen	*In vitro* skin permeation and *in vivo* absorption	[43]
PAMAM G1–G5 NH$_2$	Diflunisal	*In vitro* skin permeation and *in vivo* absorption	[43]
PAMAM G4 NH$_2$ and G4 OH	Indomethacin	*In vitro* skin permeation and *in vivo* absorption	[44]
PAMAM G2 and G3 NH$_2$	Riboflavin	*In vitro* skin permeation	[45]
PAMAM G4 NH$_2$	5-Fluorouracil	*In vitro* skin permeation	[46]
Benzyloxycarbonyl, Nitromethane, methylamine	5-Aminolaevulinic acid	*In vitro* cell culture and tissue skin explant studies	[47]
PAMAM G5, G7 and G9 NH$_2$ dendrimers	pEGFP1	*In vitro* cell culture, *in vivo* transfection in hairless mice	[48]
Pan-DR-binding epitope (PADRE)-derivatised dendrimers	Liposomal amphotericin	*In vivo* absorption and immunological study	[49]
Peptide dendrimers	Peptide dendrimers	Iontophoresis-mediated *in vitro* skin permeation studies	[50]

penetration enhancement for transdermal applications. Recently, Kalhapure *et al.* used Janus-type dendrimers with oleic acid moiety as a lipophilic side chain for transdermal penetration enhancement of diclofenac sodium. The results from *in vitro* skin permeability studies have shown that both G1 and G2 oleodendrimers have shown enhanced penetration of diclofenac sodium [51].

6.2.1.2 Poly α-Esters

Poly α-esters are one of the most commonly used synthetic polymers for fabricating nano-particles for topical and transdermal delivery. These polymers contain hydrolytically labile aliphatic ester linkages in their backbone and are thermoplastic. These polymers are bio-compatible and biodegradable because they undergo disintegration at the ester bonds by hydrolysis [52]. The most common polymers of this type include polyglycolide (PGA), poly(L-lactide) (PLA), polylactide-*co*-glycolide (PLGA) and polycaprolactone (PCL).

Li *et al.* reported that methoxy poly (ethylene glycol)-block-poly (DL-lactic acid) nano-particles loaded with paclitaxel were able to penetrate skin not only via appendageal routes including sweat ducts and hair follicles but also, via compact structures of stratum corneum and epidermal routes [53]. PLA NPs were also investigated for the transcutaneous delivery of the ovalbumin (OVA) antigen. PLA NPs were loaded with OVA by a double emulsion process. The average size of the OVA-loaded PLA NPs was approximately 150 nm, and they were applied onto the skin of BALB/c mice. The skin penetration studies have shown that NPs were able to penetrate into the mouse skin through the ducts of hair follicles. Furthermore, OVA-loaded NPs showed relatively high levels of IFN-γ and IL-2 secreted from splenocytes of mice immunised with the OVA-loaded NPs indicating the potential of using PLA NPs for transcutaneous antigen delivery [54].

PLGA which is a copolymer of PLA and PGA is another important class of poly α-esters which is widely used for transdermal delivery. For topical and transdermal purposes usually PLGA copolymer with molecular weight ranging from 17 to 75 has been used [55]. PLGA NPs with average diameter of 280 nm were investigated for topical delivery of protoporphyrin IX (PpIX) in photodynamic therapy of skin cancers. The results have indicated that PLGA NPs with PpIX were localised in the epidermis and dermis indicating a potential use in photodynamic therapy in skin cancer treatment [56]. In another interesting study PLGA nanoparticles with dimercaptosuccinic acid (PLGA-DMSA) were used for topical delivery of desoxycholate amphotericin B (D-AMB) for the treatment of cutaneous leishmaniasis in C57BL/6 mice. The results indicated that PLGA-DMSA NPs loaded with D-AMB showed significantly greater reductions in the number of parasites and in cell viability when compared to free D-AMB [57].

PCL is another class of polyester-type polymer which is highly hydrophobic, nontoxic and biodegradable [58]. PCL has a distinction of having excellent mechanical properties and slow biodegradability when compared to other polyesters. PCL is generally used in transdermal applications as a block copolymer of lactide or glycolide with PEG. PCL-block-PEG NPs were studies for transdermal delivery of minoxidil in hairy and hairless guinea pig skin. The results have shown that PCL-PEG NPs were able to penetrate through skin via the shunt routes such as skin appendages [59]. In another report, a self-assembled PCL polymerosomes were developed for epidermal targeting in the treatment of melanomas and basal cell carcinomas. The results have indicated that these PCL polymerosomes were able to localise into inner-corneocyte space in stratum corneum indicating a deformable nature of the NP's [60].

Overall, poly α-ester class of polymers because of their desirable properties such as biocompatibility, biodegradability and easily deformable structure have shown great promise in topical and transdermal delivery. However, their larger size usually limits their skin penetration through SC barrier. Many reports indicate that their preferable route of skin penetration is through hair follicles via transappendageal pathway.

6.2.1.3 Natural Biopolymers

Chitosan is an important natural biopolymer and is most abundant natural polymer after cellulose. Chitosan has many advantages including biocompatibility, stability, ease of sterilisation and susceptibility to enzymatic degradation. When compared with polysaccharides chitosan is relatively nontoxic and nonimmunogenic and thus is widely used of drug delivery applications [61]. Chitosan can enhance penetration of therapeutics into skin either by altering the secondary structure of keratin, increasing the water content in stratum corneum and cell membrane fluidity or by depolarising the negatively charged cell membrane resulting in decreasing the membrane potential driving the drug into skin [62, 63]. Because of its ability to form stable complexes with plasmid DNA chitosan was used for topical gene delivery. A study by Mumper group showed that pDNA condensed chitosan nanoparticles resulted in both detectable and quantifiable levels of luciferase expression in mouse skin 24 h after topical application, indicating the potential of chitosan for transdermal delivery of NPs [64].

The other natural biopolymers and polysaccharides used for topical and transdermal drug delivery include sodium alginate [65, 66], pectin [67, 68], lecithin [69, 70], collagen [71, 72] and hyaluronic acid [73–75].

6.2.2   Lipid Based Nanocarriers

Because of their lipophilic characteristics, lipid based nanocarriers were most sought after devices for topical and transdermal delivery applications. The lipid based nanocarriers include liposomes, niosomes, ethosomes and Transferosomes®. All these nanocarriers were extensively studied in the past 30 years for transdermal delivery of drugs. The mode of action of these nanocarriers in enhanced skin permeation has been widely investigated and reviewed [5, 76–78].

Liposomes are micro- or nano-sized vesicles that contain amphipathic phospholipids arranged in concentric bilayers enclosing an aqueous environment. Depending on the physicochemical characteristics of the drug, molecules can either be encapsulated in the aqueous space (hydrophilic compounds) or intercalated into the lipid bilayer (lipophilic compounds) [79, 80]. Because of their similarity to biological membranes, ability to interact with similar lipids in skin, and decreased systemic absorption, liposomes have been extensively studied for transdermal applications [81]. Initially, liposomes were used in cosmetics such as moisturising and restorative creams. Later, they have been used for penetration of drug through skin. Liposomes were investigated for the treatment of xeroderma pigmentosum or skin cancer with a liposomal system containing a DNA repair enzyme. The results have shown that liposomal formulation showed reduction in DNA damage in the first few hours after the treatment [82].

Niosomes represent another category of lipid-based nanocarriers which have structure and properties similar to that of liposomes. Niosomes are fabricated using synthetic non-ionic surfactants and have better chemical stability and are relatively cheaper to produce [83]. Niosomes have been widely investigated for their topical and transdermal delivery of therapeutics. Niosomes were used to encapsulate various drugs such as enoxacin [84], anti-inflammatory drugs [85] and gene and nucleic acids [86].

Ethosomes were designed to improve skin permeation by enhancing the elasticity of the lipid vesicle, and contain up to 20–45% of ethanol [5]. Ethosomes were used to improve skin penetration of drugs such as melatonin which have very low skin permeability [87]. In a comparative study by Dayan *et al.* skin permeability of an anti-muscarinic drug (THP-Hcl) was evaluated using ethosomes and classic liposomes. The results indicated that the permeability of THP through skin of nude mice was significantly greater by ethosomal formulation when compared with liposomal formulation [88, 89].

Transferosomes are ultra-deformable vesicles made of phospholipids with surfactants that act as edge activators to promote vesicle elasticity and deformability [77, 90]. Transferosomes are 10^5 times more deformable than traditional liposomes. The elastic and ultra-deformable properties of Transferosomes facilitate a rapid penetration through the intercellular lipids of the SC. The hydration gradient that exists naturally in the skin is also thought to be the driving force for the transdermal permeation of Transferosomes [7]. The transdermal permeation of insulin containing Transferosomes was investigated in intact surface of rat and human skin *in vivo*. The results indicated that when these Transferosomes were applied epicutaneously, blood sugar concentration was kept constant for at least 10 h, suggesting hypoglycemic activity of this formulation [91]. In a recent report by Cevc G group reported the utility of Transferosomes in immunisation against tetanus. A tetanus toxoid formulated with Transferosomes was applied to mice epicutaneously that were challenged with a lethal (50× LD50) parenteral tetanus toxin. The results have shown that Transferosomes formulation with tetanus toxoid was able to protect 100% of the tested mice, indicating a specific Th2 cellular response [92].

Many other novel liposomes were developed and investigated for their ability to penetrate through skin. These include Pharmacosomes®, which are a colloidal dispersion of drugs covalently linked to phospholipids and are mainly used for the entrapment of polar drug molecules [93]. Vesosomes® are vesicles fused with other vesicles and were shown to be promising systems for transcutaneous immunisation [94]. Proliposomes® are formulations that contain dry lipid and drug which form liposomal dispersion immediately after hydration. These were developed for improved physical and chemical stability [95]. Similarly, Proniosomes® are dry form of niosomes required to be hydrated immediately before use. They were developed to overcome the disadvantage of limited shelf-life usually seen in niosomes [96].

Finally, solid lipid nanoparticles (SLNs) and nanostructured lipid carriers (NLCs) are other class of lipid based nanocarriers widely investigated for their use in topical and transdermal delivery. SLNs are usually made of one lipid that is solid at room and body temperatures, and are characterised by a perfect crystalline lattice matrix. SLNs are stabilised in the dispersion by layering their surface with a surfactant [97]. By contrast, NLCs are made of both solid and liquid lipids that are macroscopically solid at room and body temperatures, and their matrix may be either a distorted crystalline lattice, or an amorphous lipid blend, or a solid lipid matrix entrapping liquid lipid based nanocompartments [5, 98]. Similar to SLNs, NLCs are also stabilised by covering their surface with surfactants. Both SLNs and NLCs can have either positive or negative surface [98]. Both SLNs and NLCs are rigid and have been widely investigated for their cosmetic and dermatological use [98, 99]. It is summarised that both these carriers form an occlusive film and enhanced drug permeation by these carriers is due to occlusive properties of the particles [5]. Table 6.2 below is a compilation of some of the lipid-based nanocarriers currently investigated for enhanced skin delivery of drugs.

6.2.3 Metallic and Mineral Nanoparticles

6.2.3.1 Magnetic Nanoparticles

These are usually made of iron and iron derivatives and could be magnetic, paramagnetic, or superparamagnetic in nature. In dermatology, magnetic nanoparticles can be used for early diagnosis of skin diseases, treatment of skin diseases by drug delivery or targeting, hyperthermia, or magnetofection. Furthermore, they can be used for skin wound healing by tissue engineering approach. Moreover, because of the magnetic properties of these NPs, they can be directed or localised in a particular layer of the skin where it is desired to act [5, 17]. In a report by Baroli *et al.* it was shown that iron oxide magnetic nanoparticles

Table 6.2   *Lipid-based nanocarriers for transdermal delivery of therapeutics*

Lipid-carrier	Drug	Indication	Reference
Liposomes	Tretinoin	Acne treatment	[100]
Liposomes	Lidocaine	Local anaesthesia	[101]
Liposomes	rhEGF	Wound healing	[102]
Liposomes	Curcumin	Anti-inflammatory	[103]
Ethosomes	Psoralen	Vitiligo and psoriasis	[104]
Ethosomes	Tamoxifen citrate	Burn, scars and breast cancer	[105]
Ethosomes	Testosterone propionate	Testosterone replacement therapy	[106]
Transfersomes	5-Fluorouracil	Melanoma	[107]
Transfersomes	Buspirone Hcl	Generalised anxiety disorder	[108]
Transfersomes	Itraconazole	Antifungal	[109]
Niosomes	Lopinavir	Anti-viral	[110]
Niosomes	Ellagic acid	Anti-oxidant	[111]
SLNs	Griseofulvin	Antifungal	[112]
SLNs	Isotretinoin	Acne	[113]
SLNs	Terbinafine	Antifungal	[114]
NLCs	Minoxidil & Finasteride	Anti-alopecia	[115]
NLCs	Docetaxel-nicotinamide complex	Anti-cancer	[116]
NLCs	Enoxaparin	Anti-coagulant	[117]

rhEGF, recombinant human epidermal growth factor.

when applied to skin were found in all layers of SC, at the stratum granulosum–SC interface, in hair follicles, and in some rare cases in the viable epidermis. However, none of these particles crossed the skin [17].

6.2.3.2   Quantum Dots

Quantum dots (QDs) used in transdermal delivery are core-shell nanomaterials comprised of elements in groups II/VI, III/V, and IV/VI. They exhibit unique photoluminescence behaviour and are currently being used as fluorescent labels. The 'shells' which are surface coatings can be customised based on the applications of QD. For instance, hydrophilic coatings can be used to increase the solubility in a biocompatible medium, or use coatings that reduce leaching of metals from the core, or use surface coatings with reactive groups that facilitate conjugation to antibodies, targeting ligands, therapeutics and diagnostic molecules [118]. Because of their success as diagnostic and imaging agents in biomedicine they have been investigated for their use in skin permeation. Prow *et al.* reported that QDs with various surface modifications such as polyethylene glycol (PEG), PEG-amine and PEG-carboxyl (PEG–COOH) were able to penetrate through the viable epidermis in human skin that was previously tape-stripped 30 strips of SC [119]. In another investigation the influence of shape and size of QDs on penetration through intact skin were reported. The results indicated that both spherical QDs with 4.6 nm core-shell diameter and ellipsoid with core-shell diameter 12 nm (major axis) by 6 nm (minor axis) were localised within the epidermal and dermal layers by 8 h [120].

6.2.3.3   Titanium Oxide- and Zinc Oxide Based Nanoparticles

Titanium and zinc oxides are widely used in the cosmetic industry as sunscreens because of their ability to scatter the UV light. They are also coated with aluminium oxide, silicon dioxide, or silicon oils for increasing dispersion stability. Usually, the size of titanium dioxide is up to several microns and that of zinc oxide is between 30 and 200 nm [4]. Nohynek *et al.* have reported that both titanium and zinc oxide NPs penetrate only in the outermost layer of SC indicating that they are non-toxic to humans [4]. Another report by Kimura *et al.* also showed that zinc and titanium oxide nanoparticles did not migrate from the surface of the skin, grooves or follicles into viable epidermis and dermis [121].

6.2.4   Carbon-Based Nanomaterials

Carbon nanotubes both single- or multi-walled, carbon hollow spheres and fullerenes (C_{60}) are some of the carbon based nanomaterials used for skin penetration. In a recent report, single-walled carbon nanotubes were functionalised with succinated polyethyleimine and their ability to transdermally deliver siRNA to mouse melanoma was investigated. The results have shown that topical application of nanotubes with siRNA resulted in attenuation of tumour growth over a period of 25 days indicating their potential for transdermal therapeutic delivery [122]. Amino acid derivatised fullerenes were investigated for their penetration ability through dermatomed porcine skin that was fixed with a flexing device. The results have indicated that fullerene NPs were able to penetrate faster (within 8 h) when skin was flexed for 60 and 90 min when compared with unflexed skin (after 24 h) [123].

6.3   Interactions of Nanoparticles with the Skin

The study of NP interaction with skin is important because it provides us with knowledge about the possible pathways of NP absorption through skin. This knowledge can be extrapolated either to develop better nanoparticulate systems for skin penetration or to identify the reasons for health hazards posed by NPs in the form of environmental pollutants which may be absorbed through the skin. As discussed in Section 6.1.2 NPs are absorbed through skin mainly by three major pathways: through SC, through opening of hair follicles and through skin furrows or dermatoglyphs [6, 8]. In this section the interaction mechanism of various classes of NPs discussed in Section 6.2 will be reviewed. This section is also intended to provide information about interaction of nanoparticles with proteins which may alter behaviour of both NPs and proteins [124]. Please refer to reviews by Schneider *et al.* [125], Hansen and Lehr [8] and Desai *et al.* [126] for a comprehensive review of the topic.

The most promising drug delivery carriers which are widely investigated for their skin penetration are polymeric NPs. Shim *et al.* investigated the skin penetration of polymeric NPs in guinea pig skin and reported that the penetration of NPs is dependent on size. However, their investigation with hairless animals did not show any penetration of NPs indicating that the pathway through hair follicles might be an important pathway for entry of polymeric NPs [59]. In another interesting study by Stracke *et al.* using multiphoton laser scanning microscopy, it was reported that intact PLGA NPs do not permeate the SC [55]. Luengo *et al.* reported similar findings where flufenamic acid-loaded PLGA NPs did not permeate into viable epidermis [127]. Overall, it is imperative that intact polymeric nanoparticles do not penetrate through SC unless they are deformed on skin surface and penetrate through hair follicles.

Liposomes belong to the class of NPs which were studied extensively for their interaction with dermal barriers. Depending on their structure liposomes can enter skin by several mechanisms, including intact vesicular skin penetration, due to trans-epidermal osmotic gradient and hydration force, lateral diffusion in the SC and transappendageal pathway [126, 128, 129]. It is well established in scientific literature that skin penetration of liposomes is dependent on composition of phospholipid, size and surface charge. Shanmugam *et al.* reported that the order of steady state flux of drug through skin from liposomes followed the order cationic > anionic > neutral. The enhanced skin penetration of cationic liposomes was explained by 'Donnan exclusion effect' which suggests permeation selectivity of skin [130]. Moreover, it was also reported that the positive charge of liposomes assists in binding of liposomes to negatively charged skin cells and hair follicles in an ion-exchange manner [131]. Transferosomes are deformable liposomes which have the capability to enter into the skin by virtue of their ability to squeeze between the cells of SC [132]. Finally, SLNs and NLCs penetrate through skin by eliciting an occlusive action on the skin which can result in reduction of corneocyte packing and opening of gaps between corneocytes that facilitate penetration of drug into deeper layers of skin [133].

Metallic nanoparticles such as magnetic nanoparticles which are made of iron and iron oxides can passively penetrate through skin provided their size is smaller than 10 nm. Baroli *et al.* has shown that rigid magnetic nanoparticles with size less than 10 nm were able to penetrate through the lipid matrix of SC and hair follicles and reach up to the stratum granulosum. They also reported that NPs aggregate in SC and upper layers of dermis [17]. QDs on the other hand penetrate the skin in a size and coating dependent manner. Zhang *et al.* showed that nail shaped QDs that are coated with PEG that are less than 40 nm in diameter were able to penetrate into uppermost layers of SC and outer root sheath of hair follicles [134]. Mortensen *et al.* reported that QDs can penetrate the skin through SC intracellular lipid lamellae along the edges of differentiated corneocytes [135]. Titanium oxide NPs failed to penetrate into deeper layers of SC even after repetitive application [136]. Similarly, the epidermal penetration of zinc oxide nanoparticles was negligible after topical application to intact human skin [137, 138].

6.4   Limitations of Nanotechnology for Skin Delivery

Even though applications of nanotechnology in topical and transdermal drug delivery were widely investigated; several limitations are associated with this strategy confounding clinical success in skin delivery of therapeutics. The limitations may include but not limited to are design and characterisation of materials used in nanotechnology, systemic toxicity of nanoparticles and dependence on external mechanical skin abrasion techniques for skin penetration. The limitations related to design and characterisation of NPs include ability to reproducibly engineer NPs with specific shapes and sizes, attain optimal drug loading of NPs for effective skin delivery, control and sustain release pattern of therapeutics from NPs and ensure stability of NPs to avoid harmful degradation products [6]. Some NPs used for skin delivery purposes have shown to permeate through skin via hair follicles or SC reach viable epidermis. Once they permeate into viable epidermis NPs can elicit inflammatory responses, interact with cells resulting in increased oxidative stress, and accumulate in major organs and tissues due to systemic absorption leading to overall systemic toxicity. Therefore, it is necessary to develop more methods to understand the behaviour of NPs after absorption into skin leading to risk assessment of these NPs for potential toxicity. It is also important to develop new NP design strategies involving size, shape and surface charge that would avoid potential toxicity issues [139]. Many NPs do not permeate through skin unless the absorption site of the skin in altered using external techniques such as flexing, massaging, microneedle-assisted delivery, tape-stripping of SC layers, thermal ablation, laser radiation, iontophoresis, sonophoresis, magnetophoresis and high-velocity (ballistic) injection techniques [50, 140–142]. The dependence on external invasive mechanical methods for enhanced skin permeation of NPs may reduce the patient compliance due to high cost associated with those methods and eventually affect the clinical usefulness of NPs for skin delivery.

6.5 Conclusions

Overall nanotechnology is an attractive strategy for improving skin penetration of therapeutics. The most important and well-researched factor which governs the penetration of NPs through skin is their size. Other properties such as shape and surface charge also influence skin penetration of NPs; however, more systematic and collaborative studies are needed to establish a relationship between these properties and nanoparticle absorption. It is also important to carefully consider the use of animal models for studying skin penetration of NPs in order to establish their ability to mimic human skin absorption of NPs. The models designed for studying skin penetration of NPs must also include parameters related to unhealthy human skin, inter-species differences, formulation variables and intentional- and unintentional skin exposure to NP based formulations. Since its introduction to the field of skin delivery three decades ago, nanotechnology has gone through great strides in the development of new materials for designing NPs. Materials such as NLCs, Transferosomes, PLGA and dendrimers have shown great promise in transdermal delivery of drugs and other therapeutics such as vaccines. However, it is very important to establish concrete information about degradation profile, elimination pattern, length and magnitude of accumulation in major organs, and interaction with skin components for those NPs as they can have regulatory impact related to toxicity concerns. In conclusion, nanotechnology holds great promise in improving the delivery of drugs and other therapeutics including vaccines, proteins, nucleic acids, and so on, through skin, provided materials used are nontoxic, biocompatible, non-immunogenic, easy to manufacture, have intrinsic capability to permeate through skin barriers and do not depend solely on physical barrier disruption methods.

References

[1] Frequesntly Asked Questions (2010) *Nanotechnology 101*. http://www.nano.gov/nanotech-101/nanotechnology-facts (accessed 26 November 2013).
[2] Bertrand, N., Wu, J., Xu, X. *et al.* (2014) Cancer nanotechnology: the impact of passive and active targeting in the era of modern cancer biology. *Adv. Drug Deliv. Rev.*, **66**, 2–25.

[3] Cheng, Y., Morshed, R.A., Auffinger, B. *et al.* (2014) Multifunctional nanoparticles for brain tumor imaging and therapy. *Adv. Drug Deliv. Rev.*, **66**, 42–57.

[4] Nohynek, G.J., Lademann, J., Ribaud, C. and Roberts, M.S. (2007) Grey goo on the skin? Nanotechnology, cosmetic and sunscreen safety. *Crit. Rev. Toxicol.*, **37** (3), 251–277.

[5] Baroli, B. (2010) Penetration of nanoparticles and nanomaterials in the skin: fiction or reality? *J. Pharm. Sci.*, **99** (1), 21–50.

[6] Prow, T.W., Grice, J.E., Lin, L.L. *et al.* (2011) Nanoparticles and microparticles for skin drug delivery. *Adv. Drug Deliv. Rev.*, **63** (6), 470–491.

[7] Pegoraro, C., MacNeil, S. and Battaglia, G. (2012) Transdermal drug delivery: from micro to nano. *Nanoscale*, **4** (6), 1881–1894.

[8] Hansen, S. and Lehr, C.M. (2012) Nanoparticles for transcutaneous vaccination. *Microb. Biotechnol.*, **5** (2), 156–167.

[9] DeLouise, L.A. (2012) Applications of nanotechnology in dermatology. *J. Invest. Dermatol.*, **132** (3 Pt 2), 964–975.

[10] Prausnitz, M.R., Mitragotri, S. and Langer, R. (2004) Current status and future potential of transdermal drug delivery. *Nat. Rev. Drug Discov.*, **3** (2), 115–124.

[11] Kierszenbaum, A.L. and Tres, L.L. (2011) Integumentary system, in *Histology and Cell Biology, An Introduction to Pathology*, 3rd edn (eds A.L. Kierszenbaum and L.L. Tres), Elsevier Saunders, Philadelphia, pp. 339–344.

[12] Nemes, Z. and Steinert, P.M. (1999) Bricks and mortar of the epidermal barrier. *Exp. Mol. Med.*, **31** (1), 5–19.

[13] Elias, P.M. (2005) Stratum corneum defensive functions: an integrated view. *J. Invest. Dermatol.*, **125** (2), 183–200.

[14] Bouwstra, J.A. and Ponec, M. (2006) The skin barrier in healthy and diseased state. *Biochim. Biophys. Acta*, **1758** (12), 2080–2095.

[15] Scheuplein, R.J. (1967) Mechanism of percutaneous absorption. II. Transient diffusion and the relative importance of various routes of skin penetration. *J. Invest. Dermatol.*, **48** (1), 79–88.

[16] Scientific Committee on Consumer Products (2008) On regulatory aspects of nanomaterials. Committee on the Environment, Public Health and Food Safety, European Parliament, Brussels.

[17] Baroli, B., Ennas, M.G., Loffredo, F. *et al.* (2007) Penetration of metallic nanoparticles in human full-thickness skin. *J. Invest. Dermatol.*, **127** (7), 1701–1712.

[18] Ryman-Rasmussen, J.P., Riviere, J.E. and Monteiro-Riviere, N.A. (2007) Variables influencing interactions of untargeted quantum dot nanoparticles with skin cells and identification of biochemical modulators. *Nano Lett.*, **7** (5), 1344–1348.

[19] van der Merwe, D., Brooks, J.D., Gehring, R. *et al.* (2006) A physiologically based pharmacokinetic model of organophosphate dermal absorption. *Toxicol. Sci.*, **89** (1), 188–204.

[20] Reddy, M.B., Guy, R.H. and Bunge, A.L. (2000) Does epidermal turnover reduce percutaneous penetration? *Pharm. Res.*, **17** (11), 1414–1419.

[21] Marks, R. (2004) The stratum corneum barrier: the final frontier. *J. Nutr.*, **134** (Suppl. 8), 2017S–2021S.

[22] Tso, C.P., Zhung, C.M., Shih, Y.H. *et al.* (2010) Stability of metal oxide nanoparticles in aqueous solutions. *Water Sci. Technol.*, **61** (1), 127–133.

[23] Nitsche, J.M. and Kasting, G.B. (2013) A microscopic multiphase diffusion model of viable epidermis permeability. *Biophys. J.*, **104** (10), 2307–2320.

[24] Oesch, F., Fabian, E., Oesch-Bartlomowicz, B. *et al.* (2007) Drug-metabolizing enzymes in the skin of man, rat, and pig. *Drug Metab. Rev.*, **39** (4), 659–698.

[25] Brandner, J.M., Kief, S., Grund, C. *et al.* (2002) Organization and formation of the tight junction system in human epidermis and cultured keratinocytes. *Eur. J. Cell Biol.*, **81** (5), 253–263.

[26] Langbein, L., Grund, C., Kuhn, C. *et al.* (2002) Tight junctions and compositionally related junctional structures in mammalian stratified epithelia and cell cultures derived therefrom. *Eur. J. Cell Biol.*, **81** (8), 419–435.

[27] Brandner, J.M., McIntyre, M., Kief, S. *et al.* (2003) Expression and localization of tight junction-associated proteins in human hair follicles. *Arch. Dermatol. Res.*, **295** (5), 211–221.

[28] Baroli, B. (2010) Skin absorption and potential toxicity of nanoparticulate nanomaterials. *J. Biomed. Nanotechnol.*, **6** (5), 485–496.

[29] Gillet, A., Lecomte, F., Hubert, P. *et al.* (2011) Skin penetration behaviour of liposomes as a function of their composition. *Eur. J. Pharm. Biopharm.*, **79** (1), 43–53.

[30] Labouta, H.I., Liu, D.C., Lin, L.L. *et al.* (2011) Gold nanoparticle penetration and reduced metabolism in human skin by toluene. *Pharm. Res.*, **28** (11), 2931–2944.

[31] Singh, P. and Nanda, A. (2014) Enhanced sun protection of nano-sized metal oxide particles over conventional metal oxide particles: an *in vitro* comparative study. *Int. J. Cosmet. Sci.*, **36** (3), 273–283.

[32] Yellepeddi, V.K., Kumar, A. and Palakurthi, S. (2009) Surface modified poly(amido)amine dendrimers as diverse nanomolecules for biomedical applications. *Expert Opin. Drug Deliv.*, **6** (8), 835–850.

[33] Yellepeddi, V.K., Kumar, A., Maher, D.M. *et al.* (2011) Biotinylated PAMAM dendrimers for intracellular delivery of cisplatin to ovarian cancer: role of SMVT. *Anticancer Res*, **31** (3), 897–906.

[34] Thiagarajan, G., Ray, A., Malugin, A. and Ghandehari, H. (2010) PAMAM-camptothecin conjugate inhibits proliferation and induces nuclear fragmentation in colorectal carcinoma cells. *Pharm. Res.*, **27** (11), 2307–2316.

[35] El-Sayed, M., Ginski, M., Rhodes, C. and Ghandehari, H. (2002) Transepithelial transport of poly(amidoamine) dendrimers across Caco-2 cell monolayers. *J. Control. Release*, **81** (3), 355–365.

[36] Kitchens, K.M., Kolhatkar, R.B., Swaan, P.W. *et al.* (2006) Transport of poly(amidoamine) dendrimers across Caco-2 cell monolayers: influence of size, charge and fluorescent labeling. *Pharm. Res.*, **23** (12), 2818–2826.

[37] Mignani, S., El Kazzouli, S., Bousmina, M. and Majoral, J.P. (2013) Expand classical drug administration ways by emerging routes using dendrimer drug delivery systems: a concise overview. *Adv. Drug Deliv. Rev.*, **65** (10), 1316–1330.

[38] Sun, M.J., Fan, A.P., Wang, Z. and Zhao, Y.J. (2012) Dendrimer-mediated drug delivery to the skin. *Soft Matter.*, **8** (16), 4301–4305.

[39] Venuganti, V.V., Sahdev, P., Hildreth, M. *et al.* (2011) Structure-skin permeability relationship of dendrimers. *Pharm. Res.*, **28** (9), 2246–2260.

[40] Yang, Y., Sunoqrot, S., Stowell, C. *et al.* (2012) Effect of size, surface charge, and hydrophobicity of poly(amidoamine) dendrimers on their skin penetration. *Biomacromolecules*, **13** (7), 2154–2162.

[41] Agrawal, U., Mehra, N.K., Gupta, U. and Jain, N.K. (2013) Hyperbranched dendritic nanocarriers for topical delivery of dithranol. *J. Drug Target.*, **21** (5), 497–506.

[42] Borowska, K., Wolowiec, S., Glowniak, K. *et al.* (2012) Transdermal delivery of 8-methoxypsoralene mediated by polyamidoamine dendrimer G2.5 and G3.5 – *in vitro* and *in vivo* study. *Int. J. Pharm.*, **436** (1–2), 764–770.

[43] Cheng, Y., Man, N., Xu, T. *et al.* (2007) Transdermal delivery of nonsteroidal anti-inflammatory drugs mediated by polyamidoamine (PAMAM) dendrimers. *J. Pharm. Sci.*, **96** (3), 595–602.

[44] Chauhan, A.S., Sridevi, S., Chalasani, K.B. *et al.* (2003) Dendrimer-mediated transdermal delivery: enhanced bioavailability of indomethacin. *J. Control. Release*, **90** (3), 335–343.

[45] Filipowicz, A. and Wolowiec, S. (2011) Solubility and *in vitro* transdermal diffusion of riboflavin assisted by PAMAM dendrimers. *Int. J. Pharm.*, **408** (1–2), 152–156.

[46] Venuganti, V.V. and Perumal, O.P. (2008) Effect of poly(amidoamine) (PAMAM) dendrimer on skin permeation of 5-fluorouracil. *Int. J. Pharm.*, **361** (1–2), 230–238.

[47] Battah, S., O'Neill, S., Edwards, C. *et al.* (2006) Enhanced porphyrin accumulation using dendritic derivatives of 5-aminolaevulinic acid for photodynamic therapy: an *in vitro* study. *Int. J. Biochem. Cell Biol.*, **38** (8), 1382–1392.

[48] Bielinska, A.U., Yen, A., Wu, H.L. *et al.* (2000) Application of membrane-based dendrimer/DNA complexes for solid phase transfection *in vitro* and *in vivo*. *Biomaterials*, **21** (9), 877–887.

[49] Daftarian, P.M., Stone, G.W., Kovalski, L. *et al.* (2013) A targeted and adjuvanted nanocarrier lowers the effective dose of liposomal amphotericin B and enhances adaptive immunity in murine cutaneous leishmaniasis. *J. Infect. Dis.*, **208** (11), 1914–1922.

[50] Mutalik, S., Parekh, H.S., Anissimov, Y.G. *et al.* (2013) Iontophoresis-mediated transdermal permeation of peptide dendrimers across human epidermis. *Skin Pharmacol. Physiol.*, **26** (3), 127–138.

[51] Kalhapure, R.S. and Akamanchi, K.G. (2013) Oleodendrimers: a novel class of multicephalous heterolipids as chemical penetration enhancers for transdermal drug delivery. *Int. J. Pharm.*, **454** (1), 158–166.

[52] Pawar, K.R. and Babu, R.J. (2010) Polymeric and lipid-based materials for topical nanoparticle delivery systems. *Crit. Rev. Ther. Drug Carrier Syst.*, **27** (5), 419–459.

[53] Li, J., Zhai, Y.L., Zhang, B. *et al.* (2008) Methoxy poly(ethylene glycol)-block-poly(d,l-lactic acid) copolymer nanoparticles as carriers for transdermal drug delivery. *Polym. Int.*, **57** (2), 268–274.

[54] Mattheolabakis, G., Lagoumintzis, G., Panagi, Z. *et al.* (2010) Transcutaneous delivery of a nanoencapsulated antigen: induction of immune responses. *Int. J. Pharm.*, **385** (1–2), 187–193.

[55] Stracke, F., Weiss, B., Lehr, C.M. *et al.* (2006) Multiphoton microscopy for the investigation of dermal penetration of nanoparticle-borne drugs. *J. Invest. Dermatol.*, **126** (10), 2224–2233.

[56] da Silva, C.L., Del Ciampo, J.O., Rossetti, F.C. *et al.* (2013) PLGA nanoparticles as delivery systems for protoporphyrin IX in topical PDT: cutaneous penetration of photosensitizer observed by fluorescence microscopy. *J. Nanosci. Nanotechnol.*, **13** (10), 6533–6540.

[57] de Carvalho, R.F., Ribeiro, I.F., Miranda-Vilela, A.L. *et al.* (2013) Leishmanicidal activity of amphotericin B encapsulated in PLGA-DMSA nanoparticles to treat cutaneous leishmaniasis in C57BL/6 mice. *Exp. Parasitol.*, **135** (2), 217–222.

[58] Ponsart, S., Coudane, J. and Vert, M. (2000) A novel route to poly(epsilon-caprolactone)-based copolymers via anionic derivatization. *Biomacromolecules*, **1** (2), 275–281.

[59] Shim, J., Seok Kang, H., Park, W.S. *et al.* (2004) Transdermal delivery of mixnoxidil with block copolymer nanoparticles. *J. Control. Release*, **97** (3), 477–484.

[60] Rastogi, R., Anand, S. and Koul, V. (2009) Flexible polymerosomes – an alternative vehicle for topical delivery. *Colloids Surf. B Biointerfaces*, **72** (1), 161–166.

[61] Smith, A., Perelman, M. and Hinchcliffe, M. (2013) Chitosan: a promising safe and immune-enhancing adjuvant for intranasal vaccines. *Hum. Vaccin. Immunother.*, **10** (3), 797–807.

[62] He, W., Guo, X., Xiao, L. and Feng, M. (2009) Study on the mechanisms of chitosan and its derivatives used as transdermal penetration enhancers. *Int. J. Pharm.*, **382** (1–2), 234–243.

[63] Hasanovic, A., Zehl, M., Reznicek, G. and Valenta, C. (2009) Chitosan-tripolyphosphate nanoparticles as a possible skin drug delivery system for aciclovir with enhanced stability. *J. Pharm. Pharmacol.*, **61** (12), 1609–1616.

[64] Cui, Z. and Mumper, R.J. (2001) Chitosan-based nanoparticles for topical genetic immunization. *J. Control. Release*, **75** (3), 409–419.

[65] Kataria, K., Gupta, A., Rath, G. *et al.* (2014) *In vivo* wound healing performance of drug loaded electrospun composite nanofibers transdermal patch. *Int. J. Pharm.*, **469**, 102–110.

[66] Demir, Y.K., Akan, Z. and Kerimoglu, O. (2013) Characterization of polymeric microneedle arrays for transdermal drug delivery. *PLoS One*, **8** (10), e77289.

[67] Mazzitelli, S., Pagano, C., Giusepponi, D. *et al.* (2013) Hydrogel blends with adjustable properties as patches for transdermal delivery. *Int. J. Pharm.*, **454** (1), 47–57.

[68] Wong, T.W. and Nor, K.A. (2013) Physicochemical modulation of skin barrier by microwave for transdermal drug delivery. *Pharm. Res.*, **30** (1), 90–103.

[69] Wen, M.M., Farid, R.M. and Kassem, A.A. (2014) Nano-proniosomes enhancing the transdermal delivery of mefenamic acid. *J. Liposome Res.*, **24**, 280–289.

[70] Elnaggar, Y.S., El-Refaie, W.M., El-Massik, M.A. and Abdallah, O.Y. (2014) Lecithin-based nanostructured gels for skin delivery: an update on state of art and recent applications. *J. Control. Release*, **180C**, 10–24.

[71] Ghica, M.V., Albu, M.G., Leca, M. *et al.* (2011) Design and optimization of some collagen-minocycline based hydrogels potentially applicable for the treatment of cutaneous wound infections. *Pharmazie*, **66** (11), 853–861.

[72] Swatschek, D., Schatton, W., Muller, W. and Kreuter, J. (2002) Microparticles derived from marine sponge collagen (SCMPs): preparation, characterization and suitability for dermal delivery of all-trans retinol. *Eur. J. Pharm. Biopharm.*, **54** (2), 125–133.

[73] Martins, M., Azoia, N.G., Shimanovich, U. *et al.* (2014) Design of novel BSA/hyaluronic acid nanodispersions for transdermal pharma purposes. *Mol. Pharm.*, **11**, 1479–1488.

[74] Liu, S., Jin, M.N., Quan, Y.S. *et al.* (2014) Transdermal delivery of relatively high molecular weight drugs using novel self-dissolving microneedle arrays fabricated from hyaluronic acid and their characteristics and safety after application to the skin. *Eur. J. Pharm. Biopharm.*, **86** (2), 267–276.

[75] Kong, M., Park, H., Feng, C. *et al.* (2013) Construction of hyaluronic acid noisome as functional transdermal nanocarrier for tumor therapy. *Carbohydr. Polym.*, **94** (1), 634–641.

[76] Gupta, M., Agrawal, U. and Vyas, S.P. (2012) Nanocarrier-based topical drug delivery for the treatment of skin diseases. *Expert Opin. Drug Deliv.*, **9** (7), 783–804.

[77] Benson, H.A. (2006) Transfersomes for transdermal drug delivery. *Expert Opin. Drug Deliv.*, **3** (6), 727–737.

[78] El Maghraby, G.M., Williams, A.C. and Barry, B.W. (2006) Can drug-bearing liposomes penetrate intact skin? *J. Pharm. Pharmacol.*, **58** (4), 415–429.

[79] Elsayed, M.M., Abdallah, O.Y., Naggar, V.F. and Khalafallah, N.M. (2007) PG-liposomes: novel lipid vesicles for skin delivery of drugs. *J. Pharm. Pharmacol.*, **59** (10), 1447–1450.

[80] Elsayed, M.M., Abdallah, O.Y., Naggar, V.F. and Khalafallah, N.M. (2007) Lipid vesicles for skin delivery of drugs: reviewing three decades of research. *Int. J. Pharm.*, **332** (1–2), 1–16.

[81] Pierre, M.B. and Dos Santos Miranda Costa, I. (2011) Liposomal systems as drug delivery vehicles for dermal and transdermal applications. *Arch. Dermatol. Res.*, **303** (9), 607–621.

[82] Yarosh, D.B., O'Connor, A., Alas, L. *et al.* (1999) Photoprotection by topical DNA repair enzymes: molecular correlates of clinical studies. *Photochem. Photobiol.*, **69** (2), 136–140.

[83] Vora, B., Khopade, A.J. and Jain, N.K. (1998) Proniosome based transdermal delivery of levonorgestrel for effective contraception. *J. Control. Release*, **54** (2), 149–165.

[84] Fang, J.Y., Hong, C.T., Chiu, W.T. and Wang, Y.Y. (2001) Effect of liposomes and niosomes on skin permeation of enoxacin. *Int. J. Pharm.*, **219** (1–2), 61–72.

[85] Shahiwala, A. and Misra, A. (2002) Studies in topical application of niosomally entrapped Nimesulide. *J. Pharm. Pharm. Sci.*, **5** (3), 220–225.

[86] Vyas, S.P., Singh, R.P., Jain, S. *et al.* (2005) Non-ionic surfactant based vesicles (niosomes) for non-invasive topical genetic immunization against hepatitis B. *Int. J. Pharm.*, **296** (1–2), 80–86.

[87] Dubey, V., Mishra, D. and Jain, N.K. (2007) Melatonin loaded ethanolic liposomes: physico-chemical characterization and enhanced transdermal delivery. *Eur. J. Pharm. Biopharm.*, **67** (2), 398–405.

[88] Dayan, N. and Touitou, E. (2000) Carriers for skin delivery of trihexyphenidyl HCl: ethosomes vs. liposomes. *Biomaterials*, **21** (18), 1879–1885.

[89] Touitou, E., Dayan, N., Bergelson, L. *et al.* (2000) Ethosomes – novel vesicular carriers for enhanced delivery: characterization and skin penetration properties. *J. Control. Release*, **65** (3), 403–418.

[90] Cevc, G. (2004) Lipid vesicles and other colloids as drug carriers on the skin. *Adv. Drug Deliv. Rev.*, **56** (5), 675–711.

[91] Cevc, G., Gebauer, D., Stieber, J. *et al.* (1998) Ultraflexible vesicles, Transfersomes, have an extremely low pore penetration resistance and transport therapeutic amounts of insulin across the intact mammalian skin. *Biochim. Biophys. Acta*, **1368** (2), 201–215.

[92] Chopra, A. and Cevc, G. (2014) Non-invasive, epicutaneous immunisation with toxoid in deformable vesicles protects mice against tetanus, chiefly owing to a Th2 response. *Eur. J. Pharm. Sci.*, **56**, 55–64.

[93] Vaizoglu, M.O. and Speiser, P.P. (1986) Pharmacosomes – a novel drug delivery system. *Acta Pharm. Suec.*, **23** (3), 163–172.

[94] Mishra, V., Mahor, S., Rawat, A. *et al.* (2006) Development of novel fusogenic vesosomes for transcutaneous immunization. *Vaccine*, **24** (27–28), 5559–5570.

[95] Hiremath, P.S., Soppimath, K.S. and Betageri, G.V. (2009) Proliposomes of exemestane for improved oral delivery: formulation and *in vitro* evaluation using PAMPA, Caco-2 and rat intestine. *Int. J. Pharm.*, **380** (1–2), 96–104.

[96] Hu, C. and Rhodes, D.G. (1999) Proniosomes: a novel drug carrier preparation. *Int. J. Pharm.*, **185** (1), 23–35.

[97] Schafer-Korting, M., Mehnert, W. and Korting, H.C. (2007) Lipid nanoparticles for improved topical application of drugs for skin diseases. *Adv. Drug Deliv. Rev.*, **59** (6), 427–443.

[98] Muller, R.H., Radtke, M. and Wissing, S.A. (2002) Solid lipid nanoparticles (SLN) and nanostructured lipid carriers (NLC) in cosmetic and dermatological preparations. *Adv. Drug Deliv. Rev.*, **54** (Suppl. 1), S131–S155.

[99] Wissing, S.A. and Muller, R.H. (2002) Solid lipid nanoparticles as carrier for sunscreens: *in vitro* release and *in vivo* skin penetration. *J. Control. Release*, **81** (3), 225–233.

[100] Patel, V.B., Misra, A. and Marfatia, Y.S. (2000) Topical liposomal gel of tretinoin for the treatment of acne: research and clinical implications. *Pharm. Dev. Technol.*, **5** (4), 455–464.

[101] Taddio, A., Soin, H.K., Schuh, S. *et al.* (2005) Liposomal lidocaine to improve procedural success rates and reduce procedural pain among children: a randomized controlled trial. *CMAJ*, **172** (13), 1691–1695.

[102] Yin, F., Guo, S., Gan, Y. and Zhang, X. (2014) Preparation of redispersible liposomal dry powder using an ultrasonic spray freeze-drying technique for transdermal delivery of human epithelial growth factor. *Int. J. Nanomedicine*, **9**, 1665–1676.

[103] Zhao, Y.Z., Lu, C.T., Zhang, Y. *et al.* (2013) Selection of high efficient transdermal lipid vesicle for curcumin skin delivery. *Int. J. Pharm.*, **454** (1), 302–309.

[104] Zhang, Y.T., Shen, L.N., Zhao, J.H. and Feng, N.P. (2014) Evaluation of psoralen ethosomes for topical delivery in rats by using *in vivo* microdialysis. *Int. J. Nanomedicine*, **9**, 669–678.

[105] Sarwa, K.K., Suresh, P.K., Rudrapal, M. and Verma, V.K. (2014) Penetration of tamoxifen citrate loaded ethosomes and liposomes across human skin: a comparative study with confocal laser scanning microscopy. *Curr. Drug Deliv.*, **11**, 332–337.

[106] Meng, S., Chen, Z., Yang, L. *et al.* (2013) Enhanced transdermal bioavailability of testosterone propionate via surfactant-modified ethosomes. *Int. J. Nanomedicine*, **8**, 3051–3060.

[107] Khan, M.A., Pandit, J., Sultana, Y. *et al.* (2014) Novel carbopol-based transfersomal gel of 5-fluorouracil for skin cancer treatment: *in vitro* characterization and *in vivo* study. *Drug Deliv.* in press.

[108] Shamma, R.N. and Elsayed, I. (2013) Transfersomal lyophilized gel of buspirone HCl: formulation, evaluation and statistical optimization. *J. Liposome Res.*, **23** (3), 244–254.

[109] Zheng, W.S., Fang, X.Q., Wang, L.L. and Zhang, Y.J. (2012) Preparation and quality assessment of itraconazole transfersomes. *Int. J. Pharm.*, **436** (1–2), 291–298.

[110] Patel, K.K., Kumar, P. and Thakkar, H.P. (2012) Formulation of niosomal gel for enhanced transdermal lopinavir delivery and its comparative evaluation with ethosomal gel. *AAPS PharmSciTech*, **13** (4), 1502–1510.

[111] Junyaprasert, V.B., Singhsa, P., Suksiriworapong, J. and Chantasart, D. (2012) Physicochemical properties and skin permeation of Span 60/Tween 60 niosomes of ellagic acid. *Int. J. Pharm.*, **423** (2), 303–311.

[112] Aggarwal, N. and Goindi, S. (2013) Preparation and *in vivo* evaluation of solid lipid nanoparticles of griseofulvin for dermal use. *J. Biomed. Nanotechnol.*, **9** (4), 564–576.

[113] Raza, K., Singh, B., Singal, P. *et al.* (2013) Systematically optimized biocompatible isotretinoin-loaded solid lipid nanoparticles (SLNs) for topical treatment of acne. *Colloids Surf. B Biointerfaces*, **105**, 67–74.

[114] Chen, Y.C., Liu, D.Z., Liu, J.J. *et al.* (2012) Development of terbinafine solid lipid nanoparticles as a topical delivery system. *Int. J. Nanomedicine*, **7**, 4409–4418.

[115] Gomes, M.J., Martins, S., Ferreira, D. *et al.* (2014) Lipid nanoparticles for topical and transdermal application for alopecia treatment: development, physicochemical characterization, and *in vitro* release and penetration studies. *Int. J. Nanomedicine*, **9**, 1231–1242.

[116] Fan, X., Chen, J. and Shen, Q. (2013) Docetaxel-nicotinamide complex-loaded nanostructured lipid carriers for transdermal delivery. *Int. J. Pharm.*, **458** (2), 296–304.

[117] Jain, A., Mehra, N.K., Nahar, M. and Jain, N.K. (2013) Topical delivery of enoxaparin using nanostructured lipid carrier. *J. Microencapsul.*, **30** (7), 709–715.

[118] Michalet, X., Pinaud, F.F., Bentolila, L.A. *et al.* (2005) Quantum dots for live cells, *in vivo* imaging, and diagnostics. *Science*, **307** (5709), 538–544.

[119] Prow, T.W., Monteiro-Riviere, N.A., Inman, A.O. *et al.* (2012) Quantum dot penetration into viable human skin. *Nanotoxicology*, **6** (2), 173–185.

[120] Ryman-Rasmussen, J.P., Riviere, J.E. and Monteiro-Riviere, N.A. (2006) Penetration of intact skin by quantum dots with diverse physicochemical properties. *Toxicol. Sci.*, **91** (1), 159–165.

[121] Kimura, E., Kawano, Y., Todo, H. *et al.* (2012) Measurement of skin permeation/penetration of nanoparticles for their safety evaluation. *Biol. Pharm. Bull.*, **35** (9), 1476–1486.

[122] Siu, K.S., Chen, D., Zheng, X. *et al.* (2014) Non-covalently functionalized single-walled carbon nanotube for topical siRNA delivery into melanoma. *Biomaterials*, **35** (10), 3435–3442.

[123] Rouse, J.G., Yang, J., Ryman-Rasmussen, J.P. *et al.* (2007) Effects of mechanical flexion on the penetration of fullerene amino acid-derivatized peptide nanoparticles through skin. *Nano Lett.*, **7** (1), 155–160.

[124] Scientific Committee on Emerging and Newly Identified Health Risks (SCENIHR) SCoEaN-IHR (2009) Risk assessment of products of nanotechnologis. European Commission Health and Consumer Protection Directorate General, Brussels.

[125] Schneider, M., Stracke, F., Hansen, S. and Schaefer, U.F. (2009) Nanoparticles and their interactions with the dermal barrier. *Dermatoendocrinology*, **1** (4), 197–206.

[126] Desai, P., Patlolla, R.R. and Singh, M. (2010) Interaction of nanoparticles and cell-penetrating peptides with skin for transdermal drug delivery. *Mol. Membr. Biol.*, **27** (7), 247–259.

[127] Luengo, J., Weiss, B., Schneider, M. *et al.* (2006) Influence of nanoencapsulation on human skin transport of flufenamic acid. *Skin Pharmacol. Physiol.*, **19** (4), 190–197.

[128] El Maghraby, G.M., Barry, B.W. and Williams, A.C. (2008) Liposomes and skin: from drug delivery to model membranes. *Eur. J. Pharm. Sci.*, **34** (4–5), 203–222.

[129] de Leeuw, J., de Vijlder, H.C., Bjerring, P. and Neumann, H.A. (2009) Liposomes in dermatology today. *J. Eur. Acad. Dermatol. Venereol.*, **23** (5), 505–516.

[130] Shanmugam, S., Song, C.K., Nagayya-Sriraman, S. *et al.* (2009) Physicochemical characterization and skin permeation of liposome formulations containing clindamycin phosphate. *Arch. Pharm. Res.*, **32** (7), 1067–1075.

[131] Jung, S., Otberg, N., Thiede, G. *et al.* (2006) Innovative liposomes as a transfollicular drug delivery system: penetration into porcine hair follicles. *J. Invest. Dermatol.*, **126** (8), 1728–1732.

[132] Honeywell-Nguyen, P.L., Gooris, G.S. and Bouwstra, J.A. (2004) Quantitative assessment of the transport of elastic and rigid vesicle components and a model drug from these vesicle formulations into human skin *in vivo*. *J. Invest. Dermatol.*, **123** (5), 902–910.

[133] Wissing, S.A. and Muller, R.H. (2003) Cosmetic applications for solid lipid nanoparticles (SLN). *Int. J. Pharm.*, **254** (1), 65–68.

[134] Zhang, L.W., Yu, W.W., Colvin, V.L. and Monteiro-Riviere, N.A. (2008) Biological interactions of quantum dot nanoparticles in skin and in human epidermal keratinocytes. *Toxicol. Appl. Pharmacol.*, **228** (2), 200–211.

[135] Mortensen, L.J., Oberdorster, G., Pentland, A.P. and Delouise, L.A. (2008) *In vivo* skin penetration of quantum dot nanoparticles in the murine model: the effect of UVR. *Nano Lett.*, **8** (9), 2779–2787.

[136] Lademann, J., Weigmann, H., Rickmeyer, C. *et al.* (1999) Penetration of titanium dioxide microparticles in a sunscreen formulation into the horny layer and the follicular orifice. *Skin Pharmacol. Appl. Skin Physiol.*, **12** (5), 247–256.

[137] Gamer, A.O., Leibold, E. and van Ravenzwaay, B. (2006) The *in vitro* absorption of microfine zinc oxide and titanium dioxide through porcine skin. *Toxicol. In Vitro*, **20** (3), 301–307.

[138] Cross, S.E., Innes, B., Roberts, M.S. *et al.* (2007) Human skin penetration of sunscreen nanoparticles: *in-vitro* assessment of a novel micronized zinc oxide formulation. *Skin Pharmacol. Physiol.*, **20** (3), 148–154.

[139] Kunzmann, A., Andersson, B., Thurnherr, T. *et al.* (2011) Toxicology of engineered nanomaterials: focus on biocompatibility, biodistribution and biodegradation. *Biochim. Biophys. Acta*, **1810** (3), 361–373.

[140] Ochoa, M., Mousoulis, C. and Ziaie, B. (2012) Polymeric microdevices for transdermal and subcutaneous drug delivery. *Adv. Drug Deliv. Rev.*, **64** (14), 1603–1616.

[141] Rizwan, M., Aqil, M., Talegaonkar, S. *et al.* (2009) Enhanced transdermal drug delivery techniques: an extensive review of patents. *Recent Pat. Drug Deliv. Formul.*, **3** (2), 105–124.

[142] Zaric, M., Lyubomska, O., Touzelet, O. *et al.* (2013) Skin dendritic cell targeting via microneedle arrays laden with antigen-encapsulated poly-D,L-lactide-co-glycolide nanoparticles induces efficient antitumor and antiviral immune responses. *ACS Nano*, **7** (3), 2042–2055.

7

Magnetophoresis and Electret-Mediated Transdermal Delivery of Drugs

Abhijeet Maurya[1], Cui Lili[2] and S. Narasimha Murthy[1]

[1] *School of Pharmacy, The University of Mississippi, University, MS, USA*
[2] *Department of Inorganic Chemistry, School of Pharmacy, Second Military Medical University, Shanghai, China*

7.1   Introduction

Skin is the largest organ of the body forming a protective covering over underlying tissues and organs. The skin guards the body from extremes of temperature, UV light and pathogens in the atmosphere. In addition to its protective properties, skin has also been a site for topical application of substances. The use of cosmetics has been an age-old practise. For example, zinc oxide which is a common ingredient in currently used creams, such as ointments and sunscreens, was widely used as a facial powder in the eighteenth century. However, application of medicaments to skin for therapeutic application was not popular as skin was largely believed to be impermeable. Nonetheless, the presumption of skin's impermeability was ruled out when Bourget and coworkers demonstrated that salicylic acid was absorbed through the skin when applied as a fatty acid ointment and was well tolerated than oral therapy for treatment of rheumatoid arthritis [1]. In the later years, topical application of medicament metamorphosed to a more perceptive therapy for treating local skin ailments. It was in the mid-twentieth century that skin was recognised as a portal for systemic drug application, and thus nitroglycerine ointment for treating angina was approved as the

Novel Delivery Systems for Transdermal and Intradermal Drug Delivery, First Edition.
Ryan F. Donnelly and Thakur Raghu Raj Singh.
© 2015 John Wiley & Sons, Ltd. Published 2015 by John Wiley & Sons, Ltd.

first commercially available topical product on the shelves. Thirty years after the arrival of first topical preparation in the market, the FDA approved the first transdermal patch manufactured by ALZA Corporation containing scopolamine for treatment of motion sickness [2, 3]. Following years brought a number of transdermal drug product approvals for treating various systemic therapeutic indications [4, 5]. The success of transdermal products can be attributed to its advantages and convenience over other routes of drug administration. Delivering drugs transdermally circumvents the stomach environment, avoids first-pass metabolism, minimises the frequency of administration and offers the advantage of immediate cessation of therapy when untoward reactions are observed in the patient. These advantages of transdermal delivery over other systemic delivery routes improve adherence to transdermal medication in patients [6]. The fact that transdermal route of drug delivery is non-invasive makes this technique more reliable and adds to the overall benefits of this route. As it is encountered with other drug delivery systems, transdermal drug delivery is also faced with challenges and limitations. The transport of drug molecules through skin is challenged by the exceptional barrier function of the stratum corneum (SC), which is the outermost layer of the skin. Consequently, Monash described the entire SC as the primary deterrent to percutaneous absorption [7]. A more detailed account on the semi-permeability and the barrier properties of SC was furnished by the revolutionary works of Scheuplein and Blank [8–10]. The structure and composition of SC has been substantially explored. The SC can be represented by a simple brick and mortar morphology where the keratin-rich corneocytes form the brick which are held together in a mortar of intercellular lipids. The intercellular lipids is a highly ordered bilayer structure which forms a rigid and continuous phase of the SC and is largely responsible for the permeability barrier of the SC to exogenous molecules. The overall arrangement of SC leads to a prejudiced selection of drugs that can be delivered transdermally forming a limitation that is governed by the physicochemical properties of the drug molecule. Typically a drug is suitable for transdermal delivery if it falls within a strict physicochemical parameter. Since the intercellular lipid route forms the bulk of the epidermis, most of the drug molecules have to penetrate through this pathway. A drug candidate with a lipophilicity ($\log P$) in the range of 1–3 and with a molecular weight of less than 500 Da is most ideal for transdermal drug delivery [11, 12]. This has been one of the main factors to be considered in transdermal product development. Therefore, most of the commercially marketed transdermal products are limited to small-molecular-weight drugs with a befitting lipophilicity [4, 13]. As treatment of diseases with conventional routes of drug delivery becomes increasingly complicated and as therapeutic intervention transitions from small molecule to macromolecules, gene-therapy and protein-/peptide-based formulations and vaccines, transdermal route can be sought as a potential alternative drug delivery strategy. However, delivering macromolecules and proteins through skin is very ambitious and the only possible way to accomplish this goal is to reversibly alter the SC barrier. In this endeavour, comprehensive research has been undertaken to develop strategies that can enhance drug transport across the skin. Transdermal drug enhancement strategies can be divided into two sections: physical enhancement techniques and chemical approach. Chemical approach involves the use of a chemical penetration enhancer to reversibly disrupt the SC barrier. While it is customary to use chemical enhancers for attaining transdermal enhancement of therapeutic significance, its application has been marred by irritation, dermatitis and formulation incompatibilities and instabilities [14–17]. Thus, alternate strategies for achieving transdermal enhancement could be sought.

7.2 Physical Permeation Enhancement Techniques

Physical approaches to permeation enhancement render improved transdermal permeation by the application of an external energy source that can bring about a reversible disruption of the SC barrier or exert a driving force on the drug molecules for active transdermal drug delivery enhancement. Physical enhancement strategies have been a subject of extensive academic and industrial research and have been successfully employed to enhance the delivery of macromolecules, proteins and peptides with some of them being in early stages of proprietary product development. These techniques can be grouped into different categories depending upon the type of external energy source applied to overcome the barrier properties (Table 7.1). Selecting an appropriate physical enhancement approach requires deliberations on the physicochemical properties of the drug molecule, patch design and the duration of treatment [4, 18, 19].

Application of physical enhancement approaches to achieve enhanced drug transport has led to an increase in number of drugs that can be delivered through skin. However, the convenience of using these techniques is often offset with uncertainties over the physiology of the skin and limitations of product design. Electrically mediated drug delivery for instance can cause irreversible damage to the area of application, requiring accurate selection of current intensity and amplitude. In addition, electrically mediated techniques are device-based systems driven by electric power which makes it an expensive product. Mechanically driven enhancement methods are devoid of any such limitations; however, these techniques are (minimally) invasive raising concerns over sterility and safety issues. Thus, commercial application of these techniques has been afflicted with interventions over regulatory approval which had led to an interest in the pursuit of other physical enhancement approaches that are devoid of such limitations. The techniques such as iontophoresis, electroporation, sonophoresis and laser-based techniques are discussed in many reports. In this chapter, we considered to discuss the two emerging technologies for enhancement of transdermal drug permeation.

Table 7.1 *Classification of physical methods of transdermal enhancement*

Physical enhancement methods	Type
Electrically facilitated techniques	Iontophoresis
	Electroporation
Mechanically assisted methods	Microneedles
	Abrasion
	Skin stretching
	Skin puncture
	Perforation
	Needless injection
Other methods	Ultrasound
	Radiofrequency
	Magnetophoresis
	Thermophoresis
	Laser and photomechanical waves

7.3   Magnetophoresis

Magnetophoresis is the migration of magnetically susceptible materials under the influence of an external magnetic field. Materials acquire the property of magnetic susceptibility through their electronic configuration and their spin behaviour, and depending on this property, materials in nature can be grouped into three different categories. They are either diamagnetic, or paramagnetic or ferromagnetic.

Diamagnetism is a relatively feeble attribute of a material. In presence of an external magnetic field, diamagnetic materials tend to move in a direction opposite to that of the applied field undergoing repulsion. Thus, the relative magnetic susceptibility of diamagnetic materials is less than 0 and is occasionally referred to as negative magnetisation [20, 21]. Diamagnetism is not a permanent property of a material and is only exhibited on application of an external magnetic field. Paramagnetism is an attribute whereby a paramagnetic material allows the externally applied magnetic field to pass through them. These materials have magnetic susceptibility greater than or equal to 1; however, like diamagnetism, paramagnetism is not a permanent attribute of a material and is only apparent in the presence of an applied magnetic field. Ferromagnetic materials have a large magnetic susceptibility and exhibit strong attraction in the presence of an external magnetic field. Unlike diamagnets and paramagnets, ferromagnetic materials tend to retain their behaviour even when the magnetic field is removed. Ferromagnetic materials can be described as having magnetic properties similar to that of iron, nickel or cobalt and exhibit maximum acceptability to magnetic field lines [20].

Almost all the materials in nature embrace the property of magnetism. These magnetic properties have been exploited for various applications such as mass transport, spectroscopy, imaging and drug delivery. Water is the most common diamagnetic substance. Diamagnetism although being a weak property opposes externally applied magnetic fields. This attribute of diamagnetism was exploited by Beaugnon and Tournier. The researchers demonstrated levitation of water droplets in the presence of an inhomogeneous magnetic field [22]. This theory of diamagnetic resistance can also be extended to biological systems. Geim *et al.* recognised the presence of diamagnetic properties in living systems (water, proteins and bones) and applied this knowledge in demonstrating levitation of frog inside a solenoid under the influence of a strong 16 T magnetic field [23–25]. Earlier studies by Pauling and Coryell described oxygenated haemoglobin and deoxygenated haemoglobin to be diamagnetic and paramagnetic, respectively [26]. This led to studies involving magnetic separation of red blood cells from whole blood cells under the influence of an external inhomogeneous magnetic field [27, 28]. Diamagnetic substances tend to move away from the region of strong magnetic fields. Kuznetsov and Hasenstein investigated the effect of high-gradient magnetic field on coleoptile of barely plant. The researchers found that the plant curved away from the magnetic field because of the displacement of intracellular plant amyloplasts which are diamagnetic in nature (Figure 7.1) [29]. These studies demonstrate that magnetism is an inherent property of a living system which can be exploited to accomplish various objectives.

The principle of magnetophoresis has been used for separation of cells and proteins; however, magnetism finds numerous other biomedical applications. Magnetic nanoparticles have been used as contrast agents in magnetic resonance imaging [30, 31]. These nanoparticles also have anti-tumour activity and are known to produce hyperthermia when

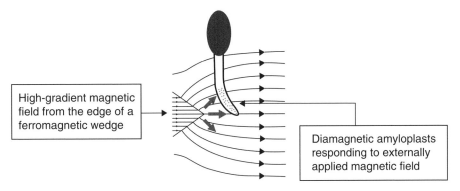

Figure 7.1 *High-gradient magnetic field–induced curvature in plant root by the virtue of displacement of amyloplasts (diamagnetic) in the direction of magnetic field. Reproduced with permission from Ref. [29]. © 1997, Oxford University Press.*

exposed to magnetic field. Thus, when injected to a tumour site, the nanoparticles produce hyperthermia in the vicinity leading to destruction of tumour cells [32, 33]. Magnetic nanoparticles have also been used for cell labelling, tissue engineering and also find application in genetic engineering where they have been used for purification and isolation of nucleic acids [34–36].

7.3.1 Drug Delivery Applications

Magnetic micro- and nanoparticles have been extensively used as drug and gene carriers which can be guided by an external magnetic field to the site of action, typically tumours, thus reducing non-specific chemotherapy [37–40]. These nanoparticles are typically paramagnetic or ferromagnetic in nature and thus possess the property to get aligned or attracted to an applied magnetic field. In this approach, Yang and coworkers were able to achieve retention of iron-oxide nanoparticles in brain tumour–induced rat model [41].

The application of magnetophoresis as a physical enhancement strategy for transdermal drug delivery is a novel and interesting prospect. In this endeavour, for the first time, Murthy and coworkers investigated the permeation of benzoic acid by baring stationary permanent magnets (field strength of ~10^{-2} T) to the drug solution in an *in vitro* permeation set-up across excised rat skin. Benzoic acid is a molecule with a diamagnetic susceptibility of -70×10^{-6} The authors reported that application of magnetic field increased the permeation of benzoic acid compared to passive. The increase in flux enhancement factor was proportional to the increasing magnetic field strength which confirmed the impact of magnetic field on drug diffusion [42]. The researchers further observed similar results for other model drugs as salbutamol sulphate and terbutaline sulphate also showed enhanced skin permeation upon application of a static magnetic field [43]. Further investigations into the magnetic field–driven skin permeation by Benson and coworkers demonstrated a four-fold increase in permeation of urea across human skin epidermis when 5% urea gel was exposed to magnetic film arrays having a field strength of 2 T/m^2 [44].

Recently, Benson and coworkers employed pulsed electromagnetic fields (PEMFs) of 400 μs conveyed at field strength of 5 mT to facilitate skin transport of naltrexone across human skin epidermis. The results from the study showed that the cumulative amount of naltrexone permeating from PEMF-mediated transport was $406.9 \pm 137.1 \, \mu g/cm^2/8 \, h$ whereas only $80.2 \pm 43.4 \, \mu g/cm^2/8 \, h$ of naltrexone permeated passively (enhancement ratio=5.6) [45]. The authors concluded that alternating magnetic fields of short duration and low magnetic field strength could facilitate transdermal drug delivery in a way similar to that of the stationary magnetic fields as demonstrated by Murthy [42].

7.3.2 Mechanism of Permeability Enhancement

The complementary transdermal transport of drug molecules by virtue of an external magnetic field can be attributed to the inherent magnetic property of the drug molecules. However, an absolute account on the mechanism for this enhancement is vaguely understood. Towards this approach, Murthy and coworkers performed mechanistic studies by performing *in vitro* permeation studies of lidocaine hydrochloride (LH) across porcine epidermis in the presence of static external magnetic fields of varying field strength (30, 150 and 300 mT) [46]. The *in vitro* permeation set-up consisted of two stationary neodymium magnets placed adjacent to the donor chamber, 2 mm above the epidermis of a Franz diffusion cell (Figure 7.2). The results from the studies clearly demonstrated the relative enhancement of LH in the presence of magnetic field compared to passive, and the enhancement efficacy was directly proportional to the increase in magnetic field strength (Table 7.2) [46]. The authors speculated that the increase in permeation of LH was due to the inherent diamagnetic nature of LH which undergoes repulsion from applied magnetic

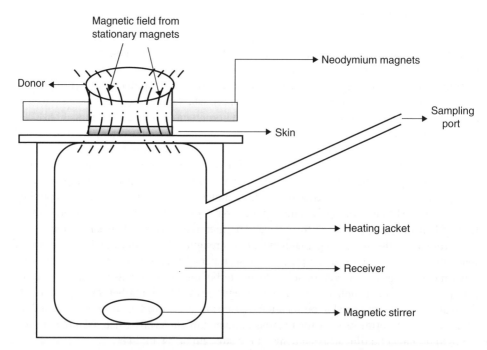

Figure 7.2 In vitro *experimental set-up for magnetophoretically assisted transdermal drug delivery. Adapted from Ref. [46], with permission from Elsevier.*

Table 7.2 *Permeation flux and flux enhancement factor of LH across porcine epidermis*

Magnetic field (mT)	Permeation flux (μg/cm^2/h)	Flux enhancement factor
0	0.18 ± 0.06	1
30	0.53 ± 0.09	2.9
150	1.01 ± 0.17	5.6
300	1.61 ± 0.12	8.9

Selected data from Ref. [46].

Table 7.3 *Transport flux of [^{3}H] Water and [1-^{14}C] Mannitol across porcine epidermis*

Magnetic field (mT)	Transport flux	
	[^{3}H] Water (nl/cm^2/h)	[1-^{14}C] Mannitol (ng/cm^2/h)
0	32.63 ± 5.25	1.89 ± 0.49
300	57.21 ± 8.63	5.18 ± 1.88

Selected data from Ref. [46].

field. *In vitro* permeation studies across 1000 Da dialysis membrane in the presence of 300 mT magnetic field strength showed similar results to that of the carried out across the epidermis. The results from these studies indicated that magnetoreplusion is clearly the predominant mechanism responsible for effecting enhanced drug delivery [46]. Since water is diamagnetic in nature, it is susceptible as well to magnetoreplusion. Murthy and coworkers performed *in vitro* transport studies with tritiated water [^{3}H] and C-14 labelled mannitol [1-^{14}C] in the presence and absence of 300 mT magnetic field strength across porcine epidermis. The results showed that transport flux was higher for water under applied magnetic field speculating a possibility of magnetophoretically induced hydrokinesis (Table 7.3) [46]. Additionally, there was a 2.7-fold increase in the transport of mannitol in presence of the magnetic field. The results from the transport study imply that magnetic field leads to movement of water molecule facilitating the transport of water-soluble components across the biological membrane. It can thus be inferred that magnetohydrokinesis could be a significant auxiliary mechanism to the overall magnetophoretic transdermal drug transport [46]. Murthy and coworkers also investigated the effect of magnetic fields on barrier property of the epidermis by performing FTIR, electrical impedance and TEWL studies. The data from these studies revealed that magnetic field did not alter the barrier properties of the epidermis and thus does not contribute to the mechanism of enhanced drug transport [46].

Contrary to the observations of Murthy and coworkers, Benson and coworkers reported reorientation of the skin due to thermal transition of lipids in the epidermis as the predominant mechanism of pulsed electromagnetic field (PEMF)-induced drug enhancement. Their observation was based on the fact that PEMF did not enhance transport of naltrexone from a polydimethyl siloxane membrane (PDMS) (Table 7.4) [45]. Moreover, Benson and coworkers studied deposition of gold nanoparticles (10 nm) across PEMF treated and untreated human skin epidermis and found that epidermis exposed to PEMF had 200 times

Table 7.4   *Permeation of naltrexone from human epidermis and PDMS membrane under passive and PEMF treatment*

Parameters	Human epidermis		PDMS	
	Mean cumulative amount ($\mu g/cm^2/8\,h$)	Flux ($\mu g/cm^2$)	Mean cumulative amount ($\mu g/cm^2/8\,h$)	Flux ($\mu g/cm^2$)
Passive	80.2 ± 43.4	9.7	104.9 ± 4.5	12.9
PEMF	406.9 ± 137.1	44.1	112.8 ± 5.1	14.2

Selected data from Ref. [45].

more deposition of gold nanoparticles than the unexposed skin. The researchers concluded that PEMF treatment resulted in the formation of transient pores in the epidermis allowing gold nanoparticles of 10 nm size to permeate through these channels [45].

7.3.3   Magnetophoretic Transdermal Patch

Magnetophoretically assisted transdermal drug transport is a physical enhancement technique that involves application of an external magnetic field energy as a driving force for drug molecules to penetrate the skin. Since the ramifications of a magnetic field on the skin are negligible, magnets can be easily placed on the skin without any safety and sterility issues. Additionally, magnets can be moulded into different shape and sizes and do not require prior charging or a power source to elicit a response. Thus a system incorporating magnets along with a drug will have all the desired properties in the making of a convenient cutaneous patch. Building on this knowledge, Murthy and coworkers fabricated a transdermal gel patch by anchoring small neodymium magnets of known magnetic field strengths arranged in parallel on a backing membrane. HPMC gels of lidocaine hydrochloride or lidocaine base was then spread onto these magnets as a thin film and the whole system was then fixed onto an adhesive 3M™ foam tape (Figure 7.3) [42]. The prepared patches were then subjected to *in vivo* studies in a rat model. The results from the *in vivo* study demonstrated enhanced permeation for both lidocaine hydrochloride and lidocaine base from the magnetic patches when compared to patches without the magnet. The study demonstrates a feasible approach of assembling a magnetophoretic patch system whereby the magnets could be incorporated in the patch formulation along with the drug. The magnet could perform dual functions of serving as a backing membrane as well as acting as physical permeation enhancer.

7.3.4   Conclusion

Magnetophoresis is an emerging physical approach for dermal and transdermal drug enhancement. Magnetic field does not cause any irreversible damage to the skin and is independent of an external power supply. The magnetic backing containing transdermal drug devices are likely to deliver relatively higher amount of drug into and across the skin as compared to passive device. The magnetophoresis-enabled systems could be relatively more economical and safer than other physical technologies of drug permeation enhancement.

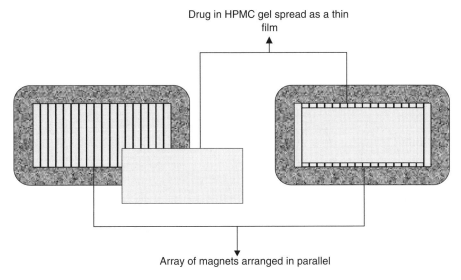

Drug in HPMC gel spread as a thin film

Array of magnets arranged in parallel

Figure 7.3 *Graphical representation of a magnetophoretic transdermal patch. Adapted from Ref. [46], with permission from Elsevier.*

7.4   Electret-Mediated Drug Delivery

An electret is a dielectric material that upon electrisation can acquire and retain the residual charge for a length of time. Electret can be prepared by using different methods and is accomplished by exposing a dielectric material to various processing conditions. These methods include thermal charging, corona charging, charging with liquid contact, radiation penetration, photoelectret process and so on [47, 48]. Electrical charges in an electret are *quasi-permanent* implying depreciation of the residual charge over the time period. Thus, analogues to magnets, electrets can generate internal and peripheral electrostatic fields which can be exploited for various applications (Figure 7.4) [49–52]. Several materials exhibit electret properties. These materials are generally polymers such as the polyethylenes (PEs), polypropylenes (PPs), polyvinylydene fluoride (PVDF), polymethyl methacrylate (PMMA), polyethylene terephthalate (PET), polytetrafluoroethylene (PTFE), polyimide (PI) and ethylene vinyl acetate (EVA). Some of the inorganic electret materials are silicon oxide, silicon nitride and aluminium oxide [53, 54]. Electrets have a very long relaxation time and thus can be used as a source of high-impedance power supply for device applications. The earliest application of electrets can be dated back to the Second World War where wax electret microphones were used in Japanese field equipments. Electret microphones are now the most common technical application of an electret [47, 55]. Electrostatic attraction properties of electret polymers have been utilised for anchoring particulate matter and domestic allergens for air cleansing process [49, 56]. Other applications of electret include measuring ion concentration in air, radiation detectors, electrophotography, acoustic transducers, electret motors and generators and electret material-based seismic detectors for security and military system [47, 57, 58].

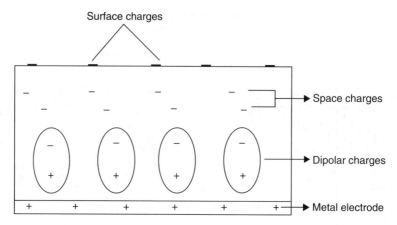

Figure 7.4 *Diagrammatic representation of charges in an electret. Adapted with permission from Ref. [49]. Reprinted with permission from Taylor & Francis Ltd.*

7.4.1 Electrets for Cutaneous Drug Delivery

Electret-mediated cutaneous drug delivery is currently at an embryonic stage of development. A typical electret employed for effecting transdermal drug enhancement was a corona-charged polypropylene (PP) or Teflon (PTFE) film with a 13–50 μm thickness and having an active surface potential of 300–3000 V (measured by a surface potentiometer). These peripheral charges perpetuate long-term electrostatic fields at the surface of the skin without precipitating permanent skin damage. Cui *et al.* described the permeation of methyl salicylate across skin treated with −1000 V electret and across untreated excised rat skin. The influence of electret was evident as the 4 h permeation study demonstrated that the cumulative amount of methyl salicylate permeated across skin treated with electret was significantly greater when compared with the untreated skin. The permeability coefficient of methyl salicylate through electret treated and untreated skin was 0.16 ± 0.02 cm/h and 0.10 ± 0.01 cm/h, respectively. The researchers also reported that increasing the surface potential of the electret from −100 to −2000 V caused an increase in the permeability coefficient of methyl salicylate (Figure 7.5). Further, Cui and coworkers studied the effect of formulation vehicles on the electret-mediated transdermal drug enhancement. The electret effect depreciated with the increase in moisture content of the formulation; however, when the electret was covered with white petroleum jelly, the surface electrostatic potential of the electret was not affected significantly suggesting that topical formulation base without moisture is appropriate for inducing electret effect, while water-based topical formulations may not be suitable due to impediment of the peripheral electrostatic field in the presence of water molecules [59].

In order to interpret the repercussions of electrets on the permeability of skin, Murthy *et al.* performed *in vitro* transport studies across porcine epidermis using salicylic acid and propofol as model test permeants. The results from the *in vitro* studies revealed that the transdermal flux for salicylic acid was significantly greater when exposed to electrets. Increasing the surface potential of the electret increased the flux enhancement factor suggesting proportionality and the enhancement efficacy of electret was independent of nature

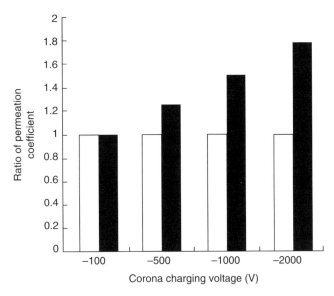

Figure 7.5   *Effect of different corona-charged electret (■) on* in vitro *transdermal enhancement of methyl salicylate as compared with control group (☐). Adapted from Ref. [59], with permission from Elsevier.*

Table 7.5   *Pharmacokinetic parameters of salicylic acid after topical application of 1% salicylic acid ointment for electrets-treated group (−3000V) vs. control group (untreated)*

Pharmacokinetic parameters	Control group	Electret group (−3000V)
$AUC_{(0-t)}$ (mg*hr/ml)	39.9 ± 10.87	97.7 ± 13.87
T_{max} (h)	6	6
C_{max} (mg/l)	5.18 ± 1.3	12.27 ± 3.07

Selected data from Ref. [60].
The treatment duration was 2 h.

of the peripheral electrostatic charge (+ or −) on the electrets. *In vivo* studies carried out with 1% salicylic acid ointment across Sprague Dawley rats showed results that were in alignment with the *in vitro* studies as the electret-treated group accounted for a superior pharmacokinetic profile than the control group (Table 7.5). However, there was a relative reduction in the propofol transport across electret-treated epidermis with increase in surface voltage [60].

Murthy *et al.* also investigated the permeation of fluorescein isothiocyanate (FITC)-labelled dextran of distinct molecular weights (1, 4 and 10 kDa) across porcine epidermis pre-treated with electret (3000V). The electret effect was apparent only in case of 1 kDa FITC dextran (fourfold enhancement compared to control), while there was no significant enhancement for higher molecular weight dextrans [60].

These studies demonstrated that electret application is a potential physical enhancement strategy for transdermal drug delivery. It can be inferred from the *in vitro* permeation experiments that the enhancement efficacy of the electrets may be limited only to hydrophilic drugs with a molecular weight not greater than 1 kDa [59, 60].

7.4.2   Electret Layer in a Patch

Application of an external electrical energy to amplify the transport of drugs across the skin has been a long sought approach in transdermal drug delivery. While their success has been fairly limited to research, products employing iontophoresis as an enhancement strategy have also found proprietary status (Lidosite™ and Zecuity™). Electret-assisted transdermal drug transport is a novel physical enhancement technique which has the potential to evolve into a stable and relatively inert patch formulation. In this endeavour, Cui *et al.* fabricated a transdermal patch of meloxicam with electret as the backing membrane [61]. Patches were prepared by solvent casting method wherein the drug along with a plasticiser was homogenously mixed in an appropriate solvent. This solution was then laminated onto the electret (−300V) and was dried subsequently at room temperature to desiccate the residual solvent. *In vitro* permeation experiments with the prepared patch showed enhanced permeation of meloxicam from electret patches than from the control patches. The cumulative amounts of meloxicam permeated from the electret-based patch and the control patch were 30.15 and 13.93 μg/cm², respectively. The researchers also carried out an *in vitro* comparative study for the enhancement efficacy of different chemical penetration enhancers with electrets on the permeation of meloxicam and found that the cumulative amount of meloxicam permeated upon electret application was significantly higher than the chemical enhancers employed in the experiment (Table 7.6) [61].

7.4.3   Mechanism of Permeability Enhancement

Electret is a material that embraces internal and external electrical fields thus forming an electrostatic environment and micro current in its vicinity. Under normal physiological conditions, most of the tissues and cells in the body function as typical bioelectrets.

The ramifications of electret exposure to transdermal drug delivery are through either direct actions on the skin or indirect actions on drug molecule. The external electric field

Table 7.6   *Cumulative permeation of meloxicam from different enhancer incorporated patches across excised rat full thickness skin*

Enhancer	Cumulative permeation of meloxicam (μg/cm², 10h)
Control (enhancer absent, electret absent)	13.93
Electret (−300V)	30.15
10% ethyl oleate	25.94
1% menthol	19.1
3% azone	18.51
5% azone	17.58
30% sulphoxides	17.38
1% azone	16.41
20% propylene glycol	11.23

Selected data from Ref. [61].

of the electrets upsets the lipid bilayer arrangements in the SC, thereby loosening up the intercellular links leading to formation of new polar pathways [59, 60]. These pathways form the framework for drug enhancement across the skin. Additionally, the internal fields of the electret induce polarisation in the drug molecules complementing enhancement efficacy of the electret [62].

Cui *et al.* studied the effect of a −300 V PTFE/LDPE/PP electret on the morphology of the skin using scanning electron microscope (SEM), transmission electron microscope (TEM) and confocal scanning light microscopy (CSLM) [63]. The researchers found that electrets caused changes in the skin structure at a molecular level. Results from SEM and TEM analysis showed the disorientation in the pecking order of SC, widening of follicular structures and desquamation of epidermis leading to decrease in resistive properties of the SC [63]. The researches further reported enhanced permeation of sodium fluorescein in the presence of electrets. CSLM analysis showed greater fluorescent intensities in the epidermis and the follicular regions of the electret-treated group than for the untreated group confirming degeneration of SC and hair follicles and thus forming pathways for enhanced transepidermal transport and deposition of sodium fluorescein [63].

In order to have a deeper understanding of the electret effect on the structure of skin, Cui *et al.* performed differential scanning calorimetric studies of rat skin after electret application. The scientists observed an increase in peak area and decrease in transition temperature of lipids in skin that were exposed to −500 V electret when compared to the DSC profile of the normal skin [64]. This can be attributed to the fact that electrets lowered the rigidity of the SC barrier thereby decreasing the thermodynamic energy associated with it. Thus, lipid transformation from the gel to the liquid state is achieved at lower energy bringing about a shift in the transition peak at a lower temperature and an increase in the peak height. A further decrease in the transition temperatures could be observed in the DSC thermograms of the skin after application of a −1000 V electret suggesting an increase in the randomness and fluidity of the SC lipids upon increasing the surface charge of the electrets. However, no changes were observed for skin exposed to −2000 V electret suggesting that the V_{max} to achieve a profound structural change in the SC was −1000 V [64]. The suppression of permeation of a lipophilic molecule propofol across the electret pre-treated epidermis was likely due to disrupted lipid domains which are otherwise predominant pathway for absorption of propofol [60].

7.4.4   Conclusion

Electret-mediated drug delivery is a promising technique for skin permeability enhancement of polar drugs. Its application is limited in case of delivery of macromolecules. Additionally, synergistic effect of electrets with other chemical and physical permeation enhancement technique needs to be explored. The electrets works better in the absence of moisture and thus further research needs to be done to identify the best possible formulation that is suitable to exploit electret-mediated delivery of drugs into skin.

References

[1] Hare, H.A. and Martin, E. (1895) Reports on therapeutic progress. *Ther. Gaz.*, **19**, 761.
[2] Zaffaroni, A. (1972) Therapeutic adhesive patch. Google Patents.
[3] Johnson, P., Hansen, D., Matarazzo, D. *et al.* (1984) Transderm Scop for prevention of motion sickness. *N. Engl. J. Med.*, **311** (7), 468–469.

[4] Perumal, O., Murthy, S.N. and Kalia, Y.N. (2013) Turning theory into practice: the development of modern transdermal drug delivery systems and future trends. *Skin Pharmacol. Physiol.*, **26** (4–6), 331–342.

[5] Guy, R.H. and Hadgraft, J. (1987) Transdermal drug delivery: A perspective. *J. Control. Release*, **4** (4), 237–251.

[6] Molinuevo, J.L. and Arranz, F.J. (2012) Impact of transdermal drug delivery on treatment adherence in patients with Alzheimer's disease. *Expert Rev. Neurother.*, **12** (1), 31–37.

[7] Monash, S. (1957) Location of the superficial epithelial barrier to skin penetration. *J. Invest. Dermatol.*, **29** (5), 367–376.

[8] Blank, I.H., Scheuplein, R.J. and MacFarlane, D.J. (1967) Mechanism of percutaneous absorption 3. The effect of temperature on the transport of non-electrolytes across the skin. *J. Invest. Dermatol.*, **49** (6), 582–589.

[9] Blank, I.H. (1965) Cutaneous barriers. *J. Invest. Dermatol.*, **45** (4), 249–256.

[10] Scheuplein, R.J. and Blank, I.H. (1971) Permeability of the skin. *Physiol. Rev.*, **51** (4), 702–747.

[11] Finnin, B.C. and Morgan, T.M. (1999) Transdermal penetration enhancers: Applications, limitations, and potential. *J. Pharm. Sci.*, **88** (10), 955–958.

[12] Bos, J.D. and Meinardi, M.M. (2000) The 500 Dalton rule for the skin penetration of chemical compounds and drugs. *Exp. Dermatol.*, **9** (3), 165–169.

[13] Watkinson, A.C. (2013) A commentary on transdermal drug delivery systems in clinical trials. *J. Pharm. Sci.*, **102** (9), 3082–3088.

[14] Walters, K.A. (2002) *Dermatological and Transdermal Formulations*, Taylor & Francis, New York.

[15] Kanikkannan, N. and Singh, M. (2002) Skin permeation enhancement effect and skin irritation of saturated fatty alcohols. *Int. J. Pharm.*, **248** (1–2), 219–228.

[16] Karande, P., Jain, A., Ergun, K. *et al.* (2005) Design principles of chemical penetration enhancers for transdermal drug delivery. *Proc. Natl. Acad. Sci. U. S. A.*, **102** (13), 4688–4693.

[17] Robinson, M.K., Parsell, K.W., Breneman, D.L. and Cruze, C.A. (1991) Evaluation of the primary skin irritation and allergic contact sensitisation potential of transdermal triprolidine. *Fundam. Appl. Toxicol.*, **17** (1), 103–119.

[18] Zorec, B., Préat, V., Miklavčič, D. and Pavšelj, N. (2013) Active enhancement methods for intra- and transdermal drug delivery: A review. *Zdrav. Vestn.*, **82** (5), 339–356.

[19] Brown, M.B., Martin, G.P., Jones, S.A. and Akomeah, F.K. (2006) Dermal and transdermal drug delivery systems: Current and future prospects. *Drug Deliv.*, **13** (3), 175–187.

[20] Thompson, R. and Oldfield, F. (1986) Magnetic properties of solids, in *Environmental Magnetism*, Springer, London, pp. 3–12.

[21] Mulay, L.N. and Boudreaux, E.A. (1976) *Theory and Applications of Molecular Diamagnetism*, Wiley, New York.

[22] Beaugnon, E. and Tournier, R. (1991) Levitation of water and organic substances in high static magnetic fields. *J. Phys. III*, **1** (8), 1423–1428.

[23] Geim, A. (1998) Everyone's magnetism. *Phys. Today*, **51** (9), 36–39.

[24] Berry, M. and Geim, A. (1997) Of flying frogs and levitrons. *Eur. J. Phys.*, **18** (4), 307.

[25] Simon, M. and Geim, A. (2000) Diamagnetic levitation: flying frogs and floating magnets. *J. Appl. Phys.*, **87** (9), 6200–6204.

[26] Pauling, L. and Coryell, C.D. (1936) The magnetic properties and structure of hemoglobin, oxyhemoglobin and carbonmonoxyhemoglobin. *Proc. Natl. Acad. Sci. U. S. A.*, **22** (4), 210–216.

[27] Melville, D. (1975) Direct magnetic separation of red cells from whole blood. *Nature*, **255** (5511), 706.

[28] Zborowski, M., Ostera, G.R., Moore, L.R. *et al.* (2003) Red blood cell magnetophoresis. *Biophys. J.*, **84** (4), 2638–2645.

[29] Kuznetsov, O.A. and Hasenstein, K.H. (1997) Magnetophoretic induction of curvature in coleoptiles and hypocotyls. *J. Exp. Bot.*, **48** (316), 1951–1957.

[30] Nitin, N., LaConte, L., Zurkiya, O. *et al.* (2004) Functionalization and peptide-based delivery of magnetic nanoparticles as an intracellular MRI contrast agent. *J. Biol. Inorg. Chem.*, **9** (6), 706–712.

[31] Xu, W., Kattel, K., Park, J.Y. *et al.* (2012) Paramagnetic nanoparticle T1 and T2 MRI contrast agents. *Phys. Chem. Chem. Phys.*, **14** (37), 12687–12700.

[32] Wada, S., Yue, L., Tazawa, K. *et al.* (2001) New local hyperthermia using dextran magnetite complex (DM) for oral cavity: experimental study in normal hamster tongue. *Oral Dis.*, **7** (3), 192–195.

[33] Wada, S., Tazawa, K., Furuta, I. and Nagae, H. (2003) Antitumor effect of new local hyperthermia using dextran magnetite complex in hamster tongue carcinoma. *Oral Dis.*, **9** (4), 218–223.

[34] Ito, A., Hayashida, M., Honda, H. *et al.* (2004) Construction and harvest of multilayered keratinocyte sheets using magnetite nanoparticles and magnetic force. *Tissue Eng.*, **10** (5–6), 873–880.

[35] Prodělalová, J., Rittich, B., Španová, A., Petrová, K. and Beneš, M.J. (2004) Isolation of genomic DNA using magnetic cobalt ferrite and silica particles. *J. Chromatogr. A*, **1056** (1), 43–48.

[36] Berensmeier, S. (2006) Magnetic particles for the separation and purification of nucleic acids. *Appl. Microbiol. Biotechnol.*, **73** (3), 495–504.

[37] Giri, S., Trewyn, B.G., Stellmaker, M.P. and Lin, V.S.Y. (2005) Stimuli-responsive controlled-release delivery system based on mesoporous silica nanorods capped with magnetic nanoparticles. *Angew. Chem. Int. Ed.*, **44** (32), 5038–5044.

[38] Chertok, B., Moffat, B.A., David, A.E. *et al.* (2008) Iron oxide nanoparticles as a drug delivery vehicle for MRI monitored magnetic targeting of brain tumors. *Biomaterials*, **29** (4), 487–496.

[39] Slowing, I.I., Vivero-Escoto, J.L., Wu, C.-W. and Lin, V.S.-Y. (2008) Mesoporous silica nanoparticles as controlled release drug delivery and gene transfection carriers. *Adv. Drug Deliv. Rev.*, **60** (11), 1278–1288.

[40] Pan, B., Cui, D., Sheng, Y. *et al.* (2007) Dendrimer-modified magnetic nanoparticles enhance efficiency of gene delivery system. *Cancer Res.*, **67** (17), 8156–8163.

[41] Chertok, B., David, A.E. and Yang, V.C. (2011) Brain tumor targeting of magnetic nanoparticles for potential drug delivery: effect of administration route and magnetic field topography. *J. Control. Release*, **155** (3), 393–399.

[42] Murthy, S.N. (1999) Magnetophoresis: an approach to enhance transdermal drug diffusion. *Pharmazie*, **54** (5), 377–379.

[43] Murthy, S.N. and Hiremath, S. (1999) Effect of magnetic field on the permeation of salbutamol sulphate. *Indian Drugs*, **36**, 663–664.

[44] Benson, H.A., Krishnan, G., Edwards, J. *et al.* (2010) Enhanced skin permeation and hydration by magnetic field array: preliminary in-vitro and in-vivo assessment. *J. Pharm. Pharmacol.*, **62** (6), 696–701.

[45] Krishnan, G., Edwards, J., Chen, Y. and Benson, H.A. (2010) Enhanced skin permeation of naltrexone by pulsed electromagnetic fields in human skin in vitro. *J. Pharm. Sci.*, **99** (6), 2724–2731.

[46] Murthy, S.N., Sammeta, S.M. and Bowers, C. (2010) Magnetophoresis for enhancing transdermal drug delivery: mechanistic studies and patch design. *J. Control. Release*, **148** (2), 197–203.

[47] Jefimenko, O.D. and Walker, D.K. (1980) Electrets. *Phys. Teach.*, **18**, 651–659.

[48] Giacometti, J.A. and Oliveira, O.N. (1992) Corona charging of polymers. *IEEE Trans. Electr. Insul.*, **27** (5), 924–943.

[49] Thakur, R., Das, D. and Das, A. (2013) Electret air filters. *Sep. Purif. Rev.*, **42** (2), 87–129.

[50] Kao, K.C. (2004) *Dielectric Phenomena in Solids*, Academic Press, San Diego, CA.

[51] Kestelman, V.N. (2000) *Electrets in Engineering: Fundamentals and Applications*, Springer, London.

[52] Ku, C.C. and Liepins, R. (1987) *Electrical Properties of Polymers*, Hanser, Munich.

[53] Panchapakesan, R. (2007) *Electret Effect in Cement*, ProQuest, New York.

[54] Nalwa, H.S. (1995) *Ferroelectric Polymers: Chemistry: Physics, and Applications*, CRC Press, New York.

[55] Sessler, G. and West, J. (2005) Electret transducers: A review. *J. Acoust. Soc. Am.*, **53** (6), 1589–1600.

[56] Gaynor, P.T. and Hughes, J.F. (1998) Dust anchoring characteristics of electret fibres with respect to Der p 1 allergen carrying particles. *Med. Biol. Eng. Comput.*, **36** (5), 615–620.

[57] Pakhomov, A., Sicignano, A. and Goldburt, T. (2004) *Lab Testing of New Seismic Sensor for Defense and Security Applications.* European Symposium on Optics and Photonics for

Defence and Security: International Society for Optics and Photonics, Bellingham, Washington, pp. 108–116.

[58] Chang, J.-S., Kelly, A.J. and Crowley, J.M. (1995) *Handbook of Electrostatic Processes*, CRC Press, New York.

[59] Cui, L., Jiang, J., Zhang, L. *et al.* (2001) Enhancing effect of electret on transdermal drug delivery. *J. Electrost.*, **51–52** (0), 153–158.

[60] Murthy, S.N., Boguda, V.A. and Payasada, K. (2008) Electret enhances transdermal drug permeation. *Biol. Pharm. Bull.*, **31** (1), 99–102.

[61] Cui, L., Hou, X., Jiang, J. *et al.* (2008) Comparative enhancing effects of electret with chemical enhancers on transdermal delivery of meloxicam in vitro. *J. Phys. Conf. Ser.*, **142**, 12–15.

[62] Guo, X., Liang, Y., Liu, H. *et al.* (2013) Study on the electrostatic and piezoelectric properties of positive polypropylene electret cyclosporine A patch. *J. Phys. Conf. Ser.*, **418**, 012148.

[63] Jiang, J., Liang, Y., Cui, L. *et al.* (2008) Influence of porous PTFE/LDPE/PP composite electret in skin ultrastructure. *J. Phys. Conf. Ser.*, **142**, 012050.

[64] Cui, L., Liang, Y., Dong, F. *et al.* (2011) Structure of rat skin after application of electret characterized by DSC. *J. Phys. Conf. Ser.*, **301**, 012027.

8

Microporation for Enhanced Transdermal Drug Delivery

Thakur Raghu Raj Singh and Chirag Gujral

School of Pharmacy, Queen's University Belfast, Belfast, UK

8.1 Introduction

Transdermal drug delivery refers to the delivery of the drug across intact, healthy skin and into the systemic circulation. This mode of delivery is advantageous for delivering therapeutic agents that do not survive ingestion and that undergo first-pass metabolism [1]. However, the outermost layer of the skin, the *stratum corneum* (SC), constitutes the principal barrier for penetration of most drugs. The ideal properties of a molecule penetrating intact SC are [2–4]:

- Molecular mass less than 600 Da, when the diffusion coefficient in SC tends to be high
- Adequate solubility in both oil and water so that the membrane concentration gradient, which is the driving force for diffusion, may be high
- High, but balanced, SC: vehicle partition coefficient such that the drug can diffuse out of the vehicle, partition into and move across the SC, without becoming sequestered within it
- Low melting point, correlating with good solubility, as predicted by ideal solubility theory

To a large extent, many drug molecules do not possess the above properties, and therefore, chemical permeation enhancers or prodrugs can only improve SC transport by an order of magnitude [2–4]. Thus, disruption or bypassing the SC using microporation methods is of great interest, as this creates aqueous pathways that significantly enhance transdermal

Novel Delivery Systems for Transdermal and Intradermal Drug Delivery, First Edition.
Ryan F. Donnelly and Thakur Raghu Raj Singh.
© 2015 John Wiley & Sons, Ltd. Published 2015 by John Wiley & Sons, Ltd.

delivery of therapeutic agents, including vaccines and macromolecules [1]. Hence, the physicochemical properties of drug molecules are no longer restricted, with a degree of water solubility now the only prerequisite.

Microporation of the skin has been described as one of the few third-generation enhancement strategies [5] such as laser or thermal ablation [6], iontophoresis [7], electroporation [8], radio frequency (RF) [9], microneedle (MN) [10], ultrasound/phonophoresis or sonophoresis [11] and high-pressure gas/powder or liquid microporation [1]. The common goal of these techniques is creating micro-sized pores in the SC, thereby altering the permeability of the skin. Furthermore, the microporation technology is considered as minimally invasive and regarded as painless that results in further benefits. In addition to delivery of drugs across the skin, microporation devices have also been proposed for the collection of biological fluid samples from the body. This chapter attempts to provide an overview of various microporation techniques that has been evaluated for enhancing transdermal drug delivery by creating micropores on skin's surface.

8.2   High-Pressure Gas or Liquid Microporation

Invented in the 1940s, the high-pressure jet injectors are perhaps the oldest devices intended to eliminate the direct mechanical piercing of the skin by needles and/or syringes. Jet injections employ a high-speed jet to puncture the skin and deliver drugs without the use of a needle [12]. It can be broadly classified as either multi-use nozzle jet injectors (MUNJIs) or disposable cartridge jet injectors (DCJIs) [12, 13]. These devices basically consist of a power source, usually a compressed gas or spring, which, on actuation, pushes a piston. This creates an impact on the drug-loaded compartment, causing a surge in pressure and release of the drug-containing vehicle through a nozzle as a jet at a speed of between 100 and 200 m/s [13]. The jet punctures the skin on contact and delivers the drug at a depth that is dependent on the characteristics of the jet, namely, orifice diameter and jet exit velocity [14]. The jet may be liquid or gaseous, with the drug in solution or as a dry powder, where the flow of gas deposits the drug particles at a desired depth into the skin, such as directly into muscles, subcutaneous or intradermal layers [14] (Figure 8.1). Liquid jet injectors have been shown to successfully deliver 5-aminolevulinic acid (ALA) [15], insulin [16], lidocaine [17, 18], vaccines [13], midazolam [19], interferon [20] and erythropoietin [21]. On the other hand, transdermal powder delivery is where the therapeutic compound is formulated as a fine powder (20–100 μm diameter) and is accelerated in a supersonic flow of helium gas to penetrate the skin [22]. The PMED® (Chiron Corporation) device, formerly known as PowderJect® [23], has been reported to successfully deliver vaccines [24], lidocaine [25] and testosterone [26]. Dry powder formulations are generally more stable than solutions and may negate the need for the 'cold chain' to be maintained when using vaccines, for example. This would be particularly advantageous for large-scale immunisation in developing countries with hot climates. Apart from drug delivery, jet injectors have also been proposed for the delivery of anti-ageing cosmetic products, such as collagen and hyaluronic acid [27].

It is claimed that needle-free injection has several potential benefits. For example, the fear of needles (belonephobia) and piercing (diatrypophobia) can be avoided [28]. In addition, several studies have shown that by adjusting injection parameters (e.g. injection volume), specific skin strata can be targeted [15]. Furthermore, the use of jet injectors

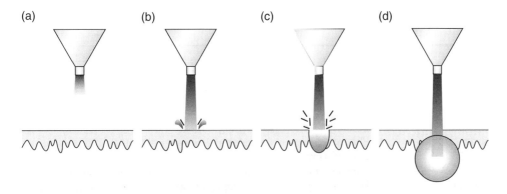

Figure 8.1 *Schematic representation of drug delivery using liquid jet injector: (a) formation of liquid jet, (b) initiation of hole formation due to impact of jet on skin surface, (c) development of hole inside skin with progress of injection and (d) deposition of drug at the end of hole in a near spherical or hemispherical pattern (spherical pattern shown). Reproduced from [14], with permission from Elsevier.*

should avoid needle-stick injuries. However, a number of limitations of needle-free injectors have been highlighted such as dosing accuracy, and location of delivery may vary due to skin variability (thickness and hydration) between patients. The long-term effect of bombarding the skin with high-speed particles or liquids is not known, and in addition, jet injection has been associated with variable adverse reactions [28, 29] and may be no less painful than conventional needle and syringe [13, 30]. In fact, needle-free liquid jet injectors have been associated with frequent bruising and pain, which offsets their advantages over the conventional needle and syringe approach [13]. In the past, these devices have been employed for mass delivery of vaccination through the skin but fell from favour when a link to spread of hepatitis B was established after vaccination with MUNJIs caused cross-contamination between patients [14]. This issue, in combination with the variability in patient response – such as occasional pain, discomfort and local reactions, inconvenience of use compared with injections and cost – is a potential barrier for the development and commercialisation of this drug delivery method [14, 31]. Despite more than 50 years of clinical use and hundreds of patents, these jet injectors have not reached their full potential in terms of replacing the routine needle-based delivery. Only very few devices have so far attracted some commercial interest following FDA clearance as shown in Table 8.1.

Growing demand in the effective delivery of macromolecules including DNA, insulin, growth hormones, vaccines and other biotechnological products [13] has led the researchers to engineer improved designs of jet injectors so as to overcome current challenges. Especially microjet injectors with low volumes, smaller nozzle diameters and pulsed small injection volumes could be of considerable benefits in reducing occasional pain and allow uniformity in dosage delivery. In addition, a better understanding of the jet injections on skin at a cellular level, variability in jet penetration depending on skin properties, mechanisms of jet injections and mechanical properties of the skin needs a special attention for effective performance. In addition to drug delivery, jet injectors can also be applied for interstitial fluid sampling.

Table 8.1   *High-pressure jet injector devices on the market or that have already received FDA clearance*

Product	Company	Energy source	Drug or formulation type
Sumavel® Dose Pro®	Zogenix	Compressed gas	Sumatriptan
Biojector® 2000	Bioject	Compressed gas	Liquid
ZetaJet™	Bioject	Spring	Liquid
Stratis®	Pharmajet	Spring	Liquid
Injex30	Injex Pharma	Spring	Insulin
Shireen PoreJet	Injex Pharma	Spring	Hyaluronic acid
PenJect® (investigational used in clinical trials)	PenJect Corporation	Compressed gas	Liquid and lyophilised drugs
Tjetr®	Antares Pharma	Spring	Human growth hormone

Reproduced from [32] with permission from Elsevier.

8.3   Ultrasound (Phonophoresis and Sonophoresis) Microporation

Ultrasound is defined as sound with a frequency ranging from 0.02 to 10.0 MHz and an intensity range of 0.0–3.0 W/cm² [33]. Sonophoresis, also known as phonophoresis, describes the effects of ultrasound on the movement of drugs through intact skin and into soft tissues [34]. Ultrasound-enhanced drug delivery has several important advantages in that it is non-invasive, can be carefully controlled and can penetrate to desired depths into the body. The early use of ultrasound as a physical enhancer in transdermal drug delivery was developed nearly 50 years ago [34]. Exposure of a biological membrane to ultrasound causes sonoporation which is the temporary, non-destructive perforation of the cell membrane. This transient state enhances the permeability of therapeutic agents into cells and tissues [35].

Ultrasound is produced by a transducer composed of a piezoelectric crystal, which defines the frequency of emitted waves and converts electric energy into mechanical energy in the form of oscillations, generating acoustic waves. During the propagation of these acoustic waves through a given medium, a wave is partially scattered and absorbed by the medium, resulting in the attenuation of the emitted wave with the lost energy being converted into heat. Ultrasound can be emitted either continuously (continuous mode) or in a sequential mode (pulsed mode) [36]. The mechanism of ultrasound effects on skin in drug delivery is not clearly understood; however, various different mechanisms have been proposed [37–39] that include:

1. Cavitation: Ultrasound generates gaseous cavities in a medium, and their subsequent collapse causes the release of shock waves, which causes structural alterations in the surrounding tissue. Cavitation leads to disordering of the lipid bilayers and the formation of aqueous channels in the skin, through which drugs can permeate.
2. Thermal effects (increasing of temperature).
3. Induction of convective transport.
4. Mechanical effects (stress occurred because of the pressure induced by ultrasound).

Applications of ultrasound differ depending upon the frequency range of ultrasound used. For example, high-frequency (3–10 MHz) ultrasound is used for diagnostic conditions in clinical imaging, medium-frequency (0.7–3.0 MHz) ultrasound for therapeutic physical therapy and low-frequency (18–100 kHz) ultrasound for lithotripsy, cataract emulsification, liposuction, cancer therapy, dental descaling and ultrasonic scalpels [40]. The experimental findings suggest that among all the ultrasound-related phenomena evaluated, cavitation has the dominant role in sonophoresis, suggesting that the application of low-frequency ultrasound should enhance transdermal transport more effectively [37]. There are innumerable applicators on the market that utilise sonophoresis technology. For instance, the ultrasonic teeth cleaning devices used by dentists have a frequency range of 25–40 kHz. Moreover, portable pocket-size sonicators for drug injection and analyte monitoring are also commercially available [33].

Ultrasound has been used in an attempt to enhance the absorption of different molecules through human skin. More recently, it was demonstrated that low-frequency ultrasound (<100 kHz), which causes cavitational disordering of SC, has been used to provide enhanced transdermal transport of low molecular weight drugs as well as high molecular weight proteins across human skin [11]. *In vitro* studies using human SC demonstrated enhanced transport (by several orders of magnitude) of the macromolecules insulin, interferon-γ and erythropoietin using low-frequency ultrasound [11]. Park and colleagues [41] used a compact, lightweight and low-frequency transducer to enhance transdermal insulin delivery. Live adult pigs were anaesthetised, and xylazine was administered to induce hyperglycaemia. The ultrasound-treated group showed a significant reduction in blood glucose compared to control. The authors proposed that the device was capable of safely reducing blood glucose to a normal range. *In vitro* and *in vivo* studies have demonstrated the efficacy of sonophoresis, with some studies reporting up to 1000-fold better penetration compared to simple topical application.

The practicality of ultrasound in health sciences as a biomedical applicator, as well as a therapeutic agent, is increasing. The use of ultrasound as an aid for increasing skin permeability depends upon the non-thermal bio-effects of its cavitation. In essence, attention should be paid to the issue of ultrasound technology's effects on the structure of the skin to develop a useful tool that takes accounts for safety issues [37]. For example, Singer *et al.* [42] demonstrated that low-intensity ultrasound induced only minor skin reactions in dogs but high-intensity ultrasound was capable of causing second-degree burns.

There is good evidence to suggest that the effect on SC is reversible in nature and also of the potential use of ultrasound within a clinical setting [43]. However, the commercial availability of ultrasound devices, especially for the transdermal drug delivery, is very limited, despite the number of patent publications that have been filed. As such, there is a need for more sophisticated devices, particularly those that can utilise the advantages of low-frequency ultrasound for its reversible effects in increasing the permeability of the skin. Additionally, the notion that a novel non-invasive ultrasonic device could be used for the extraction of clinically important analytes from the interstitial fluid of the skin may aid in further research for this area. However, the successful application of these novel devices in drug delivery or monitoring needs to be demonstrated in preclinical or clinical studies.

8.4   Iontophoresis

Iontophoresis is a century-old technique whereby an electrical potential gradient is used to drive solute molecules across the skin [32]. An electrophoretic device consists of a power source, terminating with a positive electrode (anode) and a negative electrode (cathode) (Figure 8.2). Drug transport across the skin is facilitated by two primary mechanisms, electrorepulsion and electro-osmosis. Using electrorepulsion, whereby like charges repel each other, delivery of a positively charged drug (D^+) can be achieved by dissolving the drug in a suitable vehicle in contact with an electrode of similar polarity (anode). The application of a small direct current (~0.5 mA/cm^2) causes the drug to be repelled from the anode, and it is attracted towards the oppositely charged electrode (cathode) [2]. This process is termed anodal iontophoresis. Conversely, cathodal iontophoresis occurs when anions (D^-) are repelled from the cathode towards the anode. Importantly, iontophoresis is not only reserved for charged drugs. Delivery of small neutral molecules may also be enhanced through electro-osmosis. At pH values above 4, the skin is negatively charged, due to ionisation of carboxylic acid groups within the membrane. Positively charged ions, such as Na^+, are more easily transported, as they attempt to neutralise the charge in the skin; hence, there is a flow of Na^+ to the cathode [2]. Owing to a net build-up of NaCl at the cathodal compartment, osmotic flow of water is induced from the anode to the cathode. It is this net flow of water that facilitates transfer of neutral molecules across the skin.

The transappendageal route is thought to offer the path of least electrical resistance across the skin and is suggested to be the principal pathway taken by a permeant during electrophoresis [44]. Many factors influence electrophoresis, including pH of the donor solution, electrode type, buffer concentration, current strength and current type. These parameters have been reviewed extensively elsewhere [7, 44, 45].

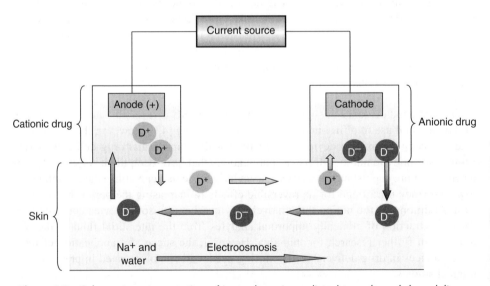

Figure 8.2   *Schematic representation of iontophoresis-mediated transdermal drug delivery.*

Iontophoresis has been used to enhance transdermal delivery of a wide range of relatively small molecules, including 5-FU [46], ALA [47], fentanyl [48], piroxicam [49], buspirone hydrochloride [50] and penbutolol sulphate [51]. Vyteris (Fair Lawn, NJ) has received approval for an iontophoretic patch containing lidocaine and adrenaline (LidoSite™) [44, 52]. Adrenaline is a vasoconstrictor and reduces lidocaine clearance from the site. Phase III clinical studies demonstrated that both children and adults experienced significantly less pain associated with venipuncture or IV cannulation when treated with the Vyteris system compared with the placebo system [7]. Other FDA-approved devices include Ionsys™ (Alza, Mountain View, CA) (single-use iontophoretic system for the systemic delivery of fentanyl). Another recently FDA-approved device is Zecuity™ (NuPathe Inc., Conshohocken, PA) that has been developed for transdermal delivery of sumatriptan.

The most widely studied macromolecule in terms of iontophoretic delivery is insulin. Monomeric human insulin has a molecular weight of approximately 6000 Da [7]. *In vivo* studies have demonstrated increased insulin delivery in animals, using iontophoresis [53]. However, achieving even the basal insulin rate (0.5–1.0 IU/h) in humans is likely to be extremely challenging [7]. Other peptides that have been successfully delivered across the skin via iontophoresis include salmon calcitonin [54], human parathyroid hormone (PTH), luteinising hormone-releasing hormone (LHRH) [55] and vasopressin [53]. Iontophoresis has traditionally been used to enhance transdermal drug delivery. However, by using 'reverse iontophoresis', substances can be extracted from the skin and analysed. This technique offers a non-invasive means of monitoring glucose and drug levels in the blood [56].

The main advantage of iontophoresis over other transdermal enhancement strategies is its ease of control. Electrical current is responsible for the increased delivery [57]. Therefore, by manipulating current density and duration, the dose may be tailored to an individual patient's needs. Safety concerns exist over the biophysical effects of iontophoresis, such as the possibility of irreversible skin damage [2]. Recent advances in the microelectronics industry have facilitated miniaturisation of iontophoretic systems. This has resulted in the development of devices which are much more convenient for the patient and have the potential to allow home use.

8.5   Electroporation

Unlike iontophoresis, which uses small voltages (<10 V), electroporation employs relatively high-voltage pulses (10–1000 V) for brief periods of time (less than a few hundred milliseconds). When applied to SC, pulses are thought to induce the formation of aqueous pores in the lipid bilayers. The aqueous pores may facilitate drug transport by passive diffusion, electro-osmosis or iontophoresis during the pulse [45, 52]. Furthermore, transdermal delivery of charged molecules may be further enhanced by iontophoretic transport through the transfollicular pathway during pulsation [58].

Electroporation has been used to deliver a number of small molecules, including lidocaine [59], nalbuphine [58] and cyclosporin [60]. Conjeevaram *et al.* [61] reported no measurable permeation of fentanyl through human epidermis under passive conditions. However, iontophoresis gave a flux of approximately 80 μg cm²/h, with a fourfold higher flux observed using electroporation. Larger molecules studied include heparin [62], PTH [63] and DNA vaccines [64]. Medi and Singh [63] examined electrically facilitated transdermal delivery of human PTH using dermatomed porcine skin. Iontophoresis significantly enhanced the flux of PTH compared to passive delivery. Electroporation pulses of 100, 200 and 300 V significantly increased PTH flux in comparison to passive as well as iontophoretic flux. Furthermore, the authors demonstrated that by following electroporation pulses with iontophoresis, flux was further increased by severalfold. Significant disruption of the SC was reported using light microscopy.

Electroporation has the potential to enhance transdermal delivery of macromolecules. However, more clinical studies are required to assess long-term safety of this technique and to gain a greater understanding of its mechanism of action [52]. Miniaturisation of such devices is essential to facilitate routine use by patients.

8.6   Laser Microporation

Microporation of skin can also be achieved by using a laser emitted at a defined wavelength that is directly absorbed by the tissue to form micropores. Laser ablation is a controlled technique which creates pores in the SC. The irradiation of laser energy causes instant tissue vaporisation due to flash evaporation of water within the irritated area following microexplosion that results in tissue ablation [65]. It is this rapid energy loss from the ablated site which protects the surrounding tissue from heat-induced damage. The two optimal wavelengths at which skin ablation can be achieved are short-wavelength ultraviolet and mid-infrared, which is absorbed by tissue proteins and tissue water, respectively. Micropores are produced by matching the energy wavelengths to the main absorption peak of water (2790 nm) and/or tissue proteins (2940 nm). The amount of SC removal can be efficiently controlled by adjusting the level of energy imparted on the skin, especially when applied at lower energy levels [65]. However, there are very few studies that demonstrated the use of laser-based devices for the enhancement of transdermal drug delivery.

Laser ablation has been successfully used to enhance skin penetration of aminolevulinic acid [66, 67], hydrocortisone [65], nalbuphine [68] and indomethacin [68]. The technique has also been successfully performed with peptides and DNA [69]. Irradiation of pigskin at 2790 nm using an erbium:yttrium–scandium–gallium–garnet (Er:YSGG) laser, resulted in a 2.1-fold increase in interferon-γ transport [65]. Fang *et al.* [66] examined the effect of molecular weight on the transdermal delivery of macromolecules using an erbium:yttrium–aluminium–garnet (Er:YAG) laser (2940 nm). The study reports transdermal delivery of model macromolecules up to 77 kDa across excised pigskin pretreated using the laser. In addition, PANTEC Biosolutions AG has described many patents [70–73] based on laser technology referred to as P.L.E.A.S.E.® (Painless Laser Epidermal System), and delivery of *in vitro* fertilisation (IVF) hormone was demonstrated in clinical trials [74]. Currently, this technology has been investigated for transcutaneous immunotherapy [75]. Controlled dermal ablation using the P.L.E.A.S.E. device generated an array of micropores of precise depth and position. Figure 8.3 shows scanning electron microscopy (SEM) of individual cell layers, allowing for high diffusion rates, no thermal damage of the tissue was observed, and full re-epithelialisation was achieved within 2 days [75].

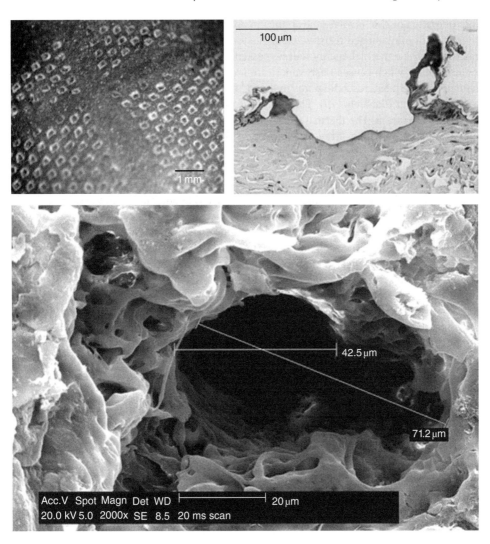

Figure 8.3 *Histological analysis of laser-microporated (using P.L.E.A.S.E. device) mouse skin. Upper left: top view of an array of micropores (500 pores/cm²). Paraffin section and SEM picture of a single micropore generated with four laser pulses delivered at 1.9 J/cm²/pulse (upper right) or eight laser pulses delivered at 0.76 J/cm²/pulse (bottom). Reproduced with permission from [40], © 2012, American Chemical Society and from [37], with permission from Elsevier.*

8.7 Thermal Microporation

It has been well documented that the flux of drugs through the skin is temperature sensitive and this factor has been utilised in transdermal and other types of delivery systems in recent times. This began with patented transdermal patch devices which used heat to formulate the patch rather than to increase the flux of drug across the skin [76–78]. Thermal microporation of skin involves the application of rapid, controlled pulses of thermal energy by means of miniaturised resistive elements to a defined site on the skin surface to create micropores. The thermal energy will be passed through the array of tiny elements for few milliseconds, which causes flash vaporisation of SC cells in an area about the width of a human hair [78]. Microscopic pores in the SC have been generated using an array of electrically resistant filaments [79]. The array was placed on rat skin and briefly heated using an electrical current. The thermal energy delivered to the skin from each filament ablates the SC, culminating in an array of micropores. This strategy was used to deliver interferon alpha-2B across hairless rat skin *in vivo*. Furthermore, the dose administered was doubled when microporation was combined with iontophoresis [80].

Eppstein [81] patented a device on behalf of Altea Therapeutics, which described a method for painlessly creating micropores (1–1000 μm) in outer layers of the skin, to facilitate the transmembrane transport of substances into the body. This was achieved through placing a heated probe on a targeted area to create micropores by vaporising water and other vaporisable substances on the skin or mucosal linings. The heat probe used is a solid, electrically or optically heated element, with the active heated probe tip no more than a few hundred microns in diameter and protruding up to a few millimetres from the supporting base. A single pulse or multiple pulses of current deliver enough thermal energy into or through the tissue to allow the ablation of the skin or biological membrane. The procedure was regarded as painless if the thermal energy was delivered to the probe in <20 ms. Based on this technology and related patents, Altea Therapeutics proprietary PassPort® transdermal delivery system has shown transdermal delivery of exenatide in clinical studies with type 2 diabetic patients (Figure 8.4). Recently, Nitto Denko acquired all assets of Altea Therapeutics for further commercial development [83].

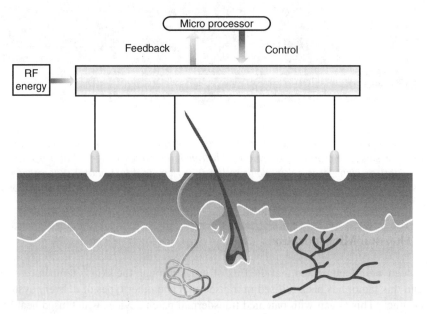

Figure 8.4 *Schematic presentation of RF-microchannels. Reproduced from [82] with permission from Elsevier.*

The earliest patent that utilised thermal energy to enhance transdermal drug delivery was patented in the early 1990s. In the last 20 years, many new devices and methods have been investigated that utilised microporation effects that are achieved when thermal energy is focused on the skin and other biological membranes. In recent years, more advanced devices have been developed, with some reaching full clinical trial and mass scale production such as Altea's PassPort™ device.

To date, thermal ablative methods rely on a two-step process, whereby the skin is first treated and then the formulation applied. For a system to be convenient for regular use, the processes of SC disruption and drug delivery need to be integrated within the same device. In addition, long-term safety data is required to determine if repeated use of such devices can cause irreversible skin damage. Furthermore, a combination of thermal microporation with other techniques has demonstrated an improved enhancement of drug delivery than with the single use of thermal microporation. The following are some of the recent patents utilising combination techniques along with thermal ablation.

8.8   RF Microporation

Percutaneous penetration can also be facilitated by the ablation of the outer layers of the skin, using an alternating electrical current at a RF of 100–500 kHz [9]. The passage of this current through cells in the upper skin strata, via an array of microelectrodes placed on the skin, propagates ionic vibrations through skin cells resulting in localised heating, liquid evaporation and the removal of cells. As a result, transient aqueous microchannels are created across SC and epidermis, called RF-microchannels, which enable or augment effective movement of water-soluble substances through the skin (Figure 8.4). Compared to other electrically assisted drug delivery techniques, such as electroporation and iontophoresis, microchannels formed using RF ablation are relatively large, enabling the transport of high molecular weight ionic, polar or neutral compounds. Furthermore, the formed microchannels do not reach underlying nerve endings and blood vessels; thus, the risk of skin trauma and neural stimulation is minimised [9, 84].

Additionally, a combination strategy of RF-microchannel generation in conjunction with iontophoresis was patented by Levin *et al.* [82]. ViaDerm™ pretreatment of the human skin *in vivo* followed by iontophoretic patch application was reported to facilitate insulin delivery by a factor of 2.5 compared to ViaDerm™ treatment only [82].

First developed for the removal of small tumours by placing a probe directly into the malignant tissues in medical electrosurgery, the RF thermal ablation has now found potential applications in the enhancement of drug delivery. The above patents demonstrated successful RF-microchannelling devices which have shown enhanced delivery of both low and high molecular weight therapeutic molecules through the pores created by the RF-based device. However, wider usage of such devices is needed to fully demonstrate the effectiveness of the technology and the suitability of devices.

8.9   Microneedles

Apart from the various different techniques mentioned above, the microporation of the skin, to desired depths, can also be achieved by the use of MNs. Although Alza Corporation appears to be the first to use MN, described in a late 1976 patent [85], the first paper to demonstrate MNs for transdermal delivery was not published until 1998 [86]. However, it was not possible to make such microstructure devices until the 1990s when high-precision microelectronics industrial tools came into existence to make such micron or submicron structural designs feasible. MNs consist of a plurality of microprojection arrays, generally ranging from 100 to 1000 µm in height, of different shapes, which are attached to a base support, and the application of such an array can create transport pathways of micron dimensions in the biological membrane. Once created, these micropores or pathways are orders of magnitude larger than molecular dimension and, therefore, should readily permit the transport of macromolecules, as well as possibly supramolecular complexes and microparticles [87]. In addition, these MNs can be used for sampling body fluids, such as for the measurement of blood glucose levels in diabetic patients. MN-based microporation is extensively discussed in our book entitled *Microneedle-Mediated Transdermal and Intradermal Drug Delivery*. Therefore, readers are requested to refer our book for further reading [88].

8.10   Conclusion

Considering the selective and limited permeation of drug molecules across the skin, it becomes imperative to develop novel technologies to overcome current limitations of transdermal drug delivery. To overcome these limitations, microporation techniques have been developed, which have successfully demonstrated delivery of small and large molecules, including vaccines, antibodies and peptides and proteins. Given the extraordinary potential of such agents as next-generation therapeutics, it is hardly surprising that a significant number of firms are actively involved in the fabrication and evaluation of a range of microporation devices. A number of patents have already been filed, of which some are already available in the market and some are undergoing preclinical and clinical development. However, before these microporation devices find widespread use, researchers must perfect the art of device and techniques for optimal usage. Ultimately, a completely integrated device for diagnostic, monitoring and drug delivery functions would be the desirable outcome of such investigations. It remains to be seen if these novel delivery devices will replace painful hypodermic injections.

References

[1] Stachowiak, J.C., Li, T.H., Arora, A. *et al.* (2009) Dynamic control of needle-free jet injection. *J. Control. Release*, **135** (2), 104–112.
[2] Barry, B.W. (2001) Novel mechanisms and devices to enable successful transdermal delivery. *Eur. J. Pharm. Sci.*, **14**, 101–114.
[3] White, S.H., Mirejovsky, D. and King, G.I. (1988) Structure of lamellar lipid domains and corneocyte envelopes of murine stratum corneum: an X-ray diffraction study. *Biochemistry*, **27**, 3725–3732.
[4] Bouwstra, J.A., Ooris, G.S., Van der spek, J.A. *et al.* (1994) The lipid and protein structure of mouse stratum corneum: a wide and small angle diffraction study. *Biochim. Biophys. Acta*, **1212**, 183–192.
[5] Prausnitz, M.R. and Langer, R. (2008) Transdermal drug delivery. *Nat. Biotechnol.*, **26** (11), 1261–1268.

[6] Lee, J.W., Gadiraju, P., Park, J.H. *et al.* (2011) Microsecond thermal ablation of skin for transdermal drug delivery. *J. Control. Release*, **154** (1), 58–68.

[7] Kalia, Y.N., Naik, A., Garrison, J. and Guy, R.H. (2004) Iontophoretic drug delivery. *Adv. Drug Deliv. Rev.*, **56**, 619–658.

[8] Becker, S., Zorec, B., Miklavčič, D. and Pavšelj, N. (2014pii: S0025-5564(14)00125-4) Transdermal transport pathway creation: electroporation pulse order. *Math. Biosci.*

[9] Kim, J., Jang, J.H., Lee, J.H. *et al.* (2012) Enhanced topical delivery of small hydrophilic or lipophilic active agents and epidermal growth factor by fractional radiofrequency micropora-tion. *Pharm. Res.*, **29** (7), 2017–2029.

[10] Donnelly, R.F., McCrudden, M.T., Zaid Alkilani, A. *et al.* (2014) Hydrogel-forming microneedles prepared from 'super Swelling' polymers combined with lyophilised wafers for transdermal drug delivery. *PLoS One*, **9** (10), e111547.

[11] Mitragotri, S., Blankschtein, D. and Langer, R. (1995) Ultrasound-mediated transdermal protein delivery. *Science*, **269** (5225), 850–853.

[12] Baxter, J. and Mitragotri, S. (2005) Jet-induced skin puncture and its impact on needle-free jet injections: experimental studies and a predictive model. *J. Control. Release*, **106**, 361–373.

[13] Mitragotri, S. (2006) Current status and future prospects of needle-free liquid jet injectors. *Nat. Rev. Drug Discov.*, **5**, 543–548.

[14] Arora, A., Prausnitz, M.R. and Mitragotri, S. (2008) Micro-scale devices for transdermal drug delivery. *Int. J. Pharm.*, **364**, 227–236.

[15] Donnelly, R.F., Morrow, D.I., McCarron, P.A. *et al.* (2007) Influence of solution viscosity and injection protocol on distribution patterns of jet injectors: application to photodynamic tumour targeting. *J. Photochem. Photobiol. B, Biol.*, **89** (2–3), 98–109.

[16] Lindmayer, I., Menassa, K., Lambert, J. *et al.* (1986) Development of new jet injector for insulin therapy. *Diabetes Care*, **9**, 294–297.

[17] Peter, D.J., Scott, J.P., Watkins, H.C. and Frasure, H.E. (2002) Subcutaneous lidocaine delivered by jet-injector for pain control before IV catheterization in the ED: the patients' perception and preference. *Am. J. Emerg. Med.*, **20**, 562–566.

[18] Zsigmond, E.K., Darby, P., Koenig, H.M. and Goll, E.F. (1999) Painless intravenous catheteri-zation by intradermal jet injection of lidocaine: a randomized trial. *J. Clin. Anesth.*, **11**, 87–94.

[19] Greenberg, R.S., Maxwell, L.G., Zahurak, M. and Yaster, M. (1995) Preanesthetic medication of children with midazolam using the biojector jet injector. *Anesthesiology*, **83**, 264–269.

[20] Brodell, R.T. and Bredle, D.L. (1995) The treatment of palmar and plantar warts using natural alpha interferon and a needleless injector. *Dermatol. Surg.*, **21**, 213–218.

[21] Suzuki, T., Takahashi, I. and Takada, G. (1995) Daily subcutaneous erythropoietin by jet injection in pediatric dialysis patients. *Nephron*, **69**, 347.

[22] Burkoth, T.L., Bellhouse, B.J., Hewson, G. *et al.* (1999) Transdermal and transmucosal powdered drug delivery. *Crit. Rev. Ther. Drug Carrier Syst.*, **16**, 331–384.

[23] Mitchell, P. (2003) UK biotech sector loses flagship PowderJect to Chiron. *Nat. Biotechnol.*, **21**, 717.

[24] Roberts, L.K., Barr, L.J., Fuller, D.H. *et al.* (2005) Clinical safety and efficacy of a powdered Hepatitis B nucleic acid vaccine delivered to the epidermis by a commercial prototype device. *Vaccine*, **23**, 4867–4878.

[25] Wolf, A.R., Stoddart, P.A., Murphy, P.J. and Sasada, M. (2002) Rapid skin anaesthesia using high velocity lignocaine particles: a prospective placebo controlled trial. *Arch. Dis. Child.*, **86**, 309–312.

[26] Longridge, D.J., Sweeney, P.A., Burkoth, T.L. and Bellhouse, B.J. (1998) Effects of particle size and cylinder pressure on dermal PowderJect delivery of testosterone to conscious rabbits. *Proc. Int. Symp. Control Rel. Bioact. Mat.*, **25**, 964.

[27] Arora, A., Mitragotri, S. and Ahmet, T. (2009) Soft tissue augmentation by needle-free injection. US Patent 20,090,030,367.

[28] Benedek, K., Walker, E., Doshier, L.A. and Stout, R. (2005) Studies on the use of needle-free injection device on proteins. *J. Chromatogr. A*, **1079**, 397–407.

[29] Houtzagers, C.M., Visser, A.P., Berntzen, P.A. *et al.* (1988) The medi-jector II: efficacy and acceptability in insulin-dependent diabetic patients with and without needle phobia. *Diabet. Med.*, **5**, 135–138.

[30] Schneider, U., Birnbacher, R. and Schober, E. (1994) Painfulness of needle and jet injection in children with diabetes mellitus. *Eur. J. Pediatr.*, **153**, 409–410.

[31] Hingson, R.A., Davis, H.S. and Rosen, M. (1963) The historical development of jet injection and envisioned uses in mass immunization and mass therapy based upon 2 decade's experience. *Mil. Med.*, **128**, 516–524.

[32] Gratieri, T., Alberti, I., Lapteva, M. and Kalia, Y.N. (2013) Next generation intra- and transdermal therapeutic systems: using non- and minimally-invasive technologies to increase drug delivery into and across the skin. *Eur. J. Pharm. Sci.*, **50** (5), 609–622.

[33] Mitragotri, S.S., Blankschtein, D. and Langer, R.S. (2000) Transdermal protein delivery or measurement using low-frequency sonophoresis. US Patent 006,018,678A.

[34] Ng, K. and Lui, Y. (2002) Therapeutic ultrasound: its application in drug delivery. *Med. Res. Rev.*, **22** (2), 204–223.

[35] Harvey, C.J., Pilcher, J.M., Eckersley, R.J. *et al.* (2002) Advances in ultrasound. *Clin. Radiol.*, **57**, 157–177.

[36] Machet, L. and Boucaud, A. (2002) Phonophoresis: efficiency, mechanisms and skin tolerance. *Int. J. Pharm.*, **243** (1–2), 1–15.

[37] Lavon, I. and Kost, J. (2004) Ultrasound and transdermal drug delivery. *Drug Discov. Today*, **9** (15), 670–676.

[38] Tachibana, K. and Tachibana, S. (1999) Application of ultrasound energy as a new drug delivery system. *Jpn. J. Appl. Phys.*, **38** (5B), 3014–3019.

[39] Joshi, A. and Raje, J. (2002) Sonicated transdermal drug transport. *J. Control. Release*, **83** (1), 13–22.

[40] Gustavo, M., Yogeshvar, N.K. and Richard, H.G. (2003) Ultrasound-enhanced transdermal transport. *J. Pharm. Sci.*, **92** (6), 1125–1137.

[41] Park, E.J., Werner, J. and Smith, N.B. (2007) Ultrasound mediated transdermal insulin delivery in pigs using a lightweight transducer. *Pharm. Res.*, **7**, 1396–1400.

[42] Singer, A., Homan, C., Church, A. and McClain, S. (1998) Low-frequency sonophoresis: pathologic and thermal effects in dogs. *Acad. Emerg. Med.*, **5**, 35–40.

[43] Mitragotri, S.S., Blankschtein, D. and Langer, R.S. (1996) Transdermal drug delivery using low-frequency sonophoresis. *Pharm. Res.*, **13**, 411–420.

[44] Batheja, P., Thakur, R. and Michniak, B. (2006) Transdermal iontophoresis. *Expert Opin. Drug Deliv.*, **3**, 127–138.

[45] Williams, A.C. (2003) *Transdermal and Topical Drug Delivery*, Pharmaceutical Press, London.

[46] Merino, V., Lopez, A., Kalia, Y.N. and Guy, R.H. (1999) Electrorepulsion versus electroosmosis: effect of pH on the iontophoretic flux of 5-fluorouracil. *Pharm. Res.*, **16**, 758–761.

[47] Lopez, R.F., Bentley, M.V., Delgado-Charro, M.B. and Guy, R.H. (2001) Iontophoretic delivery of 5-aminolevulinic acid (ALA): effect of pH. *Pharm. Res.*, **18**, 311–315.

[48] Bain, K.T. (2006) A patient-activated iontophoretic transdermal system for acute pain management with fentanyl hydrochloride: overview and applications. *J. Opioid Manag.*, **2**, 314–324.

[49] Doliwa, A., Santoyo, S. and Ygartua, P. (2001) Transdermal Iontophoresis and skin retention of piroxicam from gels containing piroxicam: hydroxypropyl-beta-cyclodextrin complexes. *Drug Dev. Ind. Pharm.*, **27**, 751–758.

[50] Al-Khalili, M., Meidan, V.M. and Michniak, B.B. (2003) Iontophoretic transdermal delivery of buspirone hydrochloride in hairless mouse skin. *AAPS PharmSci*, **5** (2), E14.

[51] Ita, K.B. and Banga, A.K. (2009) In vitro transdermal iontophoretic delivery of penbutolol sulfate. *Drug Deliv.*, **16** (1), 11–14.

[52] Fourie, L., Breytenbach, J.C., Du Plessis, J. *et al.* (2004) Percutaneous delivery of carbamazepine and selected *N*-alkyl and *N*-hydroxyalkyl analogues. *Int. J. Pharm.*, **279**, 59–66.

[53] Nair, V. and Panchagnula, R. (2003) Physicochemical considerations in the iontophoretic delivery of a small peptide: in vitro studies using arginine vasopressin as a model peptide. *Pharmacol. Res.*, **48**, 175–182.

[54] Nakamura, K., Katagai, K., Mori, K. *et al.* (2001) Transdermal administration of salmon calcitonin by pulse depolarization- iontophoresis in rats. *Int. J. Pharm.*, **218**, 93–102.

[55] Heit, M.C., Monteiro-Riviere, N.A., Jayes, F.L. and Riviere, J.E. (1994) Transdermal iontophoretic delivery of luteinizing hormone releasing hormone (LHRH): effect of repeated administration. *Pharm. Res.*, **11**, 1000–1003.

[56] Leboulanger, B., Fathi, M., Guy, R.H. and Delgado-Charro, M.B. (2004) Reverse iontophoresis as a noninvasive tool for lithium monitoring and pharmacokinetic profiling. *Pharm. Res.*, **21**, 1214–1222.

[57] Touitou, E. (2002) Drug delivery across the skin. *Expert Opin. Biol. Ther.*, **2**, 723–733.

[58] Sung, K.C., Fang, J.Y., Wang, J.J. and Hu, O.Y. (2003) Transdermal delivery of nalbuphine and its prodrugs by electroporation. *Eur. J. Pharm. Sci.*, **18**, 63–70.

[59] Wallace, M.S., Ridgeway, B., Jun, E. *et al.* (2001) Topical delivery of lidocaine in healthy volunteers by electroporation, electroincorporation, or iontophoresis: an evaluation of skin anesthesia. *Reg. Anesth. Pain Med.*, **26**, 229–238.

[60] Wang, S., Kara, M. and Krishnan, T.R. (1998) Transdermal delivery of cyclosporin-A using electroporation. *J. Control. Release*, **50**, 61–70.

[61] Conjeevaram, R., Banga, A.K. and Zhang, L. (2002) Electrically modulated transdermal delivery of fentanyl. *Pharm. Res.*, **19**, 440–444.

[62] Prausnitz, M.R., Edelman, E.R., Gimm, J.A. *et al.* (1995) Transdermal delivery of heparin by skin electroporation. *Biotechnology*, **13**, 1205–1209.

[63] Medi, B.M. and Singh, J. (2003) Electronically facilitated transdermal delivery of human parathyroid hormone (1–34). *Int. J. Pharm.*, **263**, 25–33.

[64] Drabick, J.J., Glasspool-Malone, J., King, A. and Malone, R.W. (2001) Cutaneous transfection and immune responses to intradermal nucleic acid vaccination are significantly enhanced by in vivo electropermeabilization. *Mol. Ther.*, **3**, 249–255.

[65] Nelson, J.S., McCullough, J.L., Glenn, T.C. *et al.* (1991) Midinfrared laser ablation of stratum corneum enhances in vitro percutaneous transport of drugs. *J. Invest. Dermatol.*, **97**, 874–879.

[66] Fang, J.Y., Lee, W.R., Shen, S.C. *et al.* (2004) Enhancement of topical 5-aminolaevulinic acid delivery by erbium:YAG laser and microdermabrasion: a comparison with iontophoresis and electroporation. *Br. J. Dermatol.*, **151**, 132–140.

[67] Shen, S.C., Lee, W.R., Fang, Y.P. *et al.* (2006) In vitro percutaneous absorption and in vivo protoporphyrin IX accumulation in skin and tumors after topical 5-aminolevulinic acid application with enhancement using an erbium:YAG laser. *J. Pharm. Sci.*, **95**, 929–938.

[68] Lee, W.R., Shen, S.C., Lai, H.H. *et al.* (2001b) Transdermal drug delivery enhanced and controlled by erbium:YAG laser: a comparative study of lipophilic and hydrophilic drugs. *J. Control. Release*, **75**, 155–166.

[69] Lee, W.R., Shen, S.C., Liu, C.R. *et al.* (2006) Erbium:YAG laser-mediated oligonucleotide and DNA delivery via the skin: an animal study. *J. Control. Release*, **115**, 344–353.

[70] Bragagna, T., Reinhard, B., Daniel, G. *et al.* (2006) WO2006111429.

[71] Bragagna, T., Reinhard, B., Daniel, G. and Bernhard, N. (2005) Laser microporator. EP051704.

[72] Bragagna, T., Reinhard, B., Daniel, G. and Bernhard, N. (2006) Microporator for parating a biological membran and integrated permeant administering system. WO2006111199.

[73] Bragagna, T., Reinhard, B., Daniel, G. and Bernhard, N. (2006) Microporator for creating a permeation surface. WO2006111200.

[74] http://www.please-professional.com/en/research/combination-treatments

[75] Bach, D., Weiss, R., Hessenberger, M. *et al.* (2012) Transcutaneous immunotherapy via laser-generated micropores efficiently alleviates allergic asthma in Phl p 5-sensitized mice. *Allergy*, **67** (11), 1365–1374.

[76] Konno, Y., Kawata, H., Aruga, M. *et al.* (1987) Patch. US Patent 4,685,911.

[77] Kuratomi, Y. and Miyauchi, K. (1988) Methods and instruments of moxibustion. US Patent 4,747,841.

[78] Ajay, K.B. (2006) *Therapeutic Peptides and Proteins: Formulation, Processing, and Delivery Systems*, 2nd edn, CRC Taylor & Francis, Boca Raton, pp. 264–265.

[79] Stewart, R.F. (1989) Temperature-controlled active agent dispenser. US Patent 4,830,855.

[80] Badkar, A.V., Smith, A.M., Eppstein, J.A. and Banga, A.K. (2007) Transdermal delivery of interferon alpha-2B using microporation and iontophoresis in hairless rats. *Pharm. Res.*, **24**, 1389–1395.

[81] Eppstein, J.A. (1998) Microporation of tissue for delivery of bioactive agents. WO98029134.

[82] Levin, G., Amikan, G., Meir, S. and Amir, S. (2007) Combined micro-channel generation and iontophoresis for transdermal delivery of pharmaceutical agents. US Patent 20,070,260,170.

[83] http://www.nitto.com/jp/en/press/2012/0425.jsp

[84] Sintov, A.C., Krymberk, I., Daniel, D. *et al.* (2003) Radiofrequency-driven skin microchanneling as a new way for electrically assisted transdermal delivery of hydrophilic drugs. *J. Control. Release*, **89**, 311–320.

[85] Gerstel, M.S. and Place V.A. (1976) Drug delivery device. US Patent 3,964,482.

[86] Henry, S., McAllister, D.V., Allen, M.G. and Prausnitz, M.R. (1998) Microfabricated microneedles: a novel approach to transdermal drug delivery. *J. Pharm. Sci.*, **87** (8), 922–925.

[87] Prausnitz, M.R. (2004) Microneedles for transdermal drug delivery. *Adv. Drug Deliv. Rev.*, **56** (5), 581–587.

[88] Donnelly, R.F., Thakur, R.R.S., Morrow, D.I.J. and Woolfson, A.D. (2012) *Microneedle-Mediated Transdermal and Intradermal Drug Delivery*, Wiley-Blackwell, Oxford.

9

Microneedle Technology

Helen L. Quinn, Aaron J. Courtenay, Mary-Carmel Kearney and Ryan F. Donnelly

School of Pharmacy, Queen's University Belfast, Belfast, UK

9.1   Introduction

Of the transdermal delivery enhancement technologies currently researched and discussed in this book, microneedle (MN) technology has received particular attention and is rapidly progressing towards commercialisation. MNs were first postulated as a drug delivery mechanism in the 1970s and, since manufacturing advances allowed their first realisation in 1998, have been fabricated by a variety of methods and from a range of materials. MNs are minimally invasive devices consisting of numerous micron-sized projections amassed on a baseplate. The needles puncture the outermost layer of skin, the *stratum corneum*, forming aqueous conduits through which drugs can diffuse to the dermal microcirculation. This relatively new technology marries the patient-friendly benefits of a transdermal patch with the potential delivery capabilities of a hypodermic injection. The unique attributes of MNs are that they penetrate the resilient skin barrier sufficiently to enable access to the skin's rich microcirculation yet with an average length of 50–900 μm, are short enough to avoid stimulation of dermal nerves and do not induce bleeding [1].

As a result of bypassing the *stratum corneum*, efficient transdermal transport will no longer be dependent on drug physicochemical properties and therefore may significantly increase the size of the transdermal market. Further to this, by the same mechanism, MN-based delivery systems may eliminate the potential for transdermal dosing variability, which may be at least partially due to the heterogeneous nature of skin at different sites and on different patients. Indeed, a recent report put the potential global market for MN-based drug delivery systems at just under $400 million in 2012 [2]. Since MN arrays are

Novel Delivery Systems for Transdermal and Intradermal Drug Delivery, First Edition.
Ryan F. Donnelly and Thakur Raghu Raj Singh.
© 2015 John Wiley & Sons, Ltd. Published 2015 by John Wiley & Sons, Ltd.

frequently targeted not only to the $20 billion transdermal drug delivery and $25 billion global vaccine markets but also to the $120 billion global biologics market, significant further growth can be anticipated.

It has been demonstrated that MNs can be manufactured using materials with a range of properties, including metal, silicon, carbohydrates and polymers [3]. Development strategies have been extensively investigated, with array geometries manipulated and optimised to facilitate successful perturbation of the *stratum corneum* [4]. The range of drugs, with diverse properties, amenable to delivery by MNs makes this an enticing drug delivery platform. Despite the extent and diversity of research in this field, 40 years on from the description of the first concept, there are only two marketed drug delivery MN products: MicronJet®, which is composed of four hollow silicon MNs arranged on a plastic adaptor for attachment to a standard syringe barrel, and Soluvia®, a prefillable microinjection system with a single 1.5 mm hollow silicon MN (Figure 9.1). These devices, in essence, reflect the nature of miniature hypodermic needles rather than the transdermal patch model of MNs described.

The major MN approaches employed in order to achieve facilitated drug delivery are solid, coated, hollow, dissolvable and swellable MN devices (Figure 9.2). Solid MNs require a two-step application process; the needles puncture the skin and are subsequently

(a)   (b)

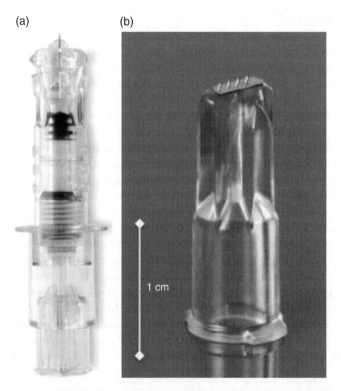

1 cm

Figure 9.1 *Commercialised MN devices for intradermal delivery: (a) BD Soluvia prefillable MN device and (b) NanoPass Technologies Ltd. MicronJet single-use, MN-based device. (a) Reproduced from Ref. [5] with permission from Elsevier and (b) Reproduced with permission from Ref. [6] with permission Nanopass Technologies Ltd. http://www.bd.com/pharmaceuticals/products/microinjection.asp.*

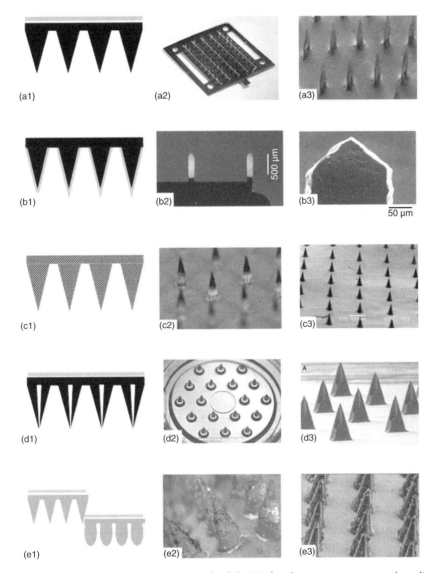

Figure 9.2 (a1) Schematic representation of solid MNs for skin pre-treatment and application of drug-loaded reservoir. (a2) Digital image of solid metallic MNs. Reproduced from Ref. [7] with permission. (a3) Scanning electron micrograph (SEM) image of stainless steel solid MNs. Reproduced from Ref. [8] with permission from Elsevier. (b1) Schematic representation of coated MNs for deposition of drug-containing layer in the skin. (b2) Digital image of dip-coated MNs containing Vitamin-B_2. Reprinted by permission from Ref. [9] copyright (2010) Macmillan Publishers Ltd. (b3) SEM image of stainless steel coated MNs. Reproduced from Ref. [10] with permission from Elsevier. (c1) Schematic representation of dissolving MNs for delivery of incorporated drug into the skin. (c2) Digital image of dissolving MNs composed of chondroitin sulphate. Reproduced from Ref. [11] with kind permission from Springer Science and Business Media. (c3) SEM image of trehalose dissolving MNs. Reproduced from Ref. [12] with permission from Elsevier. (d1) Schematic representation of hollow MNs for insertion into the skin and infusion of drug via the MN pore. (d2) Digital image of 3M hollow Microneedle Transdermal System (hMTS) polymeric MN array. Reproduced from Ref. [13] with kind permission from Springer Science and Business Media. (d3) SEM image of hollow silicon MN array. Reproduced from [14] with permission from Elsevier. (e1) Schematic representation of swelling MNs for drug delivery through hydrogel matrix in 'dry' and 'swollen' form. (e2) Digital image of swollen MNs composed of cross-linked PMVE/MA and PEG 10 000. Reproduced with permission from Ref. [15]. (e3) SEM image of MNs composed of cross-linked PMVE/MA and PEG 10 000. Reproduced with permission from Ref. [16].

removed, temporarily increasing skin permeability and facilitating the passive diffusion of drug from a reservoir, typically in the form of a patch or topical formulation placed over the site of MN insertion. Coated MNs pierce the *stratum corneum*, the drug layer dissolves, and the active drug is deposited in the skin. Dissolvable MNs are usually polymer based; the drug is incorporated into the MN formulation and released as the MN system dissolves. Hollow MNs' mechanism of action resembles that of traditional injections, using pressure to deliver a liquid formulation through the predefined MN conduit [17]. Swelling MNs, fabricated using polymers, have been developed more recently. Following insertion into the skin, they imbibe interstitial fluid and allow diffusion from a separate drug reservoir through the hydrogel matrix [18].

This chapter discusses the many aspects of MN technology, from fabrication methods and drug delivery applications to patient safety and regulatory issues. The volume of work conducted in the area is testament to the undoubted potential of MNs as a new drug delivery device.

9.2   MN Materials and Fabrication

Manufacture and fabrication of MN arrays have typically been conducted through a number of traditional industrial methods and novel fabrication strategies, the latter of which have become more common as MN design has increased in complexity. Each material employed in the manufacturing process requires significant efforts with respect to optimisation of fabrication and production conditions; however, most formulators utilise one or a combination of the following methods: micromoulding/replica moulding [19], laser cutting [20], lithography [21] and wet and dry etching [22]. Within each of these disciplines lies a wealth of engineering and manufacturing considerations, encompassing specialities involved in microelectromechanical systems (MEMS) to mass producing industrial-scale machining tools. Finding a balance between the high accuracy and precision required during MN array development and a pharmaceutically robust procedural method remains the goal of many MN research teams.

Initial MN array design centred around the employment of silicon as a raw material, which was then etched by a process known as deep reactive ion etching and photolithography (Figure 9.3) [23, 24]. This method was based on the use of an alkaline solution and ion etching, employing a veneering agent to ensure accurate MN design, followed by a photolithographic process, which uses high-intensity UV light polymerisation to construct and define the desired MN structure. Lithographically produced MN arrays have been documented extensively in academic literature with wide-ranging needle geometries and designs, utilising polymeric materials, both biopolymers and synthetics; metals such as stainless steel and titanium; glass; and silicon [25–28]. With each material comes nuance challenges with regard to fabrication of the MN array itself, and in many cases, photolithography and polymer micromachining have played a role in MN production. MN arrays can be designed to be in plane or out of plane, reflecting the manufacturing process. Ceramic MN arrays have been reported as a potential for easy mass production, whereby an alumina slurry is cast into a poly(dimethylsiloxane) MN mould. The resultant MNs are porous in nature and can allow a water-soluble active drug to be contained within the device to be delivered by an invoked capillarity effect upon insertion to the skin [29].

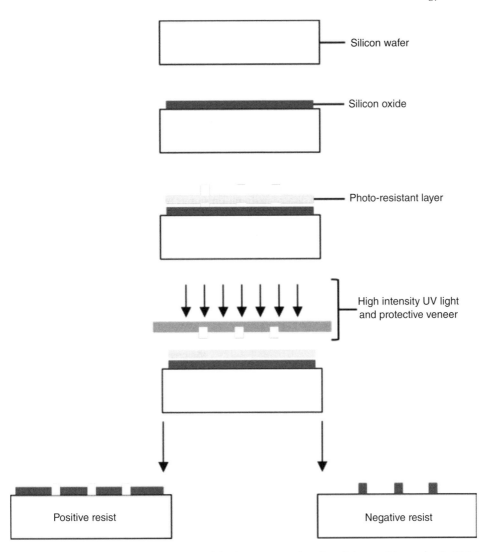

Figure 9.3 *Schematic representation of the key steps in the photolithographic method of MN fabrication: (i) silicon wafer, (ii) silicon wafer coated with silicon oxide, (iii) photo-resistive material applied by spin coating, (iv) high-intensity UV light applied in conjunction with a protective veneering agent, (v) positive resist and (vi) negative resist.*

As described, mass production of solid MNs has demonstrated real potential with regards to reproducibility, accuracy and low to moderate fabrication costs, with the first two MN-based products successfully marketed (MicronJet and Soluvia) based on silicon and metal MNs, respectively. Silicon as a material, however, is somewhat hampered by its dubious biocompatibility and lack of FDA approval, and in such instances, MN fracture and deposition in patient skin remain a significant concern [30–32]. Moreover, the possible problems resulting from inappropriate disposal of silicon or metal MNs, which both remain intact post-removal, have led to most researchers in this field focusing on MNs made from FDA-approved polymeric materials [33].

Water-soluble, inert polymers and sugars have been used in the fabrication of dissolvable MN arrays, which, on contact with skin, dissolve in the interstitial skin fluid and release their drug payload [17]. Laser-engineered poly(dimethylsiloxane) or silicon moulds are filled with a liquid polymer, whereby solvent evaporation then results in solidification of the MN array (Figure 9.4). In some instances, monomer-containing liquids have been cast followed by a polymerisation reaction to solidify the cast material and facilitate removal from the moulds. Dissolvable MNs have been manufactured from dextran [11], carboxymethyl cellulose [34], polyvinylpyrrolidone [35], polyvinyl alcohol [36], fibroin [37], poly(lactic-glycolic) acid [27] and sugars such as galactose and maltose [38]. In the majority of cases, a vacuum is applied or a centrifugation step is required to ensure the polymers occupy the appropriate portion of MN moulds in order to create reproducible MNs.

Maintaining the activity and structural integrity of biological drugs during production of MNs is an important consideration. Initially, the hot polymer and carbohydrate melts often used to make dissolvable MNs caused breakdown of biologics during processing [39]. Kim *et al.* [40] described a droplet-born air blown (DAB) method of MN fabrication, which did

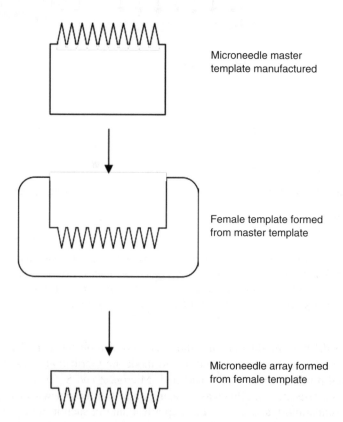

Microneedle master
template manufactured

Female template formed
from master template

Microneedle array formed
from female template

Figure 9.4 *Schematic representation of the micromoulding technique, often used for the production of polymeric MNs.*

not result in a significant loss of drug. The DAB method involves a polymeric solution shaped into a MN by passage of air across its surface. A single droplet of formulation was used for each MN and as such allowed for control of needle volume and, subsequently, control of drug loading. The process yielded reproducible MN arrays and is reported to have taken less than 10 min for fabrication of insulin-loaded dissolving MNs. A novel layer-by-layer approach, described by Iler and Kirkland in the 1960s [41, 42], was applied to MN array fabrication, involving alternate deposition of interacting species onto a substrate with intermediate rinsing steps [43]. These multilayered systems have proven to be of benefit with regard to delivery of proteinaceous compounds such as DNA and RNA vaccine components [44]. In addition, a large proportion of recent research has focused on dissolving MNs prepared from aqueous polymer blends to prevent loss of loading and maintain biological activity [45–48]. The introduction of biocompatible polymeric MN devices may herald a new era in the development of MN technology, overcoming a number of disadvantages of previous MN designs.

One possible drawback of polymeric MN systems is that some of the materials used may not inherently possess sufficient mechanical strength to pierce the *stratum corneum*. A novel approach in the development of polymeric MN arrays with enhanced mechanical strength was the combination of a metallic base with polymeric tips to achieve arrowhead MN arrays [49]. These are essentially metallic frames with projections onto which polymeric pyramid-shaped tips were fabricated. Photolithography and micromoulding techniques were employed to facilitate this combination, which when tested ensured targeted local release of the therapeutic agent to excised skin [50].

To overcome the limited dosing capacity for macromolecules loaded into dissolving MNs, Donnelly *et al.* formulated MN arrays made from hydrogel-forming polymers [18]. These MNs contain no active drug themselves. Instead, they are hard in the dry state but rapidly take up skin interstitial fluid upon insertion to form discrete hydrogel pathways between an attached patch-type drug reservoir and the dermal microcirculation. In this way, the dose of drug is not limited to what can be loaded into, or coated on the surface of, the MNs themselves. These unique swellable MNs are produced in a similar fashion to the polymeric MNs described previously, formulated from a cross-linked network of the copolymer of the free acid form of methylvinylether and maleic anhydride (PMVE/MA) and polyethylene glycol (PEG) 10 000 [18].

9.3 MN-Mediated Drug Delivery

The compounds delivered by MNs to date have typically been of high potency, meaning only a low dose is required to achieve a therapeutic effect (e.g. insulin) [45, 51] or elicit the required immune response [52, 53]. However, the majority of marketed drug substances, including many antibodies, are not low-dose, high-potency molecules. Indeed, many drugs require doses of several hundred milligrams per day in order to achieve therapeutic plasma concentrations in humans. Until now, such high doses could not be delivered transdermally from a patch of reasonable size, even for molecules whose physicochemical properties are ideal for passive diffusion across the skin's *stratum corneum* barrier. Therefore, transdermal delivery has traditionally been limited to low molecular weight, high-potency drug substances with a certain degree of lipophilicity. Since most drugs do not possess these properties, the transdermal delivery market has not expanded beyond around 20 drugs. Using MNs to enhance delivery is likely to increase this number of drugs in the coming years. However, this increase will only be maximised if high-dose molecules can also be delivered in therapeutic concentrations using MNs.

The first drug delivery via MNs was of the low molecular weight model compound, calcein, described by Henry *et al.* [54]. Multiple investigations rapidly followed, leading to the current ever-growing body of evidence for the significant drug delivery capabilities of MNs. From the precedent in 1998, solid MNs, fabricated from silicon and metal, continue to be extensively investigated for drug delivery. McAllister *et al.* reported the successful delivery of a multitude of substrates, including bovine serum albumin (BSA) and insulin using metal MNs [25]. Another group described pretreatment with solid MN, followed by application of a topical insulin solution, which resulted in the lowering of blood glucose levels in diabetic rats [55]. Pretreatment of skin with polymeric solid MNs was also found to increase topical delivery of the photosensitiser 5-aminolevulinic acid and 5-aminolevulinic acid methyl ester [56]. Stahl *et al.* used the commercially available Medik8® solid MN rollers to enhance delivery of a number of non-steroidal anti-inflammatory drugs [57]. While the conventional 'poke and patch' methodology has progressed, it has been recognised that a cumbersome two-step application process is a major drawback [18].

To create a one-step application process, solid MNs have been coated with the material to be delivered. Coated MNs have been employed for the delivery of a number of different compounds including fluorescein sodium [58], salmon calcitonin [59], desmopressin [60] and parathyroid hormone (PTH) [61], among others. Coated MNs have also found a role in the delivery of DNA. Dip coating of MN into DNA solutions, with the intradermal delivery aimed at improving novel treatment options for genetic skin diseases and cutaneous cancers, has been described [62]. Metal MNs coated in small interfering ribonucleic acid (siRNA) have been shown to silence gene expression *in vitro* and *in vivo* [63]. Aside from this, due to the limited drug loading capacity of this method, coated MNs are more frequently employed for the delivery of highly potent molecules and vaccines, the latter of which will be discussed in greater detail in Section 9.4.

Research into hollow MNs has focused mainly on array design and characterisation with several sophisticated engineering strategies presented [64]. Martanto *et al.* fabricated single-glass hollow MNs and used these to deliver sulphorhodamine solution into the skin, demonstrating the benefit of hollow design in its ability to control drug infusion rate [65]. Microinjection of insulin to diabetic rats using hollow MNs has also been successfully explored [25]. A major limitation to their use, however, is the potential blockage of the MN bore by compressed dermal tissue upon insertion, reducing potential drug release [24].

Dissolving MNs have been used to deliver a number of small molecule drugs including caffeine, lidocaine, theophylline and metronidazole [66]. Additionally, they have been used to specifically target various clinical needs by delivery of a number of biopharmaceutical molecules, including low molecular weight heparins [67], insulin [45], leuprolide acetate [68], erythropoietin [69] and human growth hormone [70]. Although successful in the delivery of these macromolecules, it was noted that, in contrast to the delivery of small water-soluble drug molecules, only the delivery of large molecular weight compounds from the MNs themselves was achievable, with the drug loaded in the baseplate remaining [45]. Another criticism of the dissolving platform was the inability to deliver therapeutically relevant concentrations of low-potency drug substances due to reduced mechanical robustness associated with high MN drug loading [3]. McCrudden *et al.* have recently taken steps to address these concerns, having successfully delivered therapeutically relevant doses of ibuprofen sodium in a rat model [51], although further work is needed in investigating this with regard to other drugs.

Hydrogel-forming MNs have been demonstrated to proficiently deliver both small molecules, such as metronidazole and theophylline, and larger molecules, such as insulin and BSA [18]. The benefit of the hydrogel system is that MN swelling rate can be controlled by altering the polymer cross-link density, thus conferring the ability to govern drug release rate, which can be tailored for specific drugs.

The vast majority of MN-based investigations involving human subjects have considered the perception, safety and practical applications of the technology [71, 72]. The field has also progressed to the delivery of drugs to humans, with naltrexone being the first drug to reach clinical studies, setting the precedent for further testing in humans [73]. Daddona *et al.* have studied the pharmacokinetic and pharmacodynamic parameters of PTH delivered using MNs in humans, with promising results to date [61].

9.3.1 Combinational Approaches

MN technology in combination with other drug delivery-enhancing methodologies has been investigated, with the aim of achieving synergistic benefits. Iontophoresis uses a low-voltage charge to force polar and ionic molecules across the skin and, when combined with MNs, enabled a threefold increase in the delivery of the macromolecule insulin [46]. This pairing of MNs and iontophoresis has been shown in other variations to produce similar benefits [74, 75]. Interestingly, another study showed that although combinational approaches involving iontophoresis did not significantly increase the delivery of low molecular weight compounds, as exemplified in that case by ibuprofen sodium, it did, however, significantly increase the delivery of protein compounds across a hydrogel MN system [76]. Sonophoresis is an alternative non-invasive method of enhancing skin permeability, using ultrasound to cause a change in the lipid arrangement of the *stratum corneum*. Sonophoresis used together with MNs has been tested with similar positive results to iontophoresis and MNs [77]. The simple technique of skin occlusion has been used in tandem with MN treatment, particularly solid MNs, to improve drug delivery by reducing the rate of skin restoration post-needle insertion and augmenting skin permeability [78]. Yan *et al.* have shown that including motorised technology in combination with MN devices can lead to improved DNA transfer into the epidermal layer of mouse skin [79]. Gene gun technology has also been used following pretreatment with solid MNs as a means of facilitating DNA-loaded microparticle delivery [80].

Modification of the drug carrier system has also been used in conjunction with MN-mediated delivery to enhance transdermal drug delivery. As the channels created in the skin are within the micrometre range, MNs can facilitate the delivery of particles that are nanometres in size. Ketoprofen-loaded nanoparticles used in combination with MNs were shown to double transdermal delivery in comparison to those with no MNs [81]. MNs have also been used to pretreat skin followed by application of docetaxel-containing liposomes. This combination was found to increase transdermal flux of docetaxel, and importantly, it caused considerable reduction in drug permeation lag time [82]. These integrated approaches reflect the versatility of MNs as a drug delivery system.

9.4   MN Vaccination

MN technology as a vaccination facilitation technique has become a great research interest in recent years. Vaccination methods have been shown to be the most effective method of reducing infectious disease, in light of both mortality and morbidity [83]. The prevalence of needle phobia and the requirement for sharps disposal and trained personnel to adminis-ter injections are further inconveniences, with the latter a particular issue in developing countries, restricting vaccination coverage [84, 85].

Delivery of vaccines to the skin has been viewed as a logical approach, as the various professional antigen-presenting immune cells contained within make the skin a natural and ideal target for vaccine delivery [86]. This transition to skin-targeted vaccine delivery was first supported by Combadière *et al.* [87]. In this study, a phase I clinical trial demonstrated that transdermal delivery of an inactivated influenza vaccine resulted in superior CD8 effector T-cell activation over administration via intramuscular (IM) injection. It has also been suggested that delivery of vaccines to the skin has a dose-sparing effect when com-pared to IM vaccination due to the significant numbers of antigen-presenting cells (APCs) present in the skin, permitting the induction of a strong immune response with lower antigen levels [86]. As a result, targeting the skin as an immunogenic organ in its own right has allowed MN vaccination to be considered in more detail, with extensive research now having been conducted in this area. The world's first MN-based vaccine device, Soluvia, was marketed in 2009 in conjunction with the intradermal vaccinations Fluzone Intradermal®, IDflu® and Intanza®.

9.4.1   Polymeric MNs and Vaccination

Dissolving MN platforms for vaccination arguably show the most promise with respect to commercialisation, owing mostly to their self-disabling nature, thereby preventing reuse. Dissolving MNs made from trehalose and carboxymethylcellulose have been utilised to deliver cell culture-derived influenza vaccine with considerable effect [88]. This capacity was further exemplified in the delivery of trivalent influenza vaccine to mice, in conjunc-tion with ovalbumin by cast carboxymethylcellulose, producing a strong immune response [34]. Superior to IM delivery methods, the use of MNs manufactured from polymerised monomeric *N*-vinylpyrrolidone and containing inactivated influenza vaccine has resulted in the initiation of a robust humoral and cellular immune response post immunisation [35]. Zaric *et al.* used poly-D,L-lactide-co-glycolide nanoparticles encasing ovalbumin as a model protein, suspended in aqueous blends of PMVE/MA to target dendritic cells (Figure 9.5). Dissolving within minutes, the dissolution strategy resulted in considerable Th1 CD4+ and potent cytotoxic CD8+ T-cell responses [52]. Matsuo *et al.* has carried out work using MicroHyala®, a novel dissolving MN array composed of hyaluronic acid, to investigate vaccine efficacy against tetanus, diphtheria, influenza and malaria, in each case producing immunisation results comparable with the parenteral route [89]. In an attempt to formulate a vaccine for Alzheimer's disease, MicroHyala was used to deliver an amyloid-beta 42-amino-acid peptide as an antigenic stimulant. Although the most recent publication did not show improved brain function in those patients treated with the active drug, it did demonstrate that efficient anti-amyloid-beta immune responses can be attained by MN application [90].

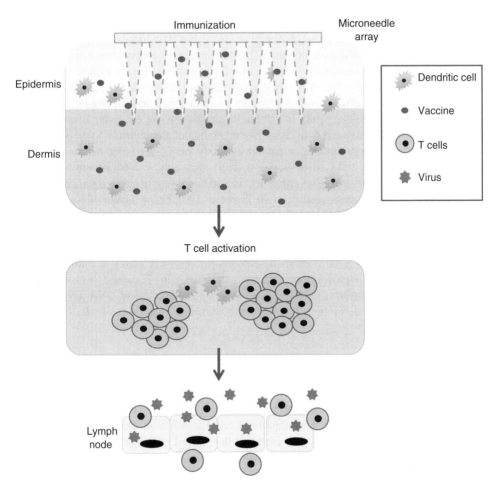

Figure 9.5 *Schematic illustration of immunisation using dissolving MN arrays loaded with vaccine particles. The vaccine particles, when delivered into the dermis, activate T cells, which are then transported to the cutaneous draining lymph nodes, resulting in clearance of virus.*

9.4.2 Solid MNs and Vaccination

Coated MN designs have also shown success with regard to stimulating appropriate immune responses following vaccine application. Coating solid MNs with antigenic material has proven to be of interest as this process requires minor alteration to long-standing techniques such as spray coating [91] and repeated dip coating to load stainless steel [92], titanium [93] and silicon [94] projections. Stainless steel MNs have been employed to deliver the vaccine against the Edmonston-Zagreb virus, demonstrating that these coated MNs have the ability to produce neutralising antibody titres equivalent to subcutaneous injection [95]. A number of groups have investigated the potential to vaccinate against influenza. Kommareddy *et al.* found that coating influenza antigen onto solid MN structures could produce comparable antibody titres to IM injection [96].

A study by Zhu *et al.* revealed that, following application of influenza vaccine-coated MNs, complete protection from the lethal challenge of H3N2 and H1N1 seasonal strains of influenza virus was afforded in mouse models [97]. Another vaccine approach against influenza, based on an influenza extracellular matrix and toll-like receptor (M2E-TLR5) fusion protein coated onto stainless steel MNs by way of micro-precision dip coating, has shown comparative protective efficacy in mice to intranasal delivery and highlights significant improvement over IM delivery [98]. In a different vaccination strategy using live virus vaccines, Vrdoljak *et al.* promoted delivery of live recombinant adenovirus and modified vaccinia virus Ankara. This produced comparable antibody and CD8+ T-cell responses to that of a conventional needle and syringe vaccine delivery approach, with equivalent doses [99].

Vaccination with DNA and RNA constructs is also beginning to show significant potential, with transfection success steadily on the increase [100]. Using a coated MN approach, DeMuth *et al.* demonstrated that a pH-dependent lipid nanocapsule layering system, for loading and delivering DNA vaccine for a model of human immunodeficiency virus, can achieve similar results to that of gene gun technology [101]. More recently, Kim *et al.* published work furthering Alzheimer's disease vaccine delivery [102]. A plasmid DNA-layered MN system has been developed, whereby the change in pH caused by insertion into skin allows coating deposition, exposing dermal APCs to the plasmid DNA payload. The study demonstrated that, comparable with subcutaneous injection, a robust humoral response could be achieved.

9.4.3   Hollow MNs and Vaccination

Vaccine delivery through a hollow MN, essentially a method of intradermal injection, is similar to that of conventional parenteral vaccine delivery in that one or more hollow needles are used to deliver a liquid formulation. Needles in this case, however, are generally 1–1.5 mm in length, and volumes delivered less than 200 µl [13]. Vaccine delivery in this way is designed to improve upon the Mantoux technique (dermal needle technique based on shallow angle insertion into patient skin), which can be difficult to perform and produces inconsistent results with respect to dermal cell response [103].

Initial and follow-on studies with anthrax vaccination in rabbit models showed up to 50-fold dose reduction when delivered by hollow MNs in comparison to IM injection. Subsequent to challenge with anthrax spores, MN vaccination resulted in equivalent survival profiles to that of the IM route [104]. Similar dose-sparing studies conducted with a rabies vaccine in healthy human adults have demonstrated adequate immuno-efficacy when MN delivery was contrasted with IM injection [105]. Multiple vaccine components combined into a single product have also received attention, particularly from Morefield *et al.* This group successfully used hollow MNs to deliver a combination vaccine, providing protection from staphylococcal toxic shock, botulism, anthrax and plague in rhesus macaque monkeys [106].

9.4.4   MN Vaccination Moving Forwards

MN vaccination has gained considerable headway within the academic community and may soon be a viable alternative to traditional vaccine delivery methods, having moved past proof of concept and into more developmental stages of research. Scale-up of the manufacturing process of MN vaccines will be required to allow mass production, alongside further endeavours to improve vaccine stability.

A key aspect of novel vaccine delivery strategies is the possibility for enhanced stability of the vaccine components crucially throughout manufacture and storage. Encapsulation into dissolving MN or, conversely, coating onto solid MN remains the major challenge to vaccine stability. As with other biologic agents, the drying stages of manufacture associated with MN loading and formulation tend to cause loss of antigenicity. Coating solid MN techniques often require the addition of suitable excipients to stabilise the antigen during the coating process and hence minimise loss of activity. Trehalose has been shown to be particularly effective in this, primarily due to the prevention of aggregation [107].

Long-term storage of injectable vaccines is a challenge, with cold chain transport and storage often required to preserve antigen activity [108]. MN approaches may hold the key to overcoming these challenges, as many of the strategies described previously result in vaccine material formulated in the dry state and, hence, are less likely to require refrigeration. Most recently, Kommareddy *et al.* has shown improvement in stability of the influenza subunit vaccine, achieving successful storage of coated MNs at room temperature in a desiccated environment for at least 8 weeks [96].

It has been suggested that scale-up of animal model testing should be explored. Rattanapak *et al.* suggested moving from rodent models to more appropriate animal models such as the miniature pig [109]. Nonhuman primates are also an important step prior to human testing. This progression in use of animal models is important to drive MN vaccination forwards so that the full potential of this novel delivery methodology can be achieved. However, results from these animal studies need to be carefully considered in terms of extrapolation to humans, in particular with respect to possible differences in immunogenicity and skin insertion factors. Ultimately, clinical studies in humans will be the determinant of vaccine efficacy, with some MN vaccines having already reached this stage.

Phase II and phase III clinical trials conducted with microinjection of a trivalent influenza vaccination observed a difference in response between healthy adult patients aged between 18 and 57 years [110] and healthy elderly patients aged greater than 60 years [111]. Ageing is known to affect the cutaneous immune response, with an age-associated reduction in number of Langerhans cell. However, recent work has highlighted no change in phenotype or function of these monocyte-derived Langerhans cells, suggesting epidermal environment may be more important [112]. Ultimately, the efficacy of MN vaccine delivery will need to be demonstrated as the various dermal components alter with age, ensuring vaccine protection across a breadth of ages.

9.5   Further MN Applications

The utility of MNs in bypassing formidable biological barriers, such as the skin's *stratum corneum*, opens up a range of additional applications beyond transdermal and intradermal drug delivery. Drug delivery into the eye [113, 114], enhanced administration of active cosmeceutical ingredients [115, 116] and therapeutic drug monitoring are currently under intense investigation in both academia and industry [117–119].

9.5.1   Therapeutic Drug Monitoring

Currently, plasma monitoring of both drug and endogenous substances for reasons such as dose optimisation, diagnostic purposes, drug adherence, and so on is primarily based on the analysis of invasively extracted blood samples. There are several well-recognised disadvantages of this process including needle phobia, high expense of analytical equipment and trained healthcare staff, risk of infection, difficulty in certain patient populations (neonates) as well as the time-consuming nature of the process. MNs have been demonstrated to have potential in addressing some of these limitations.

Similar to hypodermic needles, some MN designs can be used for the two-way movement of liquid, thereby allowing extraction of interstitial fluid from the skin [117]. This can subsequently be used for the purposes of therapeutic drug monitoring. Interstitial fluid represents a good media for analyte monitoring, as concentrations in interstitial fluid often accurately reflect those in the blood. The premise of MN-mediated therapeutic drug monitoring is that if interstitial fluid can be extracted using MNs and effectively analysed, then there is a possibility of pain-free, blood-free monitoring. If drug substances could be both monitored and delivered from the same interconnected device, then the prospect of an MN-based closed-loop delivery system could become a reality.

Research in this area has to date focused mainly on the logistics of MN-mediated therapeutic drug monitoring, for example, determining the optimum rate of fluid extraction. Initial designs tended to rely on passive fluid movement via capillary action; however, more sophisticated mechanisms have also been explored such as vacuum and osmotic pressure [117]. Glass MNs in combination with vacuum pressure have been described for the monitoring of glucose levels, a process traditionally regarded as painful and cumbersome. MNs were inserted into the skin of hairless rats and human volunteers at depths between 700 and 1500 μm. Following application of a vacuum between 2 and 10 min, interstitial fluid was extracted and glucose levels were analysed. Results from the study found that interstitial fluid glucose levels correlated well with blood glucose levels before and after insulin administration [120]. Although a glass MN device is not ideal, the study provides preliminary evidence that minimally invasive MN-mediated glucose monitoring is a possibility.

The majority of minimally invasive MN-based approaches for sampling have considered the extraction of interstitial fluid; however, there are several reports of needles on the upper end of the micron scale being used for the extraction of blood samples, albeit in a less invasive manner than conventional hypodermic needles. Li *et al.* have described metallic hollow MNs (1800 μm height, 60 μm inner diameter, 120 μm tip diameter and 15° bevel) capable of extracting 20 μl of blood from a mouse-tail vein. They claim that the sharp tip and precise dimensions of the needles has a low risk of pain inducement yet are long enough to extract blood [121]. In terms of therapeutic monitoring, blood analyte concentration is important; however, devices such as this do not enable blood-free analysis and would therefore still be associated with cross-contamination risk and require specialist disposal. More promising are MN devices capable of reliably and accurately extracting interstitial fluid, which can then be correlated with blood levels.

A variant of therapeutic monitoring via MNs has been adopted by O'Mahony *et al.* [122]. They have employed MNs to optimise signal obtainment for electrocardiography (ECG), with the potential for other applications such as electroencephalography and electromyography. The *stratum corneum* causes substantial interference in signal acquirement during these procedures due to high electrical impedance, with electrolytic gels or skin abrasion being required. These gels, however, are uncomfortable for the patient and have a relatively short duration of action. An MN patch consisting of 25 projections, 300 μm in height, connected with two electrodes was inserted into a volunteer and ECG signals were recorded. The ECG images were of comparable quality to conventionally obtained signals, demonstrating the ability of these silicon MNs to overcome the electrically impeding *stratum corneum.*

9.5.2 Cosmetic Applications

The cosmeceutical industry has shown interest in MN technology [123]. The applications by this industry are twofold; in the treatment of skin blemishes, the MNs are used to stimulate repair responses by the skin but have also been postulated for use in the delivery of compounds to the skin for cosmetic purposes.

Kim *et al.* have reported the use of hyaluronic-based dissolving MNs for the intradermal delivery of ascorbic acid and retinyl retinoate [124]. Volunteers were randomly allocated to either the ascorbic acid or the retinyl retinoate group. All subjects applied MNs to the 'crow's feet' area of the face twice daily which were left in place for 6 h. Both MN formulations displayed improved skin appearance in terms of roughness and wrinkle appearance. In addition, the MNs demonstrated sufficient mechanical robustness for insertion into the skin at drug loadings of 35% for retinyl retinoate and 60% for ascorbic acid. Application of a patch to the face for such an extended period may not be an ideal cosmetic device; however, this study provides an example of the explorative investigations being considered in the field of cosmeceutics.

In a study by Fabbrocini *et al.*, the effect of Dermaroller® treatment on acne scars was explored [125]. Patients displaying a variety of acne scar severities were treated on two occasions (8 weeks apart) with needle rollers to create between 250 and 300 microscopic holes/cm². The skin was photographed on several occasions by experienced dermatologists and assessed according to the Goldman and Barron acne classification system. In addition, a microrelief impression was made using silicone rubber on five patients to assess skin tomography before and after MN treatment. Results of the study found that there was aesthetic improvement, with a reduction in scar severity in all subjects. While the full mechanism of enhanced skin appearance as a result of MN treatment has not been elucidated, it has been found that penetrating the skin with MNs increases collagen and elastin production in the upper dermis. This study highlights one of the many potential cosmetic applications of MNs. It did also, however, identify the importance of needle length consideration; in this study, the needles used induced bleeding as they were up to 2 mm in length.

The potential cosmetic applications of MNs are extensive, with continually emerging work in this area. A recent study by Kumar *et al.* investigated skin pretreatment with MNs to enhance the local delivery of eflornithine cream, a topical product used to reduce facial hirsutism [126]. The positive results achieved in these preliminary *in vitro* and *in vivo* studies have promoted the continuation of work in this area. The combination of MNs and already established topical products must be addressed with caution however, with recent news stories reporting adverse granulomatous reactions following MN use with non-sterile topical products not intended for intradermal application [127].

9.5.3   Other Potential Applications

Beyond the original hypothesised use, the functional capabilities of MNs have been well recognised, with a plethora of research fields utilising the concept for their own particular purposes. There are ever-increasing reports of MNs being used for purposes beyond transdermal drug delivery. Recently, Yang *et al.* have demonstrated that their advanced MN system could be used in skin graft adherence [128]. Biphasic, conical shaped MNs composed of a non-swelling polystyrene core and a polystyrene-*block*-poly(acrylic acid) swellable tip were inserted into tissue with minimal force and also displayed enhanced tissue adhesion strength in comparison to staples, which are conventionally used in skin graft adherence. Furthermore, the swelling tips can be loaded with therapeutic agents to facilitate both local drug delivery and tissue adhesion.

As the capabilities of MN devices continue to be recognised, there is a corresponding increase in research applications. Thakur *et al.* have described the fabrication of hollow MNs from hypodermic needles for the delivery of thermoresponsive poloxamer formulations containing fluorescein sodium into a rabbit sclera [129]. The MNs were used to deliver *in situ* forming gels to release drug intra-sclerally in a minimally invasive manner. Optical coherence tomography (OCT) was used to confirm both penetration of the needles and recovery of the sclera post-treatment. This highlights the multi-faceted nature of the research being conducted within the field of MNs, and while it is primarily focused on intra- and transdermal drug delivery, there are many potential applications beyond this.

9.6   Patient Factors Relating to MN Use

9.6.1   Effects of MN Insertion on the Skin

The consequences of MN insertion and removal from the skin are naturally of paramount importance and must be considered to ensure both patient comfort and patient safety. This includes both the sensation experienced by the patient when MNs are applied and the effects following removal, such as closure of the created micropores and any sustained erythema. It is important to know the dimensions of the micropores they create in patients' skin and how quickly normal skin barrier function recovers. Traditionally, coloured dyes have been used to stain the pores created, and transepidermal water loss (TEWL) measurements used to quantify disturbances in skin barrier function following MN removal [130, 131]. Although these techniques confirm that the skin's *stratum corneum* barrier has been compromised, they provide no information with respect to the true depth of MN penetration. Recently, OCT, the optical analogue of ultrasound imaging, has been used to investigate MN-mediated skin puncture [132–134]. Since it is capable of imaging the skin down to depths of 2.5 mm, OCT has been used to study insertion depth and micropore width (Figure 9.6). This work indicates that MNs do not usually penetrate fully into skin. For example, approximately 80% of the shaft length of a 600 μm MN was shown to protrude beneath the *stratum corneum* in one study [134], with a micropore width of approximately 300 μm. Importantly, the technique allows the influence of different MN design and application forces on insertion depth to be studied and, if used in conjunction with transparent polymeric MNs, can be used to follow MN dissolution/

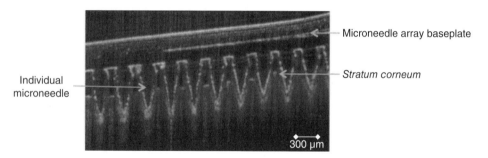

Figure 9.6 *Optical coherence tomographic image showing PMVE/MA MNs (height, 600 μm; width at base, 300 μm; spacing, 50 μm) inserted into human skin in vivo.*

swelling in real time *in vivo*. Micropore closure kinetics can also be studied. A recent study extensively investigated skin recovery in human volunteers over a period of 24 h following hydrogel-forming MN application periods of between 0.5 and 6 h [135]. Using TEWL and OCT, micropores created in the skin were monitored following MN removal to assess skin recovery over time. TEWL values showed a return to baseline in all cases at 24 h, yet OCT still highlighted the presence of micropores in the skin, exposing a discrepancy between the two techniques. Further work in this area will clarify whether indeed the skin has fully recovered at 24 h with only indentations remaining or whether skin recovery is slower than previously thought [18, 72]. Electrical impedance spectroscopy has also been used to study the micropore lifetime and their subsequent resealing in humans. Using this technique, under non-occlusive conditions, skin barrier function appears to have recovered within 2 h of treatment with stainless steel MNs, with occlusion extending the increase in skin permeability to up to 40 h, depending on MN geometry [78]. Diclofenac and fluvastatin have both recently been described as being able to delay micropore closure when used as a co-drug, allowing drug delivery for 7 days, demonstrating how the skin barrier can be further manipulated for controlled release of drug [136, 137].

With respect to skin erythema and irritation, the same study described previously found no sustained erythema following MN removal, with most cases having resolved within a few hours of removal and a few persisting cases fully resolved at 48 h [135]. This supports the findings of previous studies which have reported that although MN geometry can have an effect, skin irritation due to MN application was minimal and all signs had disappeared within 2 h, regardless of the duration of application [71]. Irritant contact dermatitis (ICD) scores, when used to measure skin irritation, have been shown to return to 0 within 1 h of MN removal [18, 138].

Volunteers generally find MN application to be painless, with only 20% of people able to differentiate between application of a flat baseplate and an array of MNs [135]. The sensation of MN application has been described as 'pressing' or 'heavy' in comparison to the 'sharp' and 'stabbing' feeling experienced with hypodermic injection [130]. In some cases, higher pain scores, determined using a visual analogue scale, were recorded with application of MN arrays of increasing needle number and length [139].

9.6.2   Patient Safety

A point worth noting when considering patient safety is that MN devices are not equivalent to conventional transdermal patches, in that they are not simply applied to the skin surface. Rather, MNs function principally by breaching the skin's protective *stratum corneum* barrier and often penetrating into the viable epidermis and dermis, typically sterile areas [140]. Accordingly, it is imperative that MNs, or products used with them, do not themselves contain microbial loads sufficient to cause skin or systemic infection. It is also important that bioburden be minimised to avoid immune stimulation, especially considering the rich immune cell population in the viable epidermis and dermis [141]. It has been shown that microbial penetration through MN-induced holes has been shown to be minimal and lower than that facilitated by conventional hypodermic needle puncture [142]. Although shown that certain microorganisms can traverse through the microconduits into the skin, none crossed the viable epidermis, demonstrating that MN application would be unlikely to cause infection in normal circumstances. Similar findings were reported by an *in vivo* study using rats treated with MNs and infected with *Staphylococcus aureus* [143].

Hydrogel-forming MNs prepared from the copolymer PMVE/MA have been shown to exhibit antimicrobial properties and are, therefore, inherently unlikely to cause infection [138]. In addition, the self-disabling nature of dissolving and hydrogel-forming MNs minimises the risk of infection that frequently arises with hypodermic needle reuse. Moreover, it will be important that no local or systemic reactions to the materials used to fabricate MNs occur. Recent reports, as highlighted previously, suggest that, when used inappropriately, MNs can indeed cause health problems, such as skin irritation and intradermal granulomas, as well as systemic hypersensitivity. In each case however, medical supervision was absent and MNs were used in conjunction with topical products not designed to be applied intradermally [127]. Current knowledge, gathered from clinical experience with MNs, supports the presumption that appropriate use is unlikely to cause skin infection, given that there have been no cases throughout the numerous animal and human studies carried out since their conception. To date, however, little is known about any long-term effects that may occur due to repeatedly penetrating the skin with MNs. The skin is replete with APCs, and so, it is vital that delivery via MNs does not elicit immune responses to the cargo being transported into the skin, potentially worth considering with biomolecules in particular.

Polymer deposition from dissolving MNs is also an issue worth considering for patient safety. While the polymers used for MN production are typically approved pharmaceutical excipients, given the novelty of the delivery mechanism of MNs, the polymers have never before been used intradermally. The amounts of polymer left behind in skin after MN removal and information on clearance rates and routes should be considered. This may be a non-issue for one-off vaccine administration but could be important if dissolving MNs were to be used regularly for insulin delivery, for example.

To provide some form of guarantee as to patient safety, it may be a regulatory requirement that MNs are produced or rendered sterile. Any contained microorganisms would have to be identified and pyrogen content minimised. Such an undertaking will require careful selection of the method. Aseptic manufacture will be expensive and will present practical challenges if MNs are to be made on a very large scale, for example, with vaccine delivery products. Terminal sterilisation using gamma irradiation, moist heat or microwave heating may damage the MNs or biomolecule loadings, while ethylene oxide may permeate polymeric MN materials, thus contaminating the delivery system.

9.6.3 Acceptability to Patients and Healthcare Providers

The success of MNs as a commercial product in widespread use will naturally depend on the ability of the device to perform its intended task to a high standard but also on its acceptability to the ultimate end user of the product. Alongside ensuring a well-performing delivery system, consideration has to be given to the patient and also the healthcare professionals who will be prescribing and, in some cases, administering the device.

A study by Birchall *et al.* made the first progress in ascertaining perceptions of MNs from healthcare professionals and members of the general public [144]. Positively, the majority of participants recruited to these focus group studies were able to identify a number of advantages to the technology, including reduced pain, tissue damage and transmission of infection on application in comparison to traditional needles. The potential for self-administration was highlighted, as was their applicability in children and needle-phobic patients. However, concerns were raised about effectiveness and how a patient would know the device had been used properly. Overall, the group reported that 100% of the public and 74% of the healthcare professional participants who participated in the study were positive in their feedback on MN technology.

As highlighted by adults involved in the Birchall study, children are a potential group who could particularly benefit from MNs. Mooney *et al.* used focus groups to explore children's views on MN use as an alternative approach to blood sampling in monitoring applications [145]. A widespread disapproval of conventional blood sampling using needles was evident, with pain, blood and traditional needle visualisation being particularly unpopular aspects. In general, MNs had greater visual acceptability and caused less fear. Children's concerns included possible allergy and potential inaccuracies with this novel approach. However, many had confidence in the judgement of healthcare professionals if deeming this technique appropriate. They considered paediatric patient education critical for acceptance of this new approach and called for an alternative name, without any reference to 'needles'. It was concluded that a proactive response to these unique insights should enable MN array design to better meet the needs of this end-user group. Further work in this area is recommended to ascertain the perspectives of a purposive sample of children with chronic conditions who require regular monitoring. Norman *et al.* used MNs to deliver insulin to children and adolescents with type 1 diabetes with 10 of 12 participants finding MN insertion less painful than subcutaneous catheter insertion and thus hypothesised that the use of such a device could lead to increased compliance [146]. Indeed, such studies, when appropriately planned to capture the required demographics, will undoubtedly aid industry in taking necessary action to address concerns and develop informative labelling and patient counselling strategies to ensure safe and effective use of MN-based devices.

9.6.4 Patient Application

Due to the inherent elasticity of the skin alongside the irregular surface topography, it has long been thought that some form of device would be important to enable consistent and reproducible insertion of MNs. This led to the development of a number of applicator devices, based upon high-impact/high-velocity insertion or rotary designs, with the objective of applying MNs in a defined way to ensure consistent therapeutic activity on each application and to minimise inter-individual variability [147]. However, in order to capitalise on preconceived positive beliefs relating to transdermal patches, MN patches must be viewed in the same way by both prescriber and patient as a convenient and easy-to-use alternative to oral and parenteral delivery. The FDA draft guidance 'Applying Human Factors and Usability Engineering to Optimize Medical Device Design' stresses the importance of an easy-to-use device, with reduced need for training and reduced reliance on a user manual [148]. Emphasis is also placed on the requirement for evidence to prove successful use of the device with minimal chance of error prior to marketing. The progression of thought has therefore now led to the belief that manual application by hand may be preferable, the simplicity of the device being considered key to its commercial success. Recent studies have provided some convincing evidence as to the validity of self-application. Donnelly *et al.* used OCT and TEWL measurements to illustrate that human volunteers can successfully insert hydrogel-forming MNs into their own skin by hand to uniform depths to yield consistent transient disturbances of skin barrier function when counselled by a pharmacist and having read a suitable patient information leaflet [72]. Considering this option of self-application further, Norman *et al.* found that the option of a self-administered MN patch, when compared to the use of a traditional hypodermic injection, increased the intent to vaccinate from 44 to 65%, highlighting the possibility that the availability of a MN patch could increase vaccine coverage [149]. This same study tested MN self-application with minimal volunteer training using thumb pressure, compared to a snap-based applicator device. In both cases, although the only measure recorded was skin staining, it was deduced that MN patches were successfully self-administered. Thumb pressure was found in some cases to require multiple attempts and/or further instruction in order to increase the pressure of application, but in general, thumb pressure was sufficient to cause 90% of MNs to penetrate the *stratum corneum* on first attempt. The inclusion of some method of feedback in response to correct application, as petitioned for by the volunteers in the Birchall and Donnelly studies, would be likely to address these manual insertion issues [72, 144]. Indeed, in order to gain acceptance from healthcare professionals, patients and, importantly, regulatory authorities, it appears likely that a suitable means of confirming that skin puncture has taken place and the MNs have been inserted properly into the skin may need to be included within the MN product itself.

9.7 The Next Steps in MN Development

Given the significant body of work published in the MN field on microfabrication and drug delivery capabilities [3], to proceed to the next stages of MN development, it is thought that the research priority lies with the scale-up of the manufacturing process alongside their application to relevant clinical needs. There has recently been a gradual migration towards a more patient-focused research model, which is encouraging as what stands between the existing MN knowledge base and the safe and effective use of a MN product by a patient is examined. Aligning engineering and product design, alongside the usability of the device, will ultimately lead to an optimised product and faster progress through the relevant regulatory channels. The MN field is now at the stage where encompassing the end user in extensive clinical trials is important while simultaneously making the necessary steps to manufacture a product on the required scale.

9.7.1 Manufacturing Considerations

The scale-up of MN production to an industrial level will require considerable thought, especially true given the plethora of small-scale production methods described in the literature [150]. Very often, a number of steps are required for preparation, a challenge to transfer these methods to an industrial setting. A further issue to address is the likely requirement that MNs will need to be manufactured aseptically or terminally sterilised in some way, a procedure that is likely to increase cost considerably. Overall, it is likely that any manufacturer wishing to develop MN products will need to make a substantial initial capital investment, given that equivalent manufacturing technologies are not currently available. Similarly, a range of new quality control tests and pharmacopoeial standards will now also become necessary. Packaging will be important in protecting MNs from moisture and microbial ingress, and suitable advice will need to be provided to avoid damage during patient handling and insertion.

9.7.2 Regulatory Considerations

The regulatory requirements pertaining to a MN product will also need to be defined in the coming years due to the innovative nature of the technology. It is likely that the regulatory requirements set for the first MN products to be approved for human use will set the standards for follow-on products. Overall, from a regulatory perspective, it seems likely that MNs will be classed as a new dosage form, rather than an adjunct technology to existing transdermal patch drug delivery systems. In summary, the key regulatory questions that may need to be addressed are as follows:

1. *Sterility of the MN dosage form.* As a MN dosage form will penetrate the *stratum corneum* into the epidermis, rather than simply adhere to its surface (as in a conventional transdermal patch), it is likely that sterility will be a regulatory requirement, although a low bioburden may be acceptable in cases where the system has inherent and demonstrable antimicrobial activity.
2. *Uniformity of content (either from the system as a whole or, possibly, in respect of individual drug-loaded MNs within an array, depending on system design).* It is likely that this pharmacopoeial requirement, which is internationally harmonised, will be applied to MN systems, as it is for transdermal patch dosage forms.
3. *Manufacturing aspects, including packaging.* The normal aspects of quality, including security of packaging (which may also require a demonstration of adequate protection from, e.g. water ingress), will apply.
4. *Potential for reuse by patients or others.* Many current MN systems, notably those made of silicon, can be removed intact from the skin and, therefore, could be reused by the patient or others. Thus, for reasons of safety, a self-disabling system ensuring single use only may be required.
5. *Disposal.* MN materials that are not dissolvable or biodegradable may be a hazard; therefore, this environmental aspect of their use may be an issue.

6. *Deposition of MN materials in the skin, particularly with respect to long-term use.* Dissolvable, polymeric MNs will deposit in the skin the materials from which they were fabricated. This could lead to long- or short-term adverse skin effects, such as granuloma formation or local erythema, particularly where repeated use is a factor. This can be mitigated by varying the application site and may be less of an issue where the use is occasional, such as in vaccination.

7. *Ease and reliability of application by patients.* As with all dosage forms, patients must be able to use the product properly, without significant inconvenience.

8. *Assurance of delivery (proper insertion).* Since there is no obvious sensation on applying a MN dosage form, some indication of correct application and delivery (particularly for vaccination applications) may be required.

9. *Potential immunological effects.* Repeated insult of the skin, an immunologically active site, by MNs may result in an immunological reaction, depending on the material involved. Regulators may therefore require some assurances regarding immunological safety.

10. *Long-term safety profile.* There are numerous studies which have involved human volunteers, both with placebo and drug-loaded MNs, forming the basis of an initial short-term safety profile. The long-term safety of MN application, from both the perspective of intermittent use and also of repeated application, for example, when considering the number of doses required for insulin administration over a sustained period, will need to be considered.

9.7.3   Commercialisation of MN Technologies

Over the past decade, there has been a substantial increase in MN technologies, demonstrated in the number of academic publications on the subject, which has more than quadrupled since 2004. While biological agents including vaccines have been the main focus, water-soluble drugs not currently suitable for passive transdermal delivery are also of great interest. A number of companies are investing heavily in the development of MN-based delivery systems; these include 3M (MN, United States), Corium (CA, United States), Zosano Pharma (CA, United States), Becton, Dickinson and Company (NJ, United States), NanoPass Technologies (Nes Ziona, Israel) and, more recently, Löhmann Therapie Systeme (LTS; NJ, United States), the world's largest transdermal patch manufacturer [6, 151–155]. Zosano has successfully conducted phase I and II clinical trials using their ZP Patch Technology, an array of titanium microprojections, originally developed at ALZA (CA, United States), used to deliver PTH to post-menopausal women. Promising results from the earlier clinical studies and positive feedback obtained in focus groups made up of the participants lead to the preliminary presumption of success in the phase III study [61].

NanoPass Technologies have shown their MicronJet device to be useful in delivery of insulin, influenza vaccines and local anaesthetics in some clinical studies [6]. Meanwhile, 3M's microstructured transdermal systems (MTS), based on either hollow or coated solid MNs, have been evaluated in a range of preclinical studies focused on the delivery of proteins, peptides and vaccines [151].

While the aforementioned MN devices have been based upon solid or hollow MN systems, it is envisaged that devices based upon FDA-approved, biodegradable/dissolving polymeric MN formulations will, in the future, receive increased attention from pharmaceutical companies due to their inherent potential advantages. Due to the self-disabling nature of these systems, there is the opportunity for reduced transmission of infection and the prevention of needle-stick injuries, alongside ease of disposal, ultimately meaning that the impact on healthcare in the developing world in particular could be significant. Further to this, the fact that most MNs formulate biomolecules, such as vaccine antigens, in the dry state means that the cold chain may be circumvented. Considering these features of MN systems, it may be hypothesised that within the next 10 years, there will be vaccination programmes in the developing world based around MNs. Such an intervention could massively improve the quality of life, life expectancy and economic productivity of developing countries. Accordingly, the potential impact of MN research and ultimate commercialisation is vast. Once the first MN-based products is accepted by regulators, healthcare providers and patients, other MN-based products for everyday patient and consumer use will become widely used, to the benefit of the pharmaceutical, medical devices and cosmetic industries and patients worldwide.

9.8   Conclusion

Transdermal drug delivery has historically been limited to only those compounds that possess the appropriate physicochemical properties, which allow for movement across the *stratum corneum* and into the viable epidermis. The advancing nature of research into MN delivery systems shows continual improvement in transdermal delivery of therapeutics, which would otherwise never passively cross the *stratum corneum*. The ever-increasing amount of fundamental knowledge appears to be feeding industrial development. As technological advances continue, MN arrays may well become one of the major pharmaceutical dosage forms and monitoring devices of the near future. However, in order for new pharmaceutical products and medical devices based upon MN arrays to realise their undoubted potential and provide benefits for patients and industry, a number of factors will need to be taken into account. These include reliable patient application, user acceptability, safety and cost-effective mass production. MN devices have the capacity to play a role in modern healthcare, and therefore, it is now the researchers' role to ensure the present work is capitalised on and subsequently translates into patient benefit.

References

[1] Donnelly, R.F., Majithiya, R., Thakur, R.R.S. *et al.* (2011) Design, optimization and characterisation of polymeric microneedle arrays prepared by a novel laser-based micromoulding technique. *Pharm. Res.*, **28** (1), 41–57.

[2] Transdermal delivery market predicted to reach $31.5 billion by 2015: PharmaLive special report. *Pharmaceutical & Medical Packaging News*. http://www.pmpnews.com/news/transdermal-delivery-market-predicted-reach-315-billion-2015-pharmalive-special-report (accessed 30 October 2014).

[3] Tuan-Mahmood, T.-M., McCrudden, M.T.C., Torrisi, B.M. *et al.* (2013) Microneedles for intradermal and transdermal drug delivery. *Eur. J. Pharm. Sci.*, **44** (5), 623–637.

[4] Olatunji, O., Das, D.B., Garland, M.J. *et al.* (2013) Influence of array interspacing on the force required for successful microneedle skin penetration: theoretical and practical approaches. *J. Pharm. Sci.*, **102** (4), 1209–1221.

[5] Frenck, R.W., Belshe, R., Brady, R.C. *et al.* (2011) Comparison of the immunogenicity and safety of a split-virion, inactivated, trivalent influenza vaccine (Fluzone®) administered by intradermal and intramuscular route in healthy adults. *Vaccine*, **29** (34), 5666–5674.

[6] Nanopass. http://www.nanopass.com/content-e.asp?cid=19 (accessed 30 October 2014).

[7] Laboratory for Drug Delivery. http://drugdelivery.chbe.gatech.edu/gallery_microneedles.html (accessed 30 October 2014).

[8] Kaur, M., Ita, K.B., Popova, I.E. *et al.* (2014) Microneedle-assisted delivery of verapamil hydrochloride and amlodipine besylate. *Eur. J. Pharm. Biopharm.*, **86** (2), 284–291.

[9] Gill, H.S., Soderholm, J., Prausnitz, M.R. *et al.* (2010) Cutaneous vaccination using microneedles coated with hepatitis C DNA vaccine. *Gene Ther.*, **17** (6), 811–814.

[10] Gill, H.S. and Prausnitz, M.R. (2007) Coated microneedles for transdermal delivery. *J. Control. Release*, **117** (2), 227–237.

[11] Fukushima, K., Ise, A., Morita, H. *et al.* (2011) Two-layered dissolving microneedles for percutaneous delivery of peptide/protein drugs in rats. *Pharm. Res.*, **28** (1), 7–21.

[12] McGrath, M.G., Vucen, S., Vrdoljak, A. *et al.* (2014) Production of dissolvable microneedles using an atomised spray process: effect of microneedle composition on skin penetration. *Eur. J. Pharm. Biopharm.*, **86** (2), 200–211.

[13] Burton, S.A., Ng, C.-Y., Simmers, R. *et al.* (2011) Rapid intradermal delivery of liquid formulations using a hollow microstructured array. *Pharm. Res.*, **28** (1), 31–40.

[14] Vazquez, P., Herzog, G., O'Mahony, C. *et al.* (2014) Microscopic gel–liquid interfaces supported by hollow microneedle array for voltammetric drug detection. *Sens. Actuators B*, **201**, 572–578.

[15] BBC. http://www.bbc.co.uk/news/uk-northern-ireland-20146132 (accessed 30 October 2014)

[16] QUB. http://www.qub.ac.uk/directorates/EnterpriseDevelopment/CommercialOpportunities/AllLicensingOpportunities/P101240/ (accessed 30 October 2014).

[17] Kim, Y.-C., Park, J.-H. and Prausnitz, M.R. (2012) Microneedles for drug and vaccine delivery. *Adv. Drug Deliv. Rev.*, **64** (14), 1547–1568.

[18] Donnelly, R.F., Thakur, R.R.S., Garland, M.J. *et al.* (2012) Hydrogel-forming microneedle arrays for enhanced transdermal drug delivery. *Adv. Funct. Mater.*, **22** (23), 4879–4890.

[19] Park, J.-H., Choi, S.-O., Kamath, R. *et al.* (2007) Polymer particle-based micromolding to fabricate novel microstructures. *Biomed. Microdevices*, **9** (2), 223–234.

[20] Aoyagi, S., Izumi, H., Isono, Y. *et al.* (2007) Laser fabrication of high aspect ratio thin holes on biodegradable polymer and its application to a microneedle. *Sens. Actuators A*, **139** (1–2), 293–302.

[21] Pérennès, F., Marmiroli, B., Matteucci, M. *et al.* (2006) Sharp beveled tip hollow microneedle arrays fabricated by LIGA and 3D soft lithography with polyvinyl alcohol. *J. Micromech. Microeng.*, **16** (3), 473–479.

[22] Jung, P.G., Lee, T.W., Oh, D.J. *et al.* (2008) Nickel microneedles fabricated by sequential copper and nickel electroless plating and copper chemical wet etching. *Sens. Mater.*, **20** (1), 45–53.

[23] Henry, S., McAllister, D.V., Allen, M.G. *et al.* (1998) Microfabricated microneedles: a novel approach to transdermal drug delivery. *J. Pharm. Sci.*, **88** (9), 948.

[24] Gardeniers, H.J.G.E., Luttge, R., Berenschot, E.J.W. *et al.* (2003) Silicon micromachined hollow microneedles for transdermal liquid transport. *J. Microelectromech. Syst.*, **12** (6), 855–862.

[25] McAllister, D.V., Wang, P.M., Davis, S.P. *et al.* (2003) Microfabricated needles for transdermal delivery of macromolecules and nanoparticles: fabrication methods and transport studies. *Proc. Natl. Acad. Sci. U. S. A.*, **100** (24), 13755–13760.

[26] Davis, S.P., Martanto, W., Allen, M.G. *et al.* (2005) Hollow metal microneedles for insulin delivery to diabetic rats. *IEEE Trans. Biomed. Eng.*, **52** (5), 909–915.

[27] Park, J.-H., Allen, M.G. and Prausnitz, M.R. (2006) Polymer microneedles for controlled-release drug delivery. *Pharm. Res.*, **23** (5), 1008–1019.

[28] Luttge, R., Berenschot, E.J.W., de Boer, M.J. *et al.* (2007) Integrated lithographic molding for microneedle-based devices. *J. Microelectromech. Syst.*, **16** (4), 872–884.

[29] Bystrova, S. and Luttge, R. (2011) Micromolding for ceramic microneedle arrays. *Microelectron. Eng.*, **88** (8), 1681–1684.

[30] Chow, A.Y., Pardue, M.T., Chow, V.Y. *et al.* (2001) Implantation of silicon chip microphotodiode arrays into the cat subretinal space. *IEEE Trans. Neural Syst. Rehabil. Eng.*, **9** (1), 86–95.

[31] Voskerician, G., Shive, M.S., Shawgo, R.S. *et al.* (2003) Biocompatibility and biofouling of MEMS drug delivery devices. *Biomaterials*, **24** (11), 1959–1967.

[32] Schmidt, S., Horch, K. and Normann, R. (1993) Biocompatibility of silicon-based electrode arrays implanted in feline cortical tissue. *J. Biomed. Mater. Res.*, **27** (11), 1393–1399.

[33] Pierre, M.B.R. and Rossetti, F.C. (2014) Microneedle-based drug delivery systems for transdermal route. *Curr. Drug Targets*, **15** (3), 281–291.

[34] Raphael, A.P., Prow, T.W., Crichton, M.L. *et al.* (2010) Targeted, needle-free vaccinations in skin using multilayered, densely-packed dissolving microprojection arrays. *Small*, **6** (16), 1785–1793.

[35] Sullivan, S.P., Koutsonanos, D.G., Del Pilar Martin, M. *et al.* (2010) Dissolving polymer microneedle patches for influenza vaccination. *Nat. Med.*, **16** (8), 915–920.

[36] Wendorf, J.R., Ghartey-Targoe, E.B., Williams, S.C. *et al.* (2011) Transdermal delivery of macromolecules using solid-state biodegradable microstructures. *Pharm. Res.*, **28** (1), 22–30.

[37] You, X., Chang, J.-H., Ju, B.-K. *et al.* (2011) Rapidly dissolving fibroin microneedles for transdermal drug delivery. *Mater. Sci. Eng. C*, **31** (8), 1632–1636.

[38] Martin, C.J., Allender, C.J., Brain, K.R. *et al.* (2012) Low temperature fabrication of biodegradable sugar glass microneedles for transdermal drug delivery applications. *J. Control. Release*, **158** (1), 93–101.

[39] Donnelly, R.F., Morrow, D.I.J., Thakur, R.R.S. *et al.* (2009) Processing difficulties and instability of carbohydrate microneedle arrays. *Drug Dev. Ind. Pharm.*, **35** (10), 1242–1254.

[40] Kim, J.D., Kim, M., Yang, H. *et al.* (2013) Droplet-born air blowing: novel dissolving microneedle fabrication. *J. Control. Release*, **170** (3), 430–436.

[41] Iler, R.K. (1966) Multilayers of colloidal particles. *J. Colloid Interface Sci.*, **21** (6), 569–594.

[42] Kirkland, J.J. (1965) Porous thin-layer modified glass bead supports for gas liquid chromatography. *Anal. Chem.*, **37** (12), 1458–1461.

[43] Ariga, K., Yamuachi, Y., Rydzek, G. *et al.* (2014) Layer-by-layer nanoarchitectonics: invention, innovation, and evolution. *Chem. Lett.*, **43** (1), 36–68.

[44] DeMuth, P.C., Min, Y., Huang, B. *et al.* (2013) Polymer multilayer tattooing for enhanced DNA vaccination. *Nat. Mater.*, **12** (4), 367–376.

[45] Migalska, K., Morrow, D.I.J., Garland, M.J. *et al.* (2011) Laser-engineered dissolving microneedle arrays for transdermal macromolecular drug delivery. *Pharm. Res.*, **28** (8), 1919–1930.

[46] Garland, M.J., Caffarel-Salvador, E., Migalska, K. *et al.* (2012) Dissolving polymeric microneedle arrays for electrically assisted transdermal drug delivery. *J. Control. Release*, **159** (1), 52–59.

[47] Demir, Y., Akan, Z. and Kerimoglu, O. (2013) Characterization of polymeric microneedle arrays for transdermal drug delivery. *PLoS One*, **8** (10).

[48] Park, J.-H. and Prausnitz, M.R. (2010) Analysis of mechanical failure of polymer microneedles by axial force. *J. Korean Phys. Soc.*, **56** (4), 1223–1227.

[49] Chu, L.Y. and Prausnitz, M.R. (2011) Separable arrowhead microneedles. *J. Control. Release*, **149** (3), 242–249.

[50] Choi, S., Rajaraman, S. and Yoon, Y. (2006) 3-D patterned microstructures using inclined UV exposure and metal transfer micromolding. Solid-State Sensor, Actuator Microsystems Workshop, South Carolina, pp. 348–351.

[51] McCrudden, M.T.C., Alkilani, A.Z., McCrudden, C.M. *et al.* (2014) Design and physicochemical characterisation of novel dissolving polymeric microneedle arrays for transdermal delivery of high dose, low molecular weight drugs. *J. Control. Release*, **180C**, 71–80.

[52] Zaric, M., Lyubomska, O., Touzelet, O. *et al.* (2013) Skin dendritic cell targeting via microneedle arrays laden with antigen-encapsulated poly-D,L-lactide-co-glycolide nanoparticles induces efficient antitumor and antiviral immune responses. *ACS Nano*, **7** (3), 2042–2055.

[53] Koutsonanos, D.G., Compans, R.W. and Skountzou, I. (2013) Targeting the skin for microneedle delivery of influenza vaccine. *Adv. Exp. Med. Biol.*, **785**, 121–132. doi: 10.1007/978-1-4614-6217-0_13.

[54] Henry, S., McAllister, D.V., Allen, M.G. *et al.* (1998) Microfabricated microneedles: a novel approach to transdermal drug delivery. *J. Pharm. Sci.*, **87** (8), 922–925.

[55] Zhou, C.-P., Liu, T.-L., Wang, H.-L. *et al.* (2010) Transdermal delivery of insulin using microneedle rollers in vivo. *Int. J. Pharm.*, **392** (1–2), 127–133.

[56] Mikolajewska, P., Donnelly, R.F., Garland, M.J. *et al.* (2010) Microneedle pre-treatment of human skin improves 5-aminolevulininc acid (ALA)- and 5-aminolevulinic acid methyl ester (MAL)-induced PpIX production for topical photodynamic therapy without increase in pain or erythema. *Pharm. Res.*, **27** (10), 2213–2220.

[57] Stahl, J., Wohlert, M. and Kietzmann, M. (2012) Microneedle pretreatment enhances the percutaneous permeation of hydrophilic compounds with high melting points. *BMC Pharmacol. Toxicol.*, **13**, 5.

[58] Chen, J., Qiu, Y. and Zhang, S. (2013) Controllable coating of microneedles for transdermal drug delivery. *Drug Dev. Ind. Pharm.* doi: 10.3109/03639045.2013.873447.

[59] Tas, C., Mansoor, S., Kalluri, H. *et al.* (2012) Delivery of salmon calcitonin using a microneedle patch. *Int. J. Pharm.*, **423** (2), 257–263.

[60] Cormier, M., Johnson, B., Ameri, M. *et al.* (2004) Transdermal delivery of desmopressin using a coated microneedle array patch system. *J. Control. Release*, **97** (3), 503–511.

[61] Daddona, P.E., Matriano, J.A., Mandema, J. *et al.* (2011) Parathyroid hormone (1-34)-coated microneedle patch system: clinical pharmacokinetics and pharmacodynamics for treatment of osteoporosis. *Pharm. Res.*, **28** (1), 159–165.

[62] Pearton, M., Saller, V., Coulman, S.A. *et al.* (2012) Microneedle delivery of plasmid DNA to living human skin: formulation coating, skin insertion and gene expression. *J. Control. Release*, **160** (3), 561–569.

[63] Chong, R.H.E., Gonzalez-Gonzalez, E., Lara, M.F. *et al.* (2013) Gene silencing following siRNA delivery to skin via coated steel microneedles: in vitro and in vivo proof-of-concept. *J. Control. Release*, **166** (3), 211–219.

[64] Wang, P.-C., Paik, S.-J., Kim, S.-H. *et al.* (2014) Hypodermic-needle-like hollow polymer microneedle array: fabrication and characterization. *J. Microelectromech. Syst.*, **23** (4), 991–998.

[65] Martanto, W., Moore, J. and Kashlan, O. (2006) Microinfusion using hollow microneedles. *Pharm. Res.*, **23** (1), 104–113.

[66] Garland, M.J., Migalska, K., Tuan-Mahmood, T. *et al.* (2012) Influence of skin model on in vitro performance of drug-loaded soluble microneedle arrays. *Int. J. Pharm.*, **434** (1–2), 80–89.

[67] Gomaa, Y.A., Garland, M.J., McInnes, F. *et al.* (2012) Laser engineered dissolving microneedles for active transdermal delivery of nadroparin calcium. *Int. J. Pharm.*, **82**, 299–307.

[68] Ito, Y., Murano, H., Hamasaki, N. *et al.* (2011) Incidence of low bioavailability of leuprolide acetate after percutaneous administration to rats by dissolving microneedles. *Int. J. Pharm.*, **407** (1–2), 126–131.

[69] Ito, Y., Hasegawa, R., Fukushima, K. *et al.* (2010) Self-dissolving micropile array chip as percutaneous delivery system of protein drug. *Biol. Pharm. Bull.*, **33** (4), 683–690.

[70] Lee, J.W., Choi, S.-O., Felner, E.I. *et al.* (2011) Dissolving microneedle patch for transdermal delivery of human growth hormone. *Small*, **7** (4), 531–539.

[71] Bal, S.M., Caussin, J., Pavel, S. *et al.* (2008) In vivo assessment of safety of microneedle arrays in human skin. *Eur. J. Pharm. Sci.*, **35** (3), 193–202.

[72] Donnelly, R.F., Moffatt, K., Alkilani, A.Z. *et al.* (2014) Hydrogel-forming microneedle arrays can be effectively inserted in skin by self-application: a pilot study centred on pharmacist intervention and a patient information leaflet. *Pharm. Res.*, **31** (8), 1989–1999.

[73] Wermeling, D.P., Banks, S.L., Hudson, D.A. *et al.* (2008) Microneedles permit transdermal delivery of a skin-impermeant medication to humans. *Proc. Natl. Acad. Sci. U. S. A.*, **105** (6), 2058–2063.

[74] Singh, N. and Banga, A. (2013) Controlled delivery of ropinirole hydrochloride through skin using modulated iontophoresis and microneedles. *J. Drug Target.*, **21** (4), 354–366.

[75] Kumar, V. and Banga, A.K. (2012) Modulated iontophoretic delivery of small and large molecules through microchannels. *Int. J. Pharm.*, **434** (1–2), 106–114.

[76] Donnelly, R.F., Garland, M.J. and Alkilani, A.Z. (2014) Microneedle-iontophoresis combinations for enhanced transdermal drug delivery. *Methods Mol. Biol.*, **1141**, 121–132.

[77] Han, T. and Das, D.B. (2013) Permeability enhancement for transdermal delivery of large molecule using low-frequency sonophoresis combined with microneedles. *J. Pharm. Sci.*, **102** (10), 3614–3622.

[78] Gupta, J., Gill, H.S., Andrews, S.N. *et al.* (2011) Kinetics of skin resealing after insertion of microneedles in human subjects. *J. Control. Release*, **154** (2), 148–155.

[79] Yan, G., Arelly, N., Farhan, N. *et al.* (2014) Enhancing DNA delivery into the skin with a motorized microneedle device. *Eur. J. Pharm. Sci.*, **52**, 215–222.

[80] Zhang, D., Das, D.B. and Rielly, C.D. (2014) Potential of microneedle-assisted micro-particle delivery by gene guns: a review. *Drug Deliv.*, **21** (8), 571–587.

[81] Vučen, S.R., Vuleta, G., Crean, A.M. *et al.* (2013) Improved percutaneous delivery of ketoprofen using combined application of nanocarriers and silicon microneedles. *J. Pharm. Pharmacol.*, **65** (10), 1451–1462.

[82] Qiu, Y., Gao, Y., Hu, K. *et al.* (2008) Enhancement of skin permeation of docetaxel: a novel approach combining microneedle and elastic liposomes. *J. Control. Release*, **129** (2), 144–150.

[83] Hegde, N.R., Kaveri, S.V. and Bayry, J. (2011) Recent advances in the administration of vaccines for infectious diseases: microneedles as painless delivery devices for mass vaccination. *Drug Discov. Today*, **16** (23–24), 1061–1068.

[84] Simonsen, L., Kane, A., Lloyd, J. *et al.* (1999) Unsafe injections in the developing world and transmission of bloodborne pathogens: a review. *Bull. World Health Organ.*, **77** (10), 789–800.

[85] Wright, S., Yelland, M., Heathcote, K. *et al.* (2009) Fear of needles-nature and prevalence in general practice. *Aust. Fam. Physician*, **38** (3), 172–176.

[86] Al-Zahrani, S., Zaric, M., McCrudden, C. *et al.* (2012) Microneedle-mediated vaccine delivery: harnessing cutaneous immunobiology to improve efficacy. *Expert Opin. Drug Deliv.*, **9** (5), 541–550.

[87] Combadière, B., Vogt, A., Mahé, B. *et al.* (2010) Preferential amplification of CD8 effector-T cells after transcutaneous application of an inactivated influenza vaccine: a randomized phase I trial. *PLoS One*, **5** (5), e10818.

[88] Kommareddy, S., Baudner, B.C., Oh, S. *et al.* (2012) Dissolvable microneedle patches for the delivery of cell-culture-derived influenza vaccine antigens. *J. Pharm. Sci.*, **101** (3), 1021–1027.

[89] Matsuo, K., Hirobe, S., Yokota, Y. *et al.* (2012) Transcutaneous immunization using a dissolving microneedle array protects against tetanus, diphtheria, malaria, and influenza. *J. Control. Release*, **160** (3), 495–501.

[90] Matsuo, K., Okamoto, H., Kawai, Y. *et al.* (2014) Vaccine efficacy of transcutaneous immunization with amyloid β using a dissolving microneedle array in a mouse model of Alzheimer's disease. *J. Neuroimmunol.*, **266** (1–2), 1–11.

[91] McGrath, M.G., Vrdoljak, A., O'Mahony, C. *et al.* (2011) Determination of parameters for successful spray coating of silicon microneedle arrays. *Int. J. Pharm.*, **415** (1–2), 140–149.

[92] Weldon, W.C., Zarnitsyn, V.G., Esser, E.S. *et al.* (2012) Effect of adjuvants on responses to skin immunization by microneedles coated with influenza subunit vaccine. *PLoS One*, **7** (7), e41501.

[93] Choi, H.-J., Bondy, B.J., Yoo, D.G. *et al.* (2013) Stability of whole inactivated influenza virus vaccine during coating onto metal microneedles. *J. Control. Release*, **166** (2), 159–171.

[94] Chen, X., Fernando, G.J., Raphael, A.P. *et al.* (2012) Rapid kinetics to peak serum antibodies is achieved following influenza vaccination by dry-coated densely packed microprojections to skin. *J. Control. Release*, **158** (1), 78–84.

[95] Edens, C., Collins, M.L., Ayers, J. *et al.* (2013) Measles vaccination using a microneedle patch. *Vaccine*, **31** (34), 3403–3409.

[96] Kommareddy, S., Baudner, B.C., Bonificio, A. *et al.* (2013) Influenza subunit vaccine coated microneedle patches elicit comparable immune responses to intramuscular injection in guinea pigs. *Vaccine*, **31** (34), 3435–3441.

[97] Zhu, Q., Zarnitsyn, V.G., Ye, L. *et al.* (2009) Immunization by vaccine-coated microneedle arrays protects against lethal influenza virus challenge. *Proc. Natl. Acad. Sci. U. S. A.*, **106** (19), 7968–7973.

[98] Wang, B.-Z., Gill, H.S., He, C. *et al.* (2014) Microneedle delivery of an M2e-TLR5 ligand fusion protein to skin confers broadly cross-protective influenza immunity. *J. Control. Release*, **178**, 1–7.

[99] Vrdoljak, A., McGrath, M.G., Carey, J.B. *et al.* (2012) Coated microneedle arrays for transcutaneous delivery of live virus vaccines. *J. Control. Release*, **159** (1), 34–42.

[100] Saade, F. and Petrovsky, N. (2012) Technologies for enhanced efficacy of DNA vaccines. *Expert Rev. Vaccines*, **11** (2), 189–209.

[101] DeMuth, P.C., Moon, J.J., Suh, H. *et al.* (2012) Releasable layer-by-layer assembly of stabilized lipid nanocapsules on microneedles for enhanced transcutaneous vaccine delivery. *ACS Nano*, **6** (9), 8041–8051.

[102] Kim, N.W., Lee, M.S., Kim, K.R. *et al.* (2014) Polyplex-releasing microneedles for enhanced cutaneous delivery of DNA vaccine. *J. Control. Release*, **179**, 11–17.

[103] Laurent, P.E., Bonnet, S., Alchas, P. *et al.* (2007) Evaluation of the clinical performance of a new intradermal vaccine administration technique and associated delivery system. *Vaccine*, **25** (52), 8833–8842.

[104] Mikszta, J.A., Dekker, J.P., Harvey, N.G. *et al.* (2006) Microneedle-based intradermal delivery of the anthrax recombinant protective antigen vaccine. *Infect. Immun.*, **74** (12), 6806–6810.

[105] Laurent, P.E., Bourhy, H., Fantino, M. *et al.* (2010) Safety and efficacy of novel dermal and epidermal microneedle delivery systems for rabies vaccination in healthy adults. *Vaccine*, **28** (36), 5850–5856.

[106] Morefield, G.L., Tammariello, R.F., Purcel, B.K. *et al.* (2008) An alternative approach to combination vaccines: intradermal administration of isolated components for control of anthrax, botulism, plague and staphylococcal toxic shock. *J. Immune Based Ther. Vaccines*, **6** (1), 5.

[107] Kim, Y.-C., Quan, F.S., Compans, R.W. *et al.* (2010) Formulation and coating of microneedles with inactivated influenza virus to improve vaccine stability and immunogenicity. *J. Control. Release*, **142** (2), 187–195.

[108] Lee, B.Y., Cakouros, B.E., Assi, T.M. *et al.* (2012) The impact of making vaccines thermostable in Niger's vaccine supply chain. *Vaccine*, **30** (38), 5637–5643.

[109] Rattanapak, T., Birchall, J.C., Young, K. *et al.* (2014) Dynamic visualization of dendritic cell-antigen interactions in the skin following transcutaneous immunization. *PLoS One*, **9** (2), e89503.

[110] Beran, J., Ambrozaitis, A., Laiskonis, A. *et al.* (2009) Intradermal influenza vaccination of healthy adults using a new microinjection system: a 3-year randomised controlled safety and immunogenicity trial. *BMC Med.*, **7**, 13.

[111] Arnou, R., Icardi, G., De Decker, M. *et al.* (2009) Intradermal influenza vaccine for older adults: a randomized controlled multicenter phase III study. *Vaccine*, **27** (52), 7304–7312.

[112] Ogden, S., Dearman, R.J., Kimber, I. *et al.* (2011) The effect of ageing on phenotype and function of monocyte-derived Langerhans cells. *Br. J. Dermatol.*, **165** (1), 184–188.

[113] Patel, S.R., Lin, A.S.P., Edelhauser, H.F. *et al.* (2011) Suprachoroidal drug delivery to the back of the eye using hollow microneedles. *Pharm. Res.*, **28** (1), 166–176.

[114] Lee, C.Y., You, Y.S., Lee, S.H. *et al.* (2013) Tower microneedle minimizes vitreal reflux in intravitreal injection. *Biomed. Microdevices*, **15** (5), 841–848.

[115] Park, K.Y., Kim, H.K., Kim, S.E. *et al.* (2012) Treatment of striae distensae using needling therapy: a pilot study. *Dermatol. Surg.*, **38** (11), 1823–1828.

[116] Seo, K.Y., Yoon, M.S., Kim, D.H. *et al.* (2012) Skin rejuvenation by microneedle fractional radiofrequency treatment in Asian skin; clinical and histological analysis. *Lasers Surg. Med.*, **44** (8), 631–636.

[117] Donnelly, R.F., Mooney, K., Caffarel-Salvador, E. *et al.* (2014) Microneedle-mediated minimally invasive patient monitoring. *Ther. Drug Monit.*, **36** (1), 10–17.

[118] Coffey, J.W., Corrie, S.R. and Kendall, M.A.F. (2013) Early circulating biomarker detection using a wearable microprojection array skin patch. *Biomaterials*, **34** (37), 9572–9583.

[119] Yeow, B., Coffey, J.W., Muller, D.A. *et al.* (2013) Surface modification and characterization of polycarbonate microdevices for capture of circulating biomarkers, both in vitro and in vivo. *Anal. Chem.*, **85** (21), 10196–10204.

[120] Wang, P.M., Cornwell, M. and Prausnitz, M.R. (2005) Minimally invasive extraction of dermal interstitial fluid for glucose monitoring using microneedles. *Diabetes Technol. Ther.*, **7** (1), 131–141.

[121] Li, C.G., Li, C.Y., Lee, K. *et al.* (2013) An optimized hollow microneedle for minimally invasive blood extraction. *Biomed. Microdevices*, **15** (1), 17–25.

[122] O'Mahony, C., Pini, F., Vereschagina, L. *et al.* (2013) Skin insertion mechanisms of microneedle-based dry electrodes for physiological signal monitoring. 2013 IEEE Biomedical Circuits and Systems Conference, 31 October 2013–2 November 2013, Rotterdam, pp. 69–72.

[123] Doddaballapur, S. (2009) Microneedling with dermaroller. *J. Cutan. Aesthet. Surg.*, **2** (2), 110.

[124] Kim, M., Yang, H., Kim, H. *et al.* (2014) Novel cosmetic patches for wrinkle improvement: retinyl retinoate- and ascorbic acid-loaded dissolving microneedles. *Int. J. Cosmet. Sci.*, **36** (3), 207–212.

[125] Fabbrocini, G., Fardella, N., Monfrecola, A. *et al.* (2009) Acne scarring treatment using skin needling. *Clin. Exp. Dermatol.*, **34** (8), 874–879.

[126] A. Kumar, Y.W. Naguib, Y.-C. Shi et al., A method to improve the efficacy of topical eflornithine hydrochloride cream, *Drug Deliv.* doi:10.3109/10717544.2014.951746 (2014).

[127] Soltani-Arabshahi, R., Wong, J.W., Duffy, K.L. *et al.* (2014) Facial allergic granulomatous reaction and systemic hypersensitivity associated with microneedle therapy for skin rejuvenation. *JAMA Dermatol.*, **150** (1), 68–72.

[128] Yang, S.Y., O'Cearbhaill, E.D., Sisk, G.C. *et al.* (2013) A bio-inspired swellable microneedle adhesive for mechanical interlocking with tissue. *Nat. Commun.*, **4**, 1702.

[129] Thakur, R.R.S., Fallows, S.J., McMillan, H.L. *et al.* (2014) Microneedle-mediated intrascleral delivery of in situ forming thermoresponsive implants for sustained ocular drug delivery. *J. Pharm. Pharmacol.*, **66** (4), 584–595.

[130] Haq, M.I., Smith, E., John, D.N. *et al.* (2009) Clinical administration of microneedles: skin puncture, pain and sensation. *Biomed. Microdevices*, **11** (1), 35–47.

[131] Verbaan, F.J., Bal, S.M., van der Berg, D.J. *et al.* (2008) Improved piercing of microneedle arrays in dermatomed human skin by an impact insertion method. *J. Control. Release*, **128** (1), 80–88.

[132] Enfield, J., O'Connell, M.-L., Lawlor, K. *et al.* (2010) In-vivo dynamic characterization of microneedle skin penetration using optical coherence tomography. *J. Biomed. Opt.*, **15** (4), 046001.

[133] Coulman, S.A., Birchall, J.C., Anesh, A. *et al.* (2011) In vivo, in situ imaging of microneedle insertion into the skin of human volunteers using optical coherence tomography. *Pharm. Res.*, **28** (1), 66–81.

[134] Donnelly, R.F., Garland, M.J., Morrow, D.I.J. *et al.* (2010) Optical coherence tomography is a valuable tool in the study of the effects of microneedle geometry on skin penetration characteristics and in-skin dissolution. *J. Control. Release*, **147** (3), 333–341.

[135] Donnelly, R.F., Mooney, K., McCrudden, M.T.C. *et al.* (2014) Hydrogel-forming microneedles increase in volume during swelling in skin, but skin barrier function recovery is unaffected. *J. Pharm. Sci.*, **103** (5), 1478–1486.

[136] Brogden, N.K., Banks, S.L., Crofford, L.J. *et al.* (2013) Diclofenac enables unprecedented week-long microneedle-enhanced delivery of a skin impermeable medication in humans. *Pharm. Res.*, **30** (8), 1947–1955.

[137] Ghosh, P., Brogden, N.K. and Stinchcomb, A.L. (2014) Fluvastatin as a micropore lifetime enhancer for sustained delivery across microneedle-treated skin. *J. Pharm. Sci.*, **103** (2), 652–660.

[138] Donnelly, R.F., Thakur, R.R.S., Alkilani, A.Z. *et al.* (2013) Hydrogel-forming microneedle arrays exhibit antimicrobial properties: potential for enhanced patient safety. *Int. J. Pharm.*, **451** (1–2), 76–91.

[139] Gill, H.S., Denson, D.D. and Burris, B.A. (2008) Effect of microneedle design on pain in human volunteers. *Clin. J. Pain*, **24** (7), 585–594.

[140] Morrow, D.I.J., McCarron, P.A., Woolfson, A.D. *et al.* (2007) Innovative strategies for enhancing topical and transdermal drug delivery. *Open Drug Deliv. J.*, **1** (1), 36–59.

[141] Roby, K.D. and Di Nardo, A. (2013) Innate immunity and the role of the antimicrobial peptide cathelicidin in inflammatory skin disease. *Drug Discov. Today Dis. Mech.*, **10** (3–4), e79–e82.

[142] Donnelly, R.F., Thakur, R.R.S., Tunney, M.M. *et al.* (2009) Microneedle arrays allow lower microbial penetration than hypodermic needles in vitro. *Pharm. Res.*, **26** (11), 2513–2522.

[143] Wei-Ze, L., Mei-Rong, H., Jian-Ping, Z. *et al.* (2010) Super-short solid silicon microneedles for transdermal drug delivery applications. *Int. J. Pharm.*, **389** (1–2), 122–129.

[144] Birchall, J.C., Clemo, R., Anstey, A. *et al.* (2011) Microneedles in clinical practice – an exploratory study into the opinions of healthcare professionals and the public. *Pharm. Res.*, **28** (1), 95–106.

[145] Mooney, K., McElnay, J.C. and Donnelly, R.F. (2013) Children's views on microneedle use as an alternative to blood sampling for patient monitoring. *Int. J. Pharm. Pract.*, **22** (5), 335–344.

[146] Norman, J.J., Brown, M.R., Raviele, N.A. *et al.* (2013) Faster pharmacokinetics and increased patient acceptance of intradermal insulin delivery using a single hollow microneedle in children and adolescents with type 1 diabetes. *Pediatr. Diabetes*, **14** (6), 459–465.

[147] Thakur, R.R.S. and Dunne, N. (2011) Review of patents on microneedle applicators. *Recent Pat. Drug Deliv. Formul.*, **5** (1), 11–23.

[148] Food and Drug Administration (2011) *Draft Guidance for Industry and Food and Drug Administration Staff : Applying Human Factors and Usability Engineering to Optimize Medical Device Design.* http://www.fda.gov/downloads/MedicalDevices/DeviceRegulationandGuidance/GuidanceDocuments/UCM259760.pdf (accessed 4 March 2015).

[149] Norman, J.J., Arya, J.M., McClain, M.A. *et al.* (2014) Microneedle patches: usability and acceptability for self-vaccination against influenza. *Vaccine*, **32** (16), 1856–1862.

[150] Indermun, S., Luttge, R., Choonara, Y.E. *et al.* (2014) Current advances in the fabrication of microneedles for transdermal delivery. *J. Control. Release*, **185C**, 130–138.

[151] 3M. http://solutions.3m.com/wps/portal/3M/en_WW/3M-DDSD/Drug-Delivery-Systems/Technologies/Microneedle (accessed 30 October 2014).

[152] Corium. http://www.coriumgroup.com/Tech_MicroCor.html (accessed 30 October 2014).

[153] Zosano. http://www.zosanopharma.com/ (accessed 30 October 2014).

[154] BD. http://www.bd.com/pharmaceuticals/products/microinjection.asp (accessed 30 October 2014).

[155] LTS. http://www.ltslohmann.de/en/innovation/technologien-von-lts.html (accessed 30 October 2014).

10

Intradermal Delivery of Active Cosmeceutical Ingredients

Andrzej M. Bugaj

College of Health, Beauty Care and Education, Poznań, Poland

10.1 Introduction

In 1962, Raymond Reed coined the term 'cosmetic pharmaceuticals' or 'cosmeceuticals' [1] to name 'basically functional cosmetic products', which contain biologically active ingredients and claim to reveal expected, drug-like bioactivity [2, 3]. Although this term is not recognised by the US Food and Drug Administration (FDA) or by the European Medicines Agency (EMA), it has been widely adopted by the cosmetic industry [4]. On the other hand, Federal Food, Drug and Cosmetic Act (FD&C Act) recognises that a product can be both cosmetic and drug if it has two intended uses [5]. Thus, such products fall under the scope of the definition of 'cosmeceuticals'.

Three major categories of cosmeceuticals have been noted including products for skin care, hair care, and so on, with the skin care segment accounting for the largest share of the market at 43% [6]. In the cosmetic field, the skin is itself the target of active substance, contrary to the pharmaceutical application, where the skin is often the main barrier to cross in order to deliver the drugs to the systemic circulation [7, 8]. The *stratum corneum* (SC), the outermost layer of the skin, that consists of the cells of corneocytes and densely packed lipid layers, creating together a structure of 'bricks and mortar' [9–11], provides high resistance for the transport of cosmetics ingredients into the skin [12]. On the other hand, SC acts as a primary barrier against pathogens, allergens and other harmful factors [13] and contributes to the maintenance of a healthy skin by hydration, cell adhesion and reduction

Novel Delivery Systems for Transdermal and Intradermal Drug Delivery, First Edition.
Ryan F. Donnelly and Thakur Raghu Raj Singh.
© 2015 John Wiley & Sons, Ltd. Published 2015 by John Wiley & Sons, Ltd.

of transepidermal water loss [14]. Hence, the development of cosmetic formulations should address not only their penetration through the SC but also protection and regeneration of SC structure and function.

Over the past few decades, there has been interest in exploring new techniques to modulate cosmetics absorption through the skin [15–17]. These investigations into skin delivery highlight the need to obtain vehicles of appropriate sizes, high stability and biocompatibility [18].

Most of the delivery systems used or proposed for use in cosmetic preparations are particulate systems, characterised by a large number of interacting particles which differ with respect to certain physical and/or chemical properties [19], reflecting molecular and supramolecular structure of these carriers. In term of this structure, these delivery systems can be divided into emulsions, vesicular systems and solid particulate systems.

10.2 Emulsions

Emulsions are mixture of liquids that do not normally blend (e.g. water and oil) with an emulsifier (surfactant) added to stabilise the dispersant droplets. The two basic types of emulsions are 'oil-in-water' (O/W) and 'water-in-oil' (W/O) emulsions. Emulsions can dissolve both hydrophilic and hydrophobic substances that are often used together in cosmetic preparations. Emulsions offer many advantages for cosmetic industry. They can dissolve both hydrophilic and hydrophobic substances which are often used together in cosmetic preparations. Moreover, they enable regulation of rheological properties of cosmetic formulation without considerable impact on efficiency of active ingredients. Furthermore, they may serve as carriers for pigments and as occlusive agents [20]. For these reasons, emulsions have been the most used in cosmetic delivery systems since antiquity [3, 21, 22].

The O/W emulsion of 3% ascorbic acid was appropriately stable and reveals a good release of the active agent *in vitro*, and its application *in vivo* resulted in a significant reduction of oxidative stress in the skin, an improvement of the epidermal–dermal microstructure and a reduction of fine lines and wrinkles in aged skin [23]. An O/W vitamin E-containing formulation was proved effective in preventing induction of erythema and reducing inflammatory damage caused by UV exposure. Vitamin E emulsion and vehicle control significantly suppressed visual scores and diminished skin colour measurement [24]. Schwarz and coworkers [25] studied the activity of antioxidants: α-tocopherol, Trolox, propyl gallate, gallic acid, methyl carnosoate and carnosic acid in two O/W and WO emulsions (droplet size 0.3–4 μm) and in bulk oil. In most emulsions, the most polar antioxidants, propyl gallate and gallic acid, exhibited either prooxidant activity or no antioxidant activity. Methyl carnosoate was the most active antioxidant in W/O emulsions but was less active than Trolox in O/W emulsions. α-Tocopherol was less active in bulk oil than in emulsions, but its activity in bulk oil was considerably increased when O/W emulsifiers, such as polyglyceryl glucose methyl distearate or cetheareth-15 and glyceryl stearate, were added to the oil. Such physicochemical parameters as interphase transport and partitioning, interactions with emulsifier molecules and surface accessibility of antioxidants are important for their activity in lipid containing systems [25].

The area of nanomaterials is currently receiving a great interest in cosmetic technology. Nanoparticles have a large application in beauty care. Many nanomaterials, such as nano-sized titanium dioxide [26], zinc oxide [27] or graphene – an innovative two-dimensional monolayer carbon material [28, 29], are applied in cosmetic formulations as active substances, for example, as sunscreens or hair conditioners [30, 31]. The nanocarriers such as micro- and nanoemulsions, nanocapsules, nanospheres or dendrimers are largely used in cosmetics as delivery systems to carry a wide range of ingredients that are intended to provide cosmetic benefits [32, 33]. According to common resolution of the International Organisation of Standardisation and of the European Standardisation Committee, nanoparticles may be defined as objects with all three external dimensions in the size range from approximately 1 to 100 nm [34]. However, in nanocosmetology, like in nanomedicine, size dimensions of 1–1000 nm are included; this is due to the fact that in medicine, nanotechnology aims to improve and optimize material properties for their interaction with cells and tissue to allow, for example, targeting of bioactive substances to specific tissues, or to improve their bioavailability. This approach makes use of nanoscale materials larger than 100 nm [33, 35]. On the other hand, in the scientific literature, there are considerable differences in classification of nanodispersion systems applied in beauty care. This terminological confusion takes place especially in the case of microemulsions and nanoemulsions [36–38].

10.2.1   Microemulsions

A major reason for the confusion between microemulsions and nanoemulsions is due to the prefixes used to denote them. The term 'micro-' means 10^{-6}, while the term 'nano-' means 10^{-9}, which would imply that nanoemulsions contain particles that are smaller than those in microemulsions. In practice, the opposite is usually the case – the particles in microemulsions are smaller than those in nanoemulsions [36]. The reason of this confusion may be a result of the historical development of the field of colloid science. The first article using the term 'microemulsion' appeared in 1961 [39], and hence, this term had become well-established among researchers well before the term 'nanoemulsion' was introduced in 1996 [40]. The use of the term 'nanoemulsion' has only become popular during the past decade, when interest in nanotechnology has been grown [36]. To date, there has not been determined in the scientific literature a critical particle size that should distinguish 'microemulsion' from 'nanoemulsion' and 'micro- (or nano-) emulsion' from 'conventional emulsion' [36, 41]. However, there are some changes in overall system properties that could be used to distinguish these particulate systems from one another.

   Microemulsions are thermodynamically stable (in contrast to nanoemulsions), low viscous and optically transparent dispersions of oil and water stabilised by surfactant, usually in combination with cosurfactant, which may be a short chain alcohol, amine or other weakly amphiphilic molecule [42–44]. The droplet size of microemulsions usually applied in cosmetic delivery is in the range of about 10–140 nm [45]. Apart from O/W and W/O microemulsions, in systems where the amounts of aqueous and oil phase are relatively equal, bicontinuous microemulsions may result. In this case, both oil and water exist as a continuous pseudophase in the presence of a continuously fluctuating surfactant-stabilized interface with a net curvature of zero [44]. In contrast to conventional emulsions, microemulsions form upon simple mixing of the components and do not require the high shear conditions generally used in the formation of ordinary emulsions.

   Applications of microemulsions in cosmetics can be found in numerous patents and research articles about formulations of skin care, hair care and personal care products. Besides general advantages, that is, good appearance, thermodynamic stability, high solubilisation power and ease of preparation, microemulsions can make cosmetic products more efficient and stable [44]. One of the main advantages of microemulsions is to enhance diffusion and penetration of beneficial actives, such as vitamins, through the skin or the membranes. On the other hand, the main drawback of these delivery systems is the possibility of disruption of the liquid crystalline structure of the SC that leads to skin irritation. However, by proper control of the microemulsion composition, particularly by use of non-irritant emulsifiers, this problem may be overcome [46].

The O/W microemulsions were prepared as transparent vehicles for sunscreens such as 4-methylbenzylidene camphor or octyl methoxycinnamate. Soya lecithin, decylpolyglucose, cyclomethicone, menthol, allantoin and stearyl methicone were used as microemulsion compositions. There was a report that the sunscreen microemulsions provided good skin feel, waterproof effect, non-stickiness and easy spread [47]. Lycopene, an antioxidant with low solubility in both water and oil, reportedly could be prepared in transparent liquid form by microemulsification. Its solubilisation efficiency in microemulsions containing jojoba oil, water, Brij 96V and hexanol was higher than that in each pure component. Although incorporation of lycopene into microemulsions altered microemulsion droplet shape from spherical to thread-like, the products were still transparent and attractive for cosmetic use [48]. Watanabe *et al.* studied bicontinuous microemulsions containing decamethyl cyclopentasiloxane as a solvent for silicon soils used as cleansing agent. The results of this study suggested that the microemulsions comprising O/W of 1:1 to 7:3 exhibited maximum detergency for silicon soils [49]. Microemulsion formulations may also improve the stability of the active ingredients. For instance, photostability of arbutin and kojic acid, used as whitening agents, to ultraviolet B irradiation was higher in O/W microemulsions comprising lecithin and alkyl glucosides as emulsifiers than in aqueous solutions [50]. Ascorbyl palmitate was more stable in W/O than in O/W microemulsions [51], whereas sodium ascorbyl phosphate was stable in both types of microemulsions [52]. However, profile release studies showed faster release of sodium ascorbyl phosphate from O/W microemulsion than from W/O microemulsion, probably because the location of sodium ascorbyl phosphate in microemulsions influences its release profiles [52]. The permeation study of α-tocopherol from various delivery systems, such as solutions, gels, emulsions and microemulsions, in *in vitro* pig skin model indicated that a microemulsion formulation was the best topical delivery system for α-tocopherol compared with the other studied systems [53]. However, the nature of microemulsions was found to be a crucial parameter for permeation of an active ingredient through *ex vivo–*separated human epidermis [54].

Hsu *et al.* [55] have reported the preparation of coenzyme Q10 (CoQ10) nanoparticles engineered from microemulsion precursors.

10.2.2   Nanoemulsions

Nanoemulsions can be defined as thermodynamically unstable O/W emulsions with a droplet size smaller than 100 nm [36, 56]. Compared with microemulsions, they are metastable and very fragile systems by nature. Unlike conventional emulsions, nanoemulsions are so fine that they can be sprayed on. They can also be sterilised by filtration [56]. Furthermore, nanoemulsions do not exhibit problems like inherent creaming, sedimentation, flocculation or coalescence, which are observed with conventional emulsions [57].

A large variety of products with different visual aspects, richness and skin feel are allowed with nanoemulsions: lotions, transparent milks and crystal-clear gels with different rheological behaviour. These products are easily valued in skin care, thanks to their good biophysical and sensorial benefits. Nanoemulsions with 15% oil content (droplet size about 50 nm) penetrated the skin considerably quicker than the corresponding macroemulsion. The skin moisturizing effect of these nanoemulsions was by far higher than that of body milk and 'body care water' that contained the same oil components. Out of 192 volunteers who tested this nanoemulsion as a body lotion, 88% appreciated its freshness, 72% its moisturizing effect and 84% the cosmetic results. Eighty per cent of the volunteers preferred the nanoemulsion compared to their usual product. Finally, 96% of them considered nanoemulsions as a new vehicle for skin care product. A cationic nanoemulsion was used for dry hair and the result showed substantial improvement with a prolonged effect (after several shampoos) – the hair was more fluid and shiny, less brittle and not at all greasy [56, 58].

In addition, nanoemulsions are known in the art which comprise an amphiphilic lipid phase composed of phospholipids, water and oil. These emulsions exhibit the disadvantage of being unstable on storage at temperatures from 0 to 45°C. They lead to yellow compositions and produce rancid smells after several days of storage. Nanoemulsions stabilised by a lamellar liquid crystal coating, obtained by the combination of a hydrophilic surfactant and of a lipophilic surfactant, are also known. However, these combinations are problematic to prepare [59]. Furthermore, the nanoemulsions obtained exhibit a waxy and film-forming feel which is not very pleasant for the user [60]. Ribier *et al.* [61] formulated nanoemulsions that are based on fluid nonionic amphiphilic lipids. However, these nanoemulsions exhibit the disadvantage of having a sticky effect during application to the skin. A need, therefore, continues to exist for nanoemulsions that have the disadvantages of neither those of the prior art nor of microemulsions [62].

Simonnet *et al.* [63] invented nanoemulsion based on phosphoric acid fatty acid esters with average droplet size less than 100 nm. This nanoemulsion is transparent and stable on storage and it can contain large amounts of oil while retaining good transparency and good cosmetic properties. The composition may be useful in applications to the skin, hair, scalp, mucous membranes and eyes. The same researchers patented nanoemulsion based on ethylene oxide and propylene oxide block copolymers. The inventors obtained a transparent nanoemulsion and the size of the globules was 44 nm. This nanoemulsion can be provided in the form of a lotion, serum, cream, milk or toilet water and can comprise adjuvants commonly used in cosmetics, dermatological and ophthalmic fields, such as gelling agents, preservatives, antioxidants and fragrances [62, 63].

10.2.3 Quick-Breaking Emulsions

Lochhead *et al.* [64] prepared a new type of O/W emulsion which is stable on prolonged storage but immediately separates into two phases upon contact with the skin. This emulsion is stabilised by a hydrophobically modified polymer of acrylic acid which is adsorbed at the oil–water interface through the hydrophobic groups, while the hydrophilic charged portion of the molecule remains in the aqueous phase. The hydrophilic moieties are powerful anchors and therefore the polyelectrolyte can be an electrostatic stabiliser, while in addition the polymer forms a 'microgel' around the oil droplets. On application of the stabilised emulsion surface, or addition of an electrolyte, the emulsion droplets instantly coalesce, thus forming a continuous film on the surface. For comparison, breaking or inversion of conventional emulsions on the skin may take 1 h or more. Surprisingly, the emulsions prepared with the modified polymer had a much larger droplet size (10–100 µm) compared to that of conventional O/W emulsions (0.1–5 µm). The cosmetic formulations which can be prepared with quick-breaking emulsions include moisturising lotions, barrier creams and lotions, cleansing creams and lotions, waterless hand cleaners, after-shave lotions and sunscreens.

10.2.4   Pickering Emulsions

Pickering emulsions are surfactant-free emulsions, stabilised by solid particles and having particle sizes less than 200 nm [65, 66]. These dispersions are characterised by good skin tolerability and are remarkably stable in the presence of electrolytes. When used in a product, they can increase the stability, shelf life, cutaneous retention and depth of penetration of the active substance [67]. Pickering emulsions exhibit higher effectiveness in sunscreen formulations such as zinc oxide or titanium dioxide coated with aluminium stearate, dimethicone or silicone dioxide, as well as moisturizers, astringents and antimicrobials [68–70]. One disadvantage of these emulsions may be dry effect on the skin, which can be overcome by the addition of cyclodextrins (CDs) [71–73].

10.2.5   Gel Emulsions

Several publications have described the formation of emulsions which contain very high concentrations of dispersed phase, up to about 90–99% by volume. These emulsions were shown to consist of polyhedral hydrocarbon cells (polyaphron) which are coated by a layer of hydrated surfactant [74, 75], and their preparation usually requires the use of two surfactants: one oil-soluble and one water-soluble [76]. The formation of highly concentrated W/O emulsions was also reported by Solans *et al.* [77], who were able to prepare W/O emulsions containing up to 99% w/w water as the dispersed phase. These emulsions are more complex than the previous O/W gel emulsions: the structure of the emulsions is similar to that of a foam, in which the gas is substituted by water and the continuous phase is a W/O microemulsion [77].

From a cosmetic point of view, gel emulsions have a great potential, since they offer simple ways to deliver very high concentrations of 'internal phase' components but still retain the 'feel' of the external phase. For example, for the O/W gel emulsions, a huge amount of oil can be applied to the skin (which normally can be achieved only with W/O emulsions), but the wetting, application and adhesion to the skin will resemble that of a simple O/W emulsion. Fueller *et al.* [78] invented refreshing O/W gel emulsion, containing hydroxyethylcellulose xanthan gum, carbomer, lipids, emulsifiers and menthol. Fry [79] patented organopolysiloxane gels that form gel emulsions with many hydrophilic solvents, which are suitable for use in cosmetic products.

10.2.6   Liquid Crystal Emulsions

Liquid crystalline phase is thermodynamically stable and represents a state of incomplete melting. Liquid crystals form multilayers around the emulsion droplets, decreasing the van der Waals' energy and increasing the viscosity that increases the emulsion stability. These multilayers act as rheological barriers to coalescence. Liquid crystals exhibit birefringence and dichromism and hence enhance the cosmetic appeal because of the coloured appearance of the preparations into which they are incorporated. Lipophilic materials such as vitamins, incorporated into liquid crystalline matrix, are protected from both thermal and photodegradation [80, 81].

Emulsions containing liquid crystals have been observed to have a rate of active release much slower than those without this stabilizing component. This effect is because of multilayer structure of liquid crystalline material around droplet, which effectively reduces the interfacial transport of the dissolved actives from within the droplet. For example, timed release of vitamin A palmitate containing liquid crystals dispersed in water-based gel [82–85].

One such liquid crystal containing anti-ageing formulation is Cosmedix Opti-Crystal Age-defying liquid crystal eye serum with liquid crystals to replenish damaged, thinning skin around the eyes, and growth factors and α-lipoic acid to dramatically diminish the appearance of eye lines and wrinkles. Opti-Crystal is the ultimate tool in anti-ageing eye care [86].

Iwai *et al.* [14] developed a new type of skin care product called Lamellar Gel, which contains synthesised pseudo-ceramide. Its structure is similar to that of ceramide found among the SC lipids, which allows it to control intramolecular interactions. Compared to regular emulsions, the Lamellar Gel demonstrated better skin care characteristics regarding permeability, skin hydration and skin occlusion.

10.2.7 Multiple Emulsions

Multiple emulsions are a type of polydisperse systems where both W/O and O/W emulsions exist simultaneously. Multiple emulsions are of two types: O/W/O-type multiple emulsions and W/O/W-type multiple emulsions. In the O/W/O emulsions, small oil droplets are dispersed in larger aqueous droplets with help of suitable emulsifier, and these aqueous droplets are again dispersed – also with help of suitable emulsifier – in a continuous oil phase, while in the W/O/W phase, small water droplets are dispersed in bigger oil droplets and these oil droplets are again dispersed in a continuous aqueous phase [87]. The main advantages of these delivery systems are high capacity of entrapment compared to other systems, protection of fragile substances, as well as possibility of combining incompatible ingredients (e.g. soluble and insoluble in water) in one product and of controlled release of active substances [88]. Apart from this, they are aesthetic and easily acceptable by consumers [46, 87]. For these reasons, multiple emulsions are found to be promising carriers for cosmetic delivery.

Yoshida *et al.* [89] examined the stability of vitamin A in O/W/O multiple emulsion formulated with organophilic clay mineral. The stability test conducted during 4 weeks at a temperature of 50°C revealed that prepared multiple emulsion is an effective carrier for stabilising vitamin A. Akhtar and Yazan [90] prepared W/O/W multiple emulsion containing two skin antiaging agents: vitamin C and Lipacide PVB, a wheat protein product. The obtained emulsions were characterised by good consistency, skin tolerability and stability during 6 months of storage at a temperature of 25°C. Mahmood and Akhtar [91] studied the stability of cosmetic multiple emulsion W/O/W loaded with extract of *Camellia sinensis* L. (green tea). During 30–90 days of storage in different conditions, no significant changes in emulsion parameter were observed. Only at a temperature of 40°C with air humidity of 75%, the viscosity of emulsion decreased in time.

10.3   Vesicular Systems

Vesicular systems are small membrane-enclosed sacs, ranging in size from 15 to 5000 nm, that are usually formed by bilayer arrangements of amphiphilic molecules with a central core [21, 92]. They are useful delivery systems for cosmetic ingredients because they offer a convenient method for solubilising active substances, and they do not disrupt the SC structure because they always form lamellar liquid crystalline structures on the skin [46]. The rationale for using vesicular systems in cosmetics delivery is many folds. Vesicles might act/serve as carriers, penetration enhancers, as well as systems for controlled release [16, 46].

10.3.1   Liposomes

Liposomes are the most widely used vesicular delivery systems in cosmetic industry [93]. These systems are produced in sizes ranging from 25 nm to several micrometres and consist of a single or multiple phospholipid and cholesterol bilayers forming unilamellar or multilamellar vesicles [46, 87, 94]. Liposomes are capable of carrying both hydrophilic and lipophilic active substances, encapsulated in the core or in the wall of bilayer, respectively [46, 94]. The concentration of active ingredients in the epidermis may be up to five times greater with liposome formulations than with formulations that use more conventional vehicles [95]. Furthermore, liposomes protect many active ingredients from degradation and may also provide controlled release of the active substances [16].

Liposome-based topical formulations were formulated and launched in cosmetics by Christian Dior (Capture™). Mehl and Zaias [96, 97] patented freeze-dried liposomes to encapsulate agents such as perfumes, moisturisers or vitamins for delivery to specific cells in the skin. They are encapsulated in liposomes specifically selected to carry the agents to the target cells thereby increasing the effectiveness and efficiency and minimising systemic absorption. Negatively charged liposomes strongly improved newborn pig skin hydration and tretinoin retention thereby suggesting that liposomes may be an interesting carrier for tretinoin [98]. Contreras *et al.* [99] designed an all-trans retinoic acid topical release system that modifies drug diffusion parameters in the vehicle and the skin in order to reduce systemic absorption and the adverse effects associated with topical application of these substances to the skin. In another study, the transdermal delivery of catechins was enhanced by incorporating anionic surfactants and ethanol. The stability and the safety of the practical use of liposomes developed in this study were indicated [100]. Cationic liposomes consisting of phosphatidylcholine and retinoic acid were found to increase the delivery of retinoic acid about twofold, suggesting the potential of the use of the cationic liposomes for the intradermal delivery of lipophilic drugs like retinoic acid [101].

Liposomes are also believed to act as penetration enhancers [16, 46]. However, recently it was shown that traditional liposomes are of little or no value as carriers for transdermal drug delivery because they do not deeply penetrate skin, but rather remain confined to upper layers of the SC [102, 103], due to alteration of SC lipid structure by individual components of these vesicles. Thus, new delivery systems such as transferosomes and ethosomes have been proposed.

10.3.1.1 Transfersomes

Cevc and Vierl introduced vesicular systems called transfersomes (IDEA AG, Munich, Germany) containing phospholipids, 10–25% edge activator (surfactant) and 3–10% ethanol [104, 105]. Due to this composition, transfersomes are characterised by at least one-order magnitude of more elastic membrane than that of conventional liposomes [87, 106, 107]. Another specific difference between Transfersomes and liposomes is the higher hydrophilicity of the former, which allows transfersome membrane to swell more than conventional lipid vesicle bilayers [108]. These properties help transfersomes to avoid aggregation and fusion, which are observed with liposomes exposed to an osmotic stress [109] and enable the former systems to squeeze through channels in the SC (in diameter up to 500 nm) to the deeper skin layer and even subcutaneous tissues [87]. However, due to the high content of added surfactants, these vesicles are very fragile in cosmetic formulations and are not as stable as conventional liposomes [87, 94].

Gallarate *et al.* [110] formulated transfersomes with α-tocopherol which significantly improved photostability and skin deposition of these antioxidants. The residual amount of α-tocopherol from deformable liposomes within the skin was almost four times more than that from undeformable ones. Saraf *et al.* [111] prepared anti-wrinkle cream with curcumin-loaded nano-transfersomes (size 200 nm) with 4 : 1 lecithin : Tween 20 ratio and maximum entrapment efficiency of 41%. The improvement in overall elasticity, biological elasticity, recovery of deformed skin, firmness and reduction in fatigue can be correlated with the anti-wrinkle properties of this cream [111].

Transfersomes formulation, like that of conventional liposomes, requires use of relatively pure components to avoid skin irritation by surface impurities. The emulsifiers used as edge activators should not exert irritable or toxic effect towards skin cells [112]. Apart from this, procedures and ingredients used for transfersomes preparation are expensive [108]. In this situation, a good alternative for these vesicular systems may be ethosomes [102].

10.3.1.2 Ethosomes

Ethosomes are vesicular systems made of phospholipids, ethanol at a high concentration (20–50%) and water [94]. The advantages of ethosomes in cosmeceuticals over conventional liposomes are not only to increase the stability of the cosmetics and decrease skin irritation from the irritating cosmetic chemicals but also to enhance transdermal permeation [113]. Increased cell membrane lipid fluidity caused by the ethanol of ethosomes results in increased skin permeability. So, the ethosomes permeates very easily inside the deep skin layers, where it gets fused with skin lipids and releases the drugs into deep layer of skin [102]. Furthermore, ethosomes have a much higher loading capacity of lipophilic drugs as compared to classic liposomes [94, 113]. The main advantage of ethosomes over transfersomes is the simple and relatively inexpensive synthesis method [114].

Topical administration of many antioxidants is one of the several approaches to diminish oxidative injury in the skin for cosmetic and cosmeceutical applications. A US company, Osmotics Inc. (Denver, CO), reported that cellulite cream called Lipoduction prepared by using ethosome technology penetrated the skin lipid barrier and delivered ingredients directly into the fat cells [115]. Other example of cosmetics using ethosomes is Noicellex (Novel Therapeutic Technologies, Inc., Wilmington, DE) [113].

10.3.1.3   Aquasomes

Aquasomes are nanoparticulate carrier systems with three-layered self-assembled structures. They comprise a central solid nanocrystalline core coated with polyhydroxy oligomers onto which biochemically active molecules are adsorbed. The solid core provides structural stability while the carbohydrate coating protects against dehydration and stabilises the biochemically active molecules. This property of maintaining the conformational integrity of bioactive molecules has led to the proposal that aquasomes have a quite versatile potential as a carrier system for delivery of peptide-based agents, such as vaccines, antigens, enzymes, peptide hormones and even genetic material [116, 117]. In cosmetic industry, aquasomes are used for delivery of pigments, whose cosmetic properties are sensitive to molecular conformation [118]. Genesphere® is an anti-wrinkle cream with aquasomes. It smoothens lines and eliminates crow's feet [86].

10.3.1.4   Ultrasomes and Photosomes

Ultrasomes are specialised liposomes encapsulating an endonuclease enzyme extracted from *Micrococcus luteus*; the enzyme recognises the UV damage to the skin and initiates removal of damaged DNA [119]. Ultrasomes also protect the immune system by repairing UV DNA damage and reducing the expression of TNF, IL-1, IL-6 and IL-8. Interestingly, ultrasomes also stimulate the production of melanin by melanocytes in the tanning response following UV exposure. A dose-dependent response has been observed *in vitro* [86, 120]. The Anti-Age Night Recovery Cream 'claims to moisturise, repair cellular damage, retain water and also boost your skin's natural collagen production'. In fact, the Rodan + Fields web site goes further, stating that the Night Recovery Cream contains patented ultrasomes that are DNA anti-ageing ingredients. These ultrasomes help repair cellular damage, fight wrinkles, sagging skin and loss of tone and elasticity [86].

Photosomes are incorporated in sun care products to protect the sun-exposed skin by releasing photolyase, a photo-reactivating enzyme extracted from a marine plant *Anacystis nidulans*. Photosomes during light irradiation reverse the cell DNA damage, reducing immune suppression and cancer induction [121]. In combination with ultrasomes, they constitute 'intelligent' DNA repair systems [86, 122].

Three hundred and sixty-five Cellular Elixir created by Lancaster protects the skin's DNA by over 30% and supports the skin's DNA repair process by up to 60% – the first time ever a cosmetic formula has been proved to protect and support the natural skin DNA repair process. This DNA action complex is made of photosomes and ultrasomes [86].

10.3.1.5   Asymmetric Oxygen Carrier System Liposomes

Asymmetric oxygen carrier system liposomes are designed to carry oxygen into the skin. These vesicles are composed of perfluorocarbon core surrounded by a monolayer of phospholipids, followed by a bilayer system. Perfluorocarbons are good carriers of oxygen and so this system is used to transport molecular oxygen into the skin [82, 119].

10.3.1.6   Yeast-Based Liposomes

Yeast cell derivatives repair, soothe and oxygenate the skin. In its liposomal form, it stimulates dermal fibroblasts and provides a feeling of well-being. Incorporation of vitamin C into the cell increases significantly when liposomes are used as a vehicle [82, 120]. In its liposomal form, it stimulates dermal fibroblasts and provides a feeling of well-being [119].

10.3.1.7   *Phytosomes and Marinosomes*

Phytosomes are created by anchoring an active phytoconstituent to a phospholipid [123, 124]. The resulting complex allows a more efficient delivery of the plant-based active compound to the tissues [125]. This leads to better absorption, lower concentration requirements and better stability of the actives. Botanical actives such as catechins, chamomile, ginkgo or quercetin are used topically in the form of phytosomes to increase bioavailability [126].

Marinosomes (Bordeaux, France) are liposomes based on a natural marine lipid extract containing high concentration of polyunsaturated fatty acids, such as eicosapentaenoic acid and docosahexaenoic acid, which are not present in normal skin epidermis. However, they are metabolised by skin epidermal enzymes into anti-inflammatory and anti-proliferative metabolites that are associated with a variety of benefits with respect to inflammatory skin disorders [127]. The study of marinosomes formulations in O/W emulsion showed that the membrane structures were mostly preserved even in the presence of surfactant. In parallel, the first toxicology file indicated a good skin and eye tolerance towards marinosomes which reveal intrinsic anti-inflammatory activity. These results allowed considering marinosomes as potential candidates for cosmeceutical use in view of the prevention and treatment of skin diseases [128, 129].

10.3.1.8   *Cubosomes*

Cubosomes are discrete, submicron, nanostructured particles of bicontinuous cubic liquid crystalline phase where 'bicontinuous' refers to two distinct (continuous but non-intersecting) hydrophilic regions separated by the bilayer [130, 131]. Cubosomes possess the same microstructure as the parent cubic phase but have much larger specific surface area and their dispersions have much lower viscosity than the bulk cubic phase. While most concentrated surfactants that form cubic liquid crystals lose these phases to micelle formation at high dilutions, a few surfactants have optimal water insolubility [131, 132]. Their cubic phases exist in equilibrium with excess water and can be dispersed to form cubosomes. Cubic phase materials can be formed by simple combination of biologically compatible lipids and water and are thus well suited for use in the treatment of skin, hair and other body tissue. Hewitt-Jones *et al.* [133] described antiperspirant preparations with cubic phase formation on contact with skin; El-Nokaly *et al.* [134] disclosed lipsticks comprising cubic phase while Gilchrest and Gordon [135] used diacylglyceride cubosomes for inducing melanin synthesis in melanocytes. With respect to liposome, cubosome possesses a larger ratio between the bilayer area and the particle volume and a larger breaking resistance [131].

10.3.1.9   *Dendrimers and Dendrosomes*

Dendrimers are unimolecular, monodisperse, micellar nanostructures around 20 nm in size, with a well-defined, regularly branched symmetrical structure and a high density of functional end groups at their periphery. They contain large number of external groups suitable for multi-functionalisation [136, 137]. Tournilhac and Simon [138] formulated cosmetic compositions capable of being applied to the skin, the keratinous fibres, the nails, the semimucous membranes and the mucous membranes, which include dendritic polyester polymers having terminal hydroxyl functional groups or the combination of such polymers with film-forming polymers. Baecker *et al.* [139] invented compositions for treating hair – shaping, cleansing, bleaching or dying – which contain dendrimers or dendrimer conjugates.

The research group of Percec [140, 141] synthesised new class of dendrimers, termed Janus-denrimers, which – similarl to the Latin god Janus, depicted with two faces – contain both hydrophilic and hydrophobic moieties, which together create amphiphilic dendrimers that can self-assemble into stable, monodisperse and mechanically superior liposomes denoted as dendrosomes [142, 143]. Dendrosomes exhibit a host of morphologies in addition to the classic spherical shape, including the less encountered tubular, dendrosomes within dendrosomes, polygonal, dendrocubosomes and other complex architectures such as disc-like, toroidal, rod-like, polygonal, spherical, ribbon-like and helical ribbon-like dendromicelles [140, 142, 143]. They are nontoxic and produce pH-sensitive membranes with controllable permeability. Apart from this, the methods of their production are simple and inexpensive. For these reasons, dendrosomes may be attractive structures for cosmetic delivery [142, 143].

10.3.1.10   Silicone Vesicles and Matrices

Silicones have been included in cosmetic and personal care products for the past 20 years. They act as emollients, water barriers and emulsifiers. Their cosmetically pleasing properties and feel and low toxicity allow them to be widely used in cosmetic, personal care and pharmaceutical products [144]. They deliver the active substances to their site of action *via* several methods. Volatile silicones are used for antiperspirants or fragrances where they release the transported active substances after they evaporate. Furthermore, silicones, such as stearyl dimethicone, improve the spread of sunscreen ingredients on the skin, therefore, raising the sun protection factor (SPF) level of the products. High capability of stearyl dimethicone to undergo thixotropy allows this product to be evenly distributed on the skin, improving sun protection by forming uniform homogenous film [144].

Silicones have a long history of use in hair care, such as improved conditioning, shine, manageability, reduced flyway and number of other benefits [144, 145]. Induction of silicone polyether into submicron-sized vesicles provides excellent stability in aqueous medium. These are called 'assembly-required' vesicles. The actives are encapsulated in the vesicle's bilayer and therefore the thickness of the bilayer rather than the size of the vesicle influences the amount of active transported. This method has allowed the stabilisation, preservation and transportation of vitamin A, E and C; humectants, for example lanolin alcohol; emollients, such as jojoba oil and polydimethyl siloxane; and hair colourants [146, 147].

The permeability of silicones makes them suitable for controlled release applications and for this reason they are used widely in transdermal delivery systems. Cross-linked silicones such as elastomers and adhesives are a relatively new class of cosmetic raw materials that have utility in delivery systems for active ingredients [148]. Studies have reported that the use of such a silicone elastomer blend to modulate the release rate of fragrance [149], while combining the active ingredient with a silicone surfactant can increase the release rate. One example of controlled release that has been used commercially is the incorporation of fragrance into a silicone elastomer that is highly swollen with silicone fluid.

Smith *et al.* [150] has patented the method utilising an acrylate copolymer of long siloxane chains to create stable dispersions of solid particles in hydrophobic solvents having a solubility parameter of approximately eight or less, such as a silicone fluid. The moisturising lotion disclosed was applied to the face and/or body to provide softening, moisturising and conditioning benefits. Klug *et al.* [151] invented cosmetic compositions, preferably emulsions, utilising one or more modified polysiloxanes as the emulsifiers and presented an example of an extra moisture night cream. Araki *et al.* [152] have patented silicone compounds for cleansing compositions with excellent foaming properties and detergency, while Ichikawa and Hata [153] have invented O/W emulsions with excellent appearance, stability and applicability, such as hair creams, hand creams and face creams, that contain 15–40% w/w non-polar oils selected from dimethylpolysiloxane, decamethylcyclopentasiloxane, dodecamethylcyclohexasiloxane and isoparaffins with less than 30 carbon atoms.

10.3.2   Niosomes

Niosomes are nonionic surfactant vesicles [154, 155] that can deliver hydrophilic, lipophilic or ampiphilic drugs. They improve the stability and availability of active ingredients as well as increase their skin penetration. Niosomes solubilised with nonionic surfactant solutions of the polyoxyethylene cetyl ether class are called discosomes. They allow active targeting of carried ingredients [156].

The first product 'Niosome' was introduced in 1987 by Lancôme [86]. The advantages of using niosomes in cosmetic and skin care applications include their ability to increase the stability of entrapped drugs, improved bioavailability of poorly absorbed ingredients and enhanced skin penetration. Example of anti-ageing cream containing niosome is Niosome daytime skin treatment by Lancôme [157].

10.3.3   Sphingosomes

Sphingosome may be defined as concentric, bilayered vesicle in which an aqueous volume is entirely enclosed by a membranous lipid bilayer mainly composed of natural or synthetic sphingolipid. We can say that sphingosome is a liposome which is composed of sphingolipid. Sphingosomes may be administered transdermally. They increase the stability and reduce toxicity of the encapsulated agent and improve its bioavailability, providing passive and active targeting to the site of action [131, 158].

10.3.4   Multiwalled Delivery Systems

Multiwalled delivery systems (MDS) are vesicles that contain five to seven bilayer walls [119]. They increase the stability of liposomes and offer properties such as delayed release and hydration as well as optimising delivery. The multiwalled delivery system is based on a combination of structured vesicle-forming materials and high shear processing. It provides exceptional long-term stability to cosmetic skin-treatment products. MDS is analogous to the structure of membrane lipid found in the intracellular matrix and made up of non-phospholipid amphiphilic molecules (oleic acid, derivatives of polyglycerols, amino acid residues). MDS gives stability to liposomes but by combining hydration and delivery. MDS also nourishes and protects the skin, bringing the formulator closer to optimising product performance [159].

Bicellar systems are discoidal aggregates formed by long and short alkyl chain phospholipids. These bicellar systems may have retardant effects on percutaneous absorption, which result in a promising strategy for future cosmetic delivery applications, such as UV filters [18, 160, 161].

For diminishing wrinkling and skin irritation and improving moisturising effect, cosmetic material containing triple-encapsulated retinol was patented [162].

Charmzone DeAGE CRD Red-Addition is the brand new CRD system that contains rich antioxidants found in red food: red grapes, red wine, pomegranate and tomato. MDS enables rapid absorption into the skin. It protects skin from damaging free radicals, pollution, stress and ageing. It firms, rejuvenates and repairs skin. It hydrates and protects skin from damaging factors, from the environment and ageing. It tightens loose pores while firming the facial contours [86].

10.4   Solid Particulate Systems

Solid particulate systems are solid matrices in which active ingredients are encapsulated or entrapped [163]. These systems are widely used when the active ingredient to be delivered is not stable in either the solubilised phase or the dispersed phase in emulsions [21]. Depending on the size, they can be classified into micro- or nanoparticles, and depending on the choice of the matrix, into polymeric or lipid particles [164]. Solid particulate systems also include CDs complexes, fullerenes and fibrous matrices [21, 119].

10.4.1   Microparticles

Microparticles are solid polymeric or lipid particles falling in the range of 0.1–1000 µm and include microcapsules and microspheres. They are used to make substances more compatible with each other, to protect them from oxidation or moisture action and to reduce odour of actives. These delivery systems help also in controlled release of carried ingredients. Microparticle systems include microcapsules, microspheres and microsponges.

10.4.1.1   Microcapsules

Microcapsules are hollow microparticles composed of a solid shell surrounding a core-forming space available to permanently or temporarily entrapped substances [165]. Some examples of the applications of microcapsules in cosmetic formulations are microcapsules containing sun filters such as octyl methoxycinnamate or octyl salicylate; depilatory pastes containing microencapsulated enzyme for protection against surface active agents, for example sodium lauryl sulphate, skin tanning agent containing dihydroxyacetone and glycerine in separate compartments within a microcapsule; microcapsules with oils such as mineral oil, vegetable oil, isopropyl myristate, isopropyl palmitate present in cleansing creams; and skin depigmentation products, containing microencapsulated anti-oxidants such as tocopherols, which will prevent lipid peroxidation in the skin [2].

Egg albumin microcapsules of size 222 µm containing vitamin A (15.7%) were used to prepare O/W creams. The *in vitro* study showed a prolonged release of vitamin A, the relative bioavailability of the microencapsulated formulation being 78.2% [166].

10.4.1.2   Microspheres

Microspheres are microparticle of spherical shape without membrane or any distinct outer layer [165]. The absence of outer layer forming a distinct phase is important to distinguish microspheres from microcapsules because it leads usually to first-order diffusion of active ingredients from these microparticles, whereas zero-order diffusion is characteristic for microcapsules [165]. In fact, release behaviours of active substance from these particles depend on a great variety of factors, such as the concentration and physicochemical characteristics of the active ingredient (particularly its solubility and oil/water partition coefficient); the nature, degradability, molecular weight and concentration of the polymer; the polymer solid microstructure when reprecipitated, the nature of the oil, particle size, the conditions of the *in vitro* release test (medium pH, temperature, contact time, among others) and the conditions of the preparation method [167].

Microspheres have been widely used in personal care and the cosmetic industry over the past decade. In skin care lotions and creams, microspheres can help deliver a soft and smooth feel, excellent lubricity and soft focus effects that reduce the appearance of fine lines and wrinkles [86]. Microspheres have numerous advantages in the personal care and cosmetics industry. Due to their unique geometry and the smallest surface-area-to-volume ratio of any shape, microspheres offer beneficial properties such as ball-bearing effect for superior texture, light scattering or 'optical blurring' for minimising lines and wrinkles, gentle but effective exfoliation, as well as the ability to house colourants and active ingredients [86]. The use of microspheres in the personal care and cosmetic industry is growing as microspheres are getting incorporated into an increasing number of anti-ageing creams [86].

Nylon microspheres are being used in cosmetic make-up and skin care products because of the feel and skin adhesion they impart, and their particle size and narrow particle size distribution [168]. Chemical inertia of these microspheres allows them to hold hydrophilic and lipophilic ingredients including vitamins, sunscreens, moisturisers, fragrances and many other bioactives. Nylon microspheres containing 40–50% water are used in moisturising lipsticks to avoid exudation observed in these products [168]. These microspheres loaded with vitamin E showed enhanced concentration of vitamin E in the epidermis because of continued contact with skin and microspheres, slow release of vitamin from the particles and protection of vitamin E from chemical interactions before absorption. The nylon microspheres can be used also in combination with either or both organic chemical and particulate mineral sunscreens to reduce their concentration while retaining effectiveness [168].

LipoPearl® represents standardised line of pearlescent beads containing emollient oils and vitamins that enhance the tactile and visual appearance of cosmetic and personal care products. The average size of particle is 1000–2800 μm [86]. Agar LipoSphere®, derived from a renewable marine source, provides a nice visual effect and leaves little or no residue upon rubout. Its average size is between 500 and 4000 μm [86]. Lipobead® is a uniform spherical semi-solid matrix of lactose, microcrystalline cellulose and hydroxypropyl methylcellulose, coloured by pigments. Lipobead is a simple carrier system for active ingredients in creams, lotions, gels, body cleansers, shampoos, conditioners, hair gels and foot care products, where an exciting visual effect is desired [86].

Botanical microspheres, such as Almondermin®, are composed of algae extract, which forms spheres containing a system of internal canals. Release of actives occurs by diffusion from sphere or by breaking when applied to the skin. They can be even coloured to achieve a pleasing visual effect [169].

10.4.1.3   *Solid Lipid Microparticles and Unispheres*

Microencapsulation is a process in which very thin coatings of inert natural or synthetic materials are deposited around microsized particles of solids or droplets of liquids. Commercial microparticles typically have a diameter between 1 and 1000 mm and contain 10–90% core [22]. Microparticles with size >1 μm are retained in the skin surface or deposited on the surface of the hair follicles, therefore preventing potential skin permeation [22]. Since microparticles are located on the skin surface forming a film, they can be used for protection against UV radiation in sunscreens [22]. Lipid microspheres, often called liposPheres, are fat-based encapsulation systems. They are composed of a solid hydrophobic fat core (triglycerides) stabilised by a layer of phospholipids molecules embedded in their surface [22]. The internal core contains the bioactive compound, dissolved or dispersed in the solid fat matrix. Lipid microparticles of cosmetic ingredients such as glycolic acid have shown decreased irritation potential [170], while incorporation of quercetin in lipid microparticles improved photo- and chemical stabilities of the flavonoid [171].

Albertini *et al.* [172] found congealing technique to be superior to the traditional melt dispersion method for rapid and solvent-free production of solid-lipid microspheres containing sunscreen agent butyl methoxydibenzoylmethane (avobenzone) with a high loading capacity.

In preparations of shampoos containing high concentration of surfactants, as an alternative to liposomes, unispheres are also used. These are small, coloured cellulose beads that hydrate and swell in aqueous media and disappear when rubbed into the skin leaving behind no shell [86].

10.4.1.4   Microsponges and Melanosponges

The most widely used polymeric particles are the microsponges and melanosponges which are also classified as porous polymetric systems [119]. They are uniform, spherical polymer particles, typically 10–25 μm in diameter, that can be loaded with an active agent [173]. These particles can release the active agent through controlled diffusion or as a response to various factors, such as rubbing, skin temperature, perspiration or moisture [21, 119]. A unique advantage of this delivery system is its large entrapment capacity, up to three times the weight of the polymer alone, while their major disadvantage is the use of harsh processing conditions in forming of such particles, thus limiting their applications to very stable active ingredients which can withstand these conditions that might damage sensitive payloads [21]. Microsponge delivery system provides sustained release technology for reducing irritation of a wide range of skin care actives, thereby providing increased patient/client compliance, enhanced formulation stability ensuring long-term product efficacy, extended shelf life and superior skin feel and exceptional product aesthetics. The characteristic feature is the capacity to adsorb or 'load' a high degree of active materials into the particle and on to its surface. Its large capacity for entrapment of actives, up to three times its weight, differentiates microsponges from other types of dermatological delivery systems [86]. In anti-acne formulations, a reduction in the skin irritation potential with increased efficacy was observed when benzoyl peroxide was entrapped in a microsponge system. Anti-inflammatory activity of hydrocortisone could be sustained with reduction in skin allergy responses and dermatoses. In anti-dandruff products, the unpleasant odour of zinc pyrithone and selenium sulphide was reduced. Reduced allerginicity was observed when insensitising ingredients are entrapped in microsponge systems, as in the case of cinnamic aldehyde [174]. Pure Dose Pearls with Microsponge Technology is very effective against wrinkles. Perlabella Retinol Facial Serum is encapsulated in a patented Microsponge system, which provides time-released delivery of retinol into the skin [86].

Melanosponge-α is used to distribute genetically engineered melanin over the skin surface to provide full-spectrum sun protection [119].

10.4.2   Solid Nanoparticles

Solid nanoparticulate systems can be defined as submicron colloidal systems having a mean particle diameter of 0.003–1 µm. They include nanospheres and nanocapsules [119]. Nanocapsules differ from nanospheres in that the former is a reservoir type of system, whereas the latter is a matrix system. Polymer composition for both is identical and includes biodegradable synthetic polymers like polyamides, cross-linked polysiloxanes or modified natural products, such as gelatin and albumin [119]. Just as in the case of microparticles, the kinetics of active substance release from nanoparticles depends on many physicochemical factors of this substance and of environment [167].

10.4.2.1   Nanocapsules

Nanocapsules are hollow nanoparticles composed of a solid shell that surrounds a core forming cavity available to entrap substances [165, 167]. This cavity can contain active ingredients in liquid or solid form or as a molecular dispersion. Nanocapsules can also carry the active substances on their surfaces or imbibed in the polymeric membrane. In the latter case, the release of active substance from nanocapsule is biphasic with a fast initial release phase followed by a slower second release phase [167]. The initial phase, called burst effect, is attributed to the desorption of the drug located on the nanocapsule surface, although degradation of tin polymeric membrane of nanocapsule may also participate in this step. The second phase corresponds to the diffusion of the active molecules from inner to outer compartment, determined by such factors as partition coefficient of active substance between the oily core and the aqueous external medium, relative volumes of inner and outer compartments, existence of active substance–polymer interactions and concentration of surfactants [167].

Nanocapsules are often used in the preparation of sun creams. The Sophi-Caps nanocapsules (700 nm of diameter) were developed to encapsulate 50% of the organic UVA-filters (avobenzone) together with UVB-filter (octocrylene). This formulation offers a photostabilization of avobenzone and prevents the UVA filter from crystallization. These capsules consist of a liquid core composed of a mixture of UV-B and UV-A filters with a ratio of 3.5 : 1. The core surrounding the inner cavity is composed of a membrane forming emulsifier together with an inulin-based polymeric coemulsifier. The polymeric structure of the core prevents the nanocapsules from degradation by formulation detergents. These nanocapsules are clear white and can be homogenously applied onto skin or hairs [175].

Soleil Soft-Touch Anti-Wrinkle Sun Cream SPF 15 by Lancôme is an example of nanocapsule containing anti-ageing formulation. This sun cream helps to preserve skin's youth. It contains exclusive ingredients to provide a long-lasting effect. It promotes a gradual, even and aesthetic tan. It is suitable for all skin types [86].

10.4.2.2   Nanospheres

Nanospheres are nanoparticles of spherical shape without membrane or any distinct outer layer. They are composed of a matrix where substances can be permanently or temporarily embedded, dissolved or covalently bound [165]. An ideal delivery system for water-based skin product would be a product that is completely washable with water;

yet the functional ingredients remain in contact with the skin to perform their pharmacological action. These two completely opposite performance criteria can be achieved by applying a delivery system that incorporates bioadhesive nanospheres. The nanospheres can be surface modified to promote adhesion and hence deposition on body surfaces in rinse-off products. High cationic charge density improves the deposition of the nanospheres onto the target site and prevents them from being diluted or washed off during the rinse process [119, 176].

Such delivery technology enhances adhesion mostly because of the fact that the functional ingredient is retained in a solid structure that has a high surface area per volume with presence of charges and moieties on the nanospheres surfaces. The skin adhesion study of free fragrance substance and of the same agent encapsulated in highly cationic nanospheres indicated that three times as much fragrance remained on the skin after 2 h, when the fragrance was encapsulated inside the nanospheres [176]. This advanced technology plays a very beneficial role in protecting against actinic ageing, which contributes to 90% of the ageing of our skin. Nanosphere technology is incorporated into some of Arbonne's anti-ageing skin care (RE9 NutriMinC) and hair care products. NutriMinC RE tackles the effects of ageing and sun damage. It incorporates nanospheres which fight free radicals with antioxidants [86]. Imai and Fukuda [177] have invented solid-powdered cosmetics with an average particle size of 0.1–200 nm, containing oils and polyols, to receive a cosmetic with good impact resistance and surface appearance, which provides moist feeling and long-lasting effects.

Kwon *et al.* [178] have also reported on poly(methyl metacrylate) nanoparticles loaded with coenzyme Q_{10} (CoQ_{10}). They demonstrated that CoQ_{10} was more stable within polymeric nanoparticles over dispersion and oil-based formulation when exposed to UV and high temperatures. Polymeric nanoparticles thus would help in increasing the drug's stability [179].

10.4.2.3   Solid Lipid Nanoparticles and Nanostructured Lipid Carriers

Solid lipid nanoparticles (SLN) are particles of solid lipid matrix with an average diameter of nanometre range, mainly 150–300 nm [180–182], although some authors give a range of 10–1000 nm [183, 184]. Topical preparations containing SLN as a new dermal carrier show distinct advantages compared to traditional formulations, such as high stability, improved penetration of ingredients through the skin and increased skin hydration due to occlusive effect [182, 185]. Aqueous dispersions of SLN have the desired semisolid consistency and are promising carrier systems for topical application [186]. SLN protects incorporated active ingredients against chemical degradation [187, 188]. This opens the perspective to the use of cosmetic ingredients which could not be utilized due to chemical stability problems when prepared in traditional formulations (e.g. retinol, vitamin C).

Due to the general adhesiveness of small particles, SLNs applied to the skin form a film. This film of ultrafine particles has an occlusive effect, which promotes penetration of active ingredients into, or specific localization of compounds in, specific skin layers, mainly the SC [189, 190]. This improved penetration can enhance the cosmetic performance of ingredients [191].

Irritant substances like tretinoin [192] turns out to be less irritating if applied encapsulated within SLN. There are few reports on lipidic and polymeric nanoparticles for

CoQ_{10}. Liu *et al.* showed avoidance of the systemic uptake of isotretinoin, a high accumulative amount of isotretinoin in skin and showed a significant skin targeting effect of isotretinoin from SLN formulation [193]. Irritation on applying particulate sunscreens can be avoided or minimized by entrapping molecular and particulate sunscreens into the SLN matrix. Surprisingly, it was found that SLN themselves have sun protective effect [194, 195].

Dingler *et al.* [187] reported that the incorporation of vitamin E into SLNs enhances the stability. The ultrafine particles possess an adhesive effect. This leads to a formation of fine adhesive film on the skin leading to occlusion and subsequent hydration. Hydration of the skin promotes penetration of actives and enhances their cosmetic efficiency.

Souto *et al.* [196] formulated a-lipoic acid into SLN as a novel approach for topical delivery. Depending on the type of SLN produced, controlled release of the active ingredients is possible. SLN with a drug enriched shell show burst release whereas SLN with drug enriched core lead to a sustained release [197, 198]. Both features are of interest for dermal application. Hence, SLNs represent a promising technology for topical cosmetic products and they possess the potential to be developed as the new generation of carrier.

Dahms *et al.* [199] invented a SLN dispersion for the controlled release of fragrances and/or aromas. In this formulation, SLN with varying melting points are dispersed in aqueous medium with use of emulsifiers that form not only monolayers around lipid particles but also lyotropic liquid-crystalline membranes within aqueous phase. The fragrances or aromas may be included both in the nanoparticles and/or in the emulsifier monolayers or the liquid-crystalline membrane layers. Oil-soluble and/or amphiphilic fragrance oils or aromas are preferably stored in the lipid particles while hydrophilic, water-soluble fragrances are incorporated into the aqueous phase. As a result, a controlled release, in particular cascaded release, of the fragrance oils and aromas present is possible. The fragrance oils and aromas soluble in the water phase are released first, whereas the substances present in the lipid particles are released subsequently. The lipid particles with the lowest melting point soften earlier and release the corresponding active ingredient. Only at a later time do the lipid particles with a higher melting point soften. Additionally, the fragrance or aroma molecules bound into liquid-crystalline lamellae that are formed in the aqueous phase by emulsifiers constitute a depot, enabling delayed release of fragrance or aroma substance [199].

Nanostructured lipid carriers (NLC) are mixtures of solid and fluid lipids, in which the fluid lipid phase is embedded into the solid lipid matrix [200]. Alpha-lipoic acid encapsulated SLN and NLC formulations demonstrated antioxidant activity at similar level of 0.01– 10 μm to pure α-lipoic acid with low cell cytotoxicity and good physical stability [201].

Commercially available products, such as NanoRepair Q10® cream and NanoRepair Q10® serum (Germany), which were introduced to the cosmetic market in October 2005, epitomise the success of NLCs. Through the first application of NLC technology for the encapsulation of lipophilic active ingredients, SLNs (Nanopearls®) containing Q10 is prepared. It is able to fully unleash its energy-supplying and antioxidant effect, to counteract premature skin ageing and ensure that collagen-and elastic-degrading enzymes are inhibited [86].

De Vringer [202] showed that the size of particles could change the occlusion factor. Lipoid microparticles are greatly inferior to lipoid nanoparticles in their occlusive effect, and the addition of lipoid microparticles in a cream lowers the cream's occlusivity, whereas the addition of lipoid nanoparticles in a cream raises the cream's occlusivity. Nanospheres containing β-carotene and a blend of UV-A and UV-B sunscreens were prepared. The results showed that the synergistic effect resulting from the combination of nanospheres and sunscreens has an inhibitory effect on tyrosinase due to the cinnamic nature of the UV-B screening agent.

10.4.2.4 Nanocrystals

Nanocrystals are aggregates comprising several hundred to thousands of atoms that combine into a cluster, used for the delivery of poorly soluble actives. Typical sizes of these aggregates are between 10 and 400 nm and they exhibit physical and chemical properties somewhere between that of bulk solids and molecules [203]. The first cosmetic products appeared on the market recently: Juvena in 2007 (rutin) and La Prairie in 2008 (hesperidin). Rutin and hesperidin are poorly soluble, plant glycoside antioxidants that could not previously be used dermally. Once formulated as nanocrystals, they became dermally available as measured by antioxidant effect. This dermal use of nanocrystals is protected by patents [204]. Juvena Juvedical Age-Decoder Fluid & Cream is an example of nanocrystal-containing anti-ageing formulation. Skin Nova Technology for optimal skin renewal is combined for the first time with an innovative NanoCrystal Technology to work effectively and entirely against the ageing process. Juvena Age-Decoder Face Fluid provides an environment for cell renewal and also works in the inner part of the cells [86].

10.4.3 Fullerenes

Fullerenes are ball-shaped molecules composed solely of an even number of carbon atoms. The so-called Buckminster fullerenes C60, which are the currently most adequately investigated molecules of that type, form a cage-like fused-ring polycyclic system with 12 five-membered rings and the rest six-membered rings. As these molecules have a high electron affinity (radical scavengers), they are expected as novel powerful antioxidants to reduce intracellular free radicals, preventing oxidative cell injury that are responsible for skin ageing [205–207]. Therefore, fullerenes are used in anti-ageing creams [86].

UNT Elixirin C60 Precious Eye Complex–Brightening Eye Cream is an example of anti-ageing formulation containing fullerenes. The anti-ageing and brightening benefits of Radical Sponge® Fullerene in ELIXIRIN C60 SERUM is proven by its high acclaims since its first launch. It is an infusion of wrinkle-fighting, elastin-strengthening and skin-brightening elements to protect, treat and illuminate the delicate eye area [64]. Takada *et al.* reported that Radical Sponge (Vitamin C60 BioResearch Corp., Tokyo, Japan) is endowed with a series of factors such as antioxidant ability and anti-melanogenesis effect for juvenile skin protection [208].

10.4.4 Cyclodextrins

Cyclodextrins are cyclic oligosaccharides containing a minimum of six D-(+)-glucopyranose units attached by α-glucosidic bonds. The three natural CDs are α, β and γ which differ in their ring size and solubility. Complexation with CDs can bring about stabilisation of the active ingredient against oxidative, photolytic and thermal degradation [209]. Most of the molecules fit into the internal CD cavity forming a complex and the resulting structure is called CD clathrates or inclusion complexes. α-CD typically forms inclusion complexes with both aliphatic hydrocarbons and gases. β-CD forms complexes with small aromatic molecules [210]. γ-CD can accept more bulky compounds like vitamin D. A recent trend in cosmetic applications is the use of modified CD molecules, such as hydroxypropylbetacyclodextrin (HP-β-CD), which can enhance the solubility of lipophilic ingredients in water without changing their intrinsic abilities to permeate lipophilic membranes. Simeoni *et al.* [211] have investigated the penetration of oxybenzone, a lipophilic sunscreen agent, on human skin, from HP-β-CD and sulfobutylether-β-cyclodextrin (SBE-β-CD). The authors showed that SBE-β-CD had greater solubilising activity on oxybenzone, a highly lipophilic sunscreen (about 1050-fold increase) when compared with the use of HP-β-CD (a 540-fold increase). The sunscreen penetration to the deeper living layers of the skin was remarkably decreased (1.0 and 2.0% of applied dose for epidermis and dermis respectively) compared with the unbound octyl methoxycinnamate formulation used as control and with OMC-loaded HP-β-CD (~5%). This result is interesting because this type of carrier can promote the solubilising and photostabilising properties of sunscreen agents while staying on top of the skin where they are intended to act [211]. Even with modified complexes, conflicting results have been found in the literature concerning their effect to promote or decrease skin penetration of drugs. But there still remain the problems of their molar mass and their limited capacity to penetrate the skin [212].

CD–retinol complexes patented by Wacker are now found in a wide variety of anti-wrinkle creams manufactured under the name CAVASOL®. Regenerative Anti-Ageing Face Cream - "Silk effect", manufacured by Kipos, Italy (Vivifarm Internazionale S.r.l.) is an example of CD-containing anti-ageing formulation. It is an anti-ageing treatment favouring cellular regeneration and fighting against the skin ageing process. Beta CDs/retinol contributes to renewal of epithelial tissues. It reduces wrinkles and supports the restoration of UV-damaged tissue [86].

10.4.5 Fibrous Matrices

In the fibrous matrices, the active ingredient is entrapped in a dissolved or dispersed state in large polymeric molecules typically arranged as non-woven mats [213]. A wide range of large polymeric substances varying in polarity from hydrophilic to hydrophobic ends of the spectrum can be used for this purpose [214]. A major advantage of these matrices is that they possess adhesive properties, the strength of which depends on the choice of polymer. Additional advantages of these systems are their capacity to stabilise the active ingredient, ability to control the release of the active ingredient to the surface of the skin and ease of application. The biggest limitation of these systems may be the harsh processing conditions [215], limited loading and delayed release of the active ingredient from the matrix. The most widely used applications of this system are in cosmetic patches intended primarily for controlled release of active ingredients over a duration from 4 h to a day [21].

10.5 Cosmetic Foams

The foams are not new inventions and their application in topical therapy can be traced back to three decades. However, foam formulations have been gaining in popularity with over 100 patents published globally in the last 10 years alone [116].

The use of foam technology to deliver a range of topical active agents has been claimed, including sun-screening compounds, corticosteroids, and anti-bacterial, anti-fungal and anti-viral agents. Although foams present distinct application advantages and improved patient compliance, the real reason for the rapid growth of topical foam technology is that foams as elegant, aesthetic and cosmetically appealing vehicles provide an alternative, promising formulation strategy in the highly competitive cosmetics market. Although there is a plethora of published data proving the safety profiles of topical foams, there is a lack of sufficient clinical evidence to demonstrate any superiority of foams over other traditional topical vehicles such as creams and ointments for cosmetics delivery [116, 216, 217].

The aerosol foams gained the increasingly popular type of topical formulation for a variety of skin conditions. The vehicle base of the foam can have liquid or semi-solid consistency which shares equal physicochemical characteristics of conventional carrier vehicle like gels, lotions and creams but it maintains desirable properties such as moisturising, quicker drying effects or high bioavailability of drug [116]. The aerosol base is dispensed through a gas-pressurised can that discharges the foam. The product characteristics like thickness, viscosity, texture, bubble size, density, persistence, stabile nature and spread ability are determined by the type of formulation and the dispensing container that are selected to suit the specific therapy needs. The foams may be preferred for application on large hairy surfaces (e.g. chest and back) or on the face as cleansers because they are easier to apply [116].

10.6   Cosmetic Patches

The influence of the pharmaceutical technology is apparent in the case of the cosmetic patches, not as simple cosmetic forms but as cosmetic delivery systems. Cosmetic patches today represent a convenient, simple, safe and effective way for cosmetic applications, using one of the most acceptable, modern and successful delivery techniques [218]. In theory, cosmetic patches can be applied in most cases for the same use as classical cosmetic products, for example, wrinkles, ageing, dark rings, acneic conditions, hydration of specific areas, spider veins and slimming. In practice, several of the aforementioned applications have been investigated with very positive results and a high degree of acceptability from the consumers [218]. There are several ways to categorise a cosmetic patch. It can be characterised from the patch form (matrix, reservoir), application for expected results (moisturising, anti-wrinkles), structural materials (synthetic, natural and hybrid) and the duration of application (overnight, half-hour patch). Categories of functional cosmetic patches are anti-blemish patch, pore cleansers, pimple patch, eye-counter patch, anti-ageing patch, anti-wrinkle patch and lifting patch [218].

10.7   Cosmeceuticals: The Future

The cosmeceutical industry is constantly seeking new and pioneering products that will combine both proven biological activity and an efficient delivery system. The global trend in the cosmetic industry is towards developing 'medicinally' active cosmetics and in the pharmaceutical industry towards 'cosmetically' oriented medicinal products as part of a current 'life-style' ideology [219]. Currently, there are over 400 suppliers and manufacturers of cosmeceutical products, and the industry is estimated to grow by 7.4% [220].

The key disciplines of recent years, which have great potential of innovations and growth in cosmetic industry, are molecular biology and material science. Formulation of many novel cosmetics is based on biotechnology-derived (e.g. stem cell-based) products [221], profiling of skin care regimes with use of genomics, proteomics and metabolomics [222, 223] as well as cell and tissue engineering for cosmetic purposes [224]. On the other hand, biotechnology companies have licensed some of their patented molecules to the cosmetic industry to enter proprietary lines of beauty products [225].

Material science, especially nanotechnology, has introduced innovative materials and other devices with different and useful characteristics in the formulation of delivery systems to modify SC permeation/penetration of active ingredients, to protect them against chemical or physical instability as well as to improve appearance and odour of cosmetic products [226–228].

An important aspect is to apply fundamental studies – colloid and surface chemistry – on cosmetic formulations to optimize the systems, on the one hand, and to understand the mechanisms of formation and stability of each system, on the other [46]. Careful preclinical and clinical evaluation of cosmeceuticals safety and efficiency is also a prerequisite in the development of innovative cosmeceuticals, because interactions between cosmeceuticals and skin are complex, depending on the specific composites in cosmeceutical products, condition of the skin or general health status of a subject, and the factors of environment, such as toxic metals (bismuth, aluminium) or radiation [229, 230]. This is especially important for nanocosmeceuticals, which in many cases are found to cause a number of considerable risk both to humans and to the environment [137, 231]. In this situation, working up of uniform legal regulations and technological standards to avoid terminological divergences and to unify requirements concerning nanomaterials used in cosmetic manufacturing is necessary [36, 232].

It is also essential to regulate the legal situation of cosmeceuticals, which have not been recognized so far by US and Europe regulations. New insights about function of the human skin, as well as development of new products for beauty care, may induce to change definitions of 'cosmetic' and 'drug' and to change legal status of 'cosmeceuticals' [233]. One of the initial steps for this regulation may be the American report entitled 'Classification and Regulation of Cosmetics and Drugs: A Legal Overview and Alternatives for Legislative Change' which includes provisions for a category of 'cosmeceuticals' to include products like sunscreens that fell in the gap between 'drugs' and 'cosmetics' [234].

References

[1] Reed, R. (1962) The definition of cosmeceutical. *J. Soc. Cosmet. Chem.*, **13**, 103–106.
[2] Arora, N., Agarwal, S. and Murthy, R.S.R. (2012) Latest technology advances in cosmaceuticals. *Int. J. Pharm. Sci. Drug Res.*, **4**, 168–182.
[3] Magdassi, S. (1997) Delivery systems in cosmetics. *Colloids Surf. A Physicochem. Eng. Asp.*, **123–124**, 671–679.
[4] Krause, J. and Tobin, G. (2013) Discovery, development, and regulation of natural products, in *Using Old Solutions to New Problems – Natural Drug Discovery in the 21st Century* (ed M. Kulka), InTech, Rijeka, pp. 3–35.

[5] Farris, P.K. (2014) Cosmeceuticals and clinical practice, in Cosmeceuticals and Cosmetic Practice (ed P.K. Farris), John Wiley & Sons, Inc., Hoboken, pp. 1–9.

[6] Dover, J. (2008) Cosmeceuticals: A practical approach. *Skin Therapy Lett.*, **3**, 1–7.

[7] Förster, M., Bolzinger, M.-A., Fessi, H. and Briançon, S. (2009) Topical delivery of cosmetics and drugs. Molecular aspects of percutaneous absorption and delivery. *Eur. J. Dermatol.*, **19**, 309–323.

[8] Padula, C., Nicoli, S., Aversa, V. *et al.* (2007) Bioadhesive film for dermal and transdermal drug delivery. *Eur. J. Dermatol.*, **17**, 309–312.

[9] Masters, B. and So, P. (2001) Confocal microscopy and multi-photon excitation microscopy of human skin in vivo. *Opt. Express*, **8**, 2–10.

[10] Touitou, E. (2002) Drug delivery across the skin. *Expert Opin. Biol. Ther.*, **2**, 723–733.

[11] Anissimov, Y.G., Jepps, O.G., Dancik, Y. and Roberts, M.S. (2013) Mathematical and pharmacokinetic modelling of epidermal and dermal transport processes. *Adv. Drug Deliv. Rev.*, **65**, 169–190.

[12] Hadgraft, J. (2001) Skin, the final frontier. *Int. J. Pharm.*, **224**, 1–18.

[13] van Smeden, J. (2013) A breached barrier: Analysis of stratum corneum lipids and their role in eczematous patients. Doctoral thesis. Leiden Academic Center for Drug Research (LACDR), Faculty of Science, Leiden University, the Netherlands, pp. 10–29.

[14] Iwai, H., Fukasawa, J. and Suzuki, T. (1998) A liquid crystal application in skin care cosmetics. *Int. J. Cosmet. Sci.*, **20**, 87–102.

[15] Barry, B.W. (2001) Novel mechanisms and devices to enable successful transdermal drug delivery. *Eur. J. Pharm. Sci.*, **14**, 101–114.

[16] Honeywell-Nguyen, P.L. and Bouwstra, J.A. (2005) Vesicles as a tool for transdermal and dermal delivery. *Drug Discov. Today Technol.*, **2**, 67–74.

[17] Williams, A.C. (2003) Transdermal and Topical Drug Delivery: From Theory to Clinical Practice, Pharm Press, London, pp. 123–136.

[18] Rubio, L., Rodríguez, G., Barbosa-Barros, L. *et al.* (2012) Bicellar systems as a new colloidal delivery strategy for skin. *Colloids Surf. B Biointerfaces*, **92**, 322–326.

[19] Henson, M.A. (2003) Distribution control of particulate systems based on population balance equation models. *Proc. Am. Control Conf.*, **4–6**, 3967–3972.

[20] Bhargava, H.N. (1987) The present status of formulation of cosmetic emulsions. *Drug Dev. Ind. Pharm.*, **13**, 2363–2387.

[21] Golubovic-Liakopoulos, N., Simon, S.R. and Shah, B. (2011) Nanotechnology use with cosmeceuticals. *Semin. Cutan. Med. Surg.*, **30**, 176–180.

[22] Pol, A. and Patravale, V. (2009) Novel lipid based systems for improved topical delivery of antioxidants. *Household Pers. Care Today*, **4**, 5–8.

[23] Raschke, T., Koop, U., Düsing, H.J. *et al.* (2004) Topical activity of ascorbic acid: From in vitro optimization to in vivo efficacy. *Skin Pharmacol. Physiol.*, **17**, 200–206.

[24] Zhai, H., Behnam, S., Villarama, C.D. *et al.* (2005) Evaluation of the antioxidant capacity and preventive effects of a topical emulsion and its vehicle control on the skin response to UV exposure. *Skin Pharmacol. Physiol.*, **18**, 288–293.

[25] Schwarz, K., Huang, S.-W., German, J.B. *et al.* (2000) Activities of antioxidants are affected by colloidal properties of oil-in-water and water-in-oil emulsions and bulk oils. *J. Agric. Food Chem.*, **48**, 4874–4882.

[26] Chen, X. and Selloni, A. (2014) Introduction: Titanium dioxide (TiO2) nanomaterials. *Chem. Rev.*, **114**, 9281–9282.

[27] Detoni, C.B., Coradini, K., Back, P. *et al.* (2014) Penetration, photo-reactivity and photoprotective properties of nanosized ZnO. *Photochem Photobiol Sci*, **13**, 1253–1260.

[28] Novoselov, K.S., Geim, A.K., Morozov, S.V. *et al.* (2004) Electric field in atomically thin carbon films. *Science*, **306**, 666–669.

[29] Goenka, S., Sant, V. and Sant, S. (2014) Graphene-based nanomaterials for drug delivery and tissue engineering. *J. Control. Release*, **173**, 75–88.

[30] Couteau, C., Alami, S., Guitton, M. *et al.* (2008) Mineral filters in sunscreen products – Comparison of the efficacy of zinc oxide and titanium dioxide by in vitro method. *Pharmazie*, **63**, 58–60.

[31] Giroud, F. and Favreau, V. (17 June 2004) Cosmetic composition for volumizing keratin fibers and cosmetic use of nanotubes volumizing keratin fibers, US Patent 2004/0115232 A1. http://worldwide.espacenet.com/publicationDetails/originalDocument?CC=US&NR=2004115232A 1&KC=A1&FT=D&ND=1&date=20040617&DB=&locale=en_EP (accessed 23 May 2015).

[32] Draelos, Z.D. (2011) Reinvigorating cosmetic dermatology with the nanoparticle revolution. *J. Cosmet. Dermatol.*, **10**, 251–252.

[33] Kaur, I.P. and Agrawal, R. (2007) Nanotechnology: A new paradigm in cosmeceuticals. *Recent Pat. Drug Deliv. Formul.*, **1**, 171–182.

[34] Lövestam, G., Rauscher, H., Roebben, G. *et al.* (2010) Considerations on a Definition of Nanomaterial for Regulatory Purposes, JRCReferenceReport, Luxembourg (EUR 24403 EN), p. 13.

[35] Wagner, V., Hüssig, B., Gaisser, S. and Bock, A.-K. (2008) Nanomedicine: Drivers for Development and Possible Impacts, JRC Scientific and Technical Reports, Luxembourg (EUR 23494 EN), p. 16.

[36] McClements, D.J. (2012) Nanoemulsions *versus* microemulsions: Terminology, differences, and similarities. *Soft Matter*, **8**, 1719–1729.

[37] Mason, T.G., Wilking, J.N., Meleson, K. *et al.* (2006) Nanoemulsions: Formation, structure, and physical properties. *J. Phys. Condens. Matter*, **18**, R635–R666.

[38] Solans, C., Izquierdo, P., Nolla, J. *et al.* (2005) Nano-emulsions. *Curr. Opin. Colloid Interface Sci.*, **10**, 102–110.

[39] Schulman, J.H. and Montagne, J.B. (1961) Formation of microemulsions by amino alkyl alcohols. *Ann. N. Y. Acad. Sci.*, **92**, 366–371.

[40] Calvo, P., Vila-Jato, J.L. and Alonso, M.J. (1996) Comparative in vitro evaluation of several colloidal systems, nanoparticles, nanocapsules, and nanoemulsions, as ocular drug carriers. *J. Pharm. Sci.*, **85**, 530–536.

[41] Sharma, S. and Sarangdevot, K. (2012) Nanoemulsions for cosmetics. *Int. J. Adv. Res. Pharm. Biosci.*, **2**, 408–415.

[42] Lawrence, M.J. (1994) Surfactant systems, microemulsions and vesicles as vehicles for drug delivery. *Eur. J. Drug Metab. Pharmacokinet.*, **3**, 257–269.

[43] Kogan, A. and Gati, N. (2006) Microemulsions as transdermal drug delivery vehicles. *Adv. Colloid Interface Sci.*, **123–126**, 369–385.

[44] Boonme, P. (2007) Applications of microemulsions in cosmetics. *J. Cosmet. Dermatol.*, **6**, 223–228.

[45] Attwood, D. and Kreuter, J. (1994) Colloidal Drug Delivery Systems, Marcel Dekker, New York, pp. 31–71.

[46] Tadros, T.F. (1992) Future developments in cosmetic formulations. *Int. J. Cosmet. Sci.*, **14**, 93–111.

[47] Carlotti, M.E., Gallarate, M. and Rossatto, V. (2003) O/W microemulsion as a vehicle for sunscreens. *J. Cosmet. Sci.*, **54**, 451–462.

[48] Garti, N., Shevachman, M. and Shani, A. (2004) Solubilization of lycopene in jojoba oil microemulsion. *J. Am. Oil Chem. Soc.*, **81**, 873–877.

[49] Watanabe, K., Noda, A., Masuda, M. *et al.* (2004) Bicontinuous microemulsion type cleansing containing silicone oil II: Characterization of the solution and its application to cleansing agent. *J. Oleo Sci.*, **53**, 547–555.

[50] Gallarate, M., Carlotti, M.E., Trotta, M. *et al.* (2004) Photostability of naturally occurring whitening agents in cosmetic microemulsions. *J. Cosmet. Sci.*, **55**, 139–148.

[51] Spiclin, P., Gasperlin, M. and Kmetec, V. (2001) Stability of ascorbyl palmitate in topical microemulsions. *Int. J. Pharm.*, **222**, 271–279.

[52] Spiclin, P., Homar, M., Zupancic-Valant, A. and Gasperlin, M. (2003) Sodium ascorbyl phosphate in topical microemulsions. *Int. J. Pharm.*, **256**, 65–73.

[53] Rangarajan, M. and Zatz, J.L. (2003) Effect of formulation on the topical delivery of alpha-tocopherol. *J. Cosmet. Sci.*, **54**, 161–174.

[54] Junyaprasert, V.B., Boonme, P., Songkro, S. *et al.* (2007) Transdermal delivery of hydrophobic and hydrophilic local anesthetics from o/w and w/o Brij 97-based microemulsions. *J. Pharm. Pharm. Sci.*, **10**, 288–298.

[55] Hsu, C., Cui, Z., Mumper, R.J. and Jay, M. (2003) Preparation and characterization of novel coenzyme Q10 nanoparticles engineered from microemulsion precursors. *AAPS PharmSciTech*, **4**, 1–12.

[56] Sonneville-Aubrun, O., Simonnet, J.-T. and L'Alloret, F. (2004) Nanoemulsions: A new vehicle for skincare products. *Adv. Colloid Interface Sci.*, **108–109**, 145–149.

[57] Rajalakshmi, R., Mahesh, K. and Ashok Kumar, C.K. (2011) A critical review on nano emulsions. *Int. J. Innov. Drug Discov.*, **1**, 1–8.

[58] Cox, B.R. and Dodd, M.T. (7 Decemeber 1999) Hair care compositions containing low melting point fatty alcohol and ethylene oxide/propylene oxide polymer. US Patent 5,997,851 A, Procter & Gamble Co., USA.

[59] Weder, H.G. and Weder, M.A. (15 July 1998) Nanodispersion cosmetic composition. EP 0852941 A1, Vesifact AG, Germany.

[60] Ribier, A., Simonnet, J.-T. and Michelet, J. (23 October 1996) Composition cosmétique ou dermatologique constituée d'une émulsion huile dans eau à base de globules huileux pourvus d'un enrobage cristal liquide lamellaire. EP 0705593 B1, L'Oréal, France.

[61] Ribier, A., Simonnet, J.-T. and Legret, S. (30 January 1996) Transparent nanoemulsions based on amphiphilic nonionic lipids and use in cosmetics or dermapharmacy. EP 0728460 B1, L'Oréal, France.

[62] Simonnet, J.-T., Sonneville, O. and Legret, S. (15 October 2002) Nanoemulsion based on ethylene oxide and propylene oxide block copolymers and its uses in the cosmetics, dermatological and/or ophthalmological fields. US Patent 6,464,990 B2, L'Oréal, France.

[63] Simonnet, J.-T., Sonneville, O. and Legret, S. (14 August 2001) Nanoemulsion based on phosphoric acid fatty acid esters and its uses in the cosmetics, dermatological, pharmaceutical, and/or ophthalmological fields. US Patent 6,274,150 B1, L'Oréal, France.

[64] Lochhead, R.Y., Castaneda, J.Y. and Hemker, W.J. (22 December 1993) Stable and quick-breaking topical skin compositions. EP 0268164 B1, The B.F. Goodrich Company. http://worldwide.espacenet.com/publicationDetails/originalDocument?CC=EP&NR=0268164B1&KC=B1&FT=D&ND=&date=19931222&DB=&&locale=en_EP (accessed 24 May 2015).

[65] Ramsden, W. (1903) Separation of solids in the surface-layers of solutions and 'suspensions' (observations on surface-membranes, bubbles, emulsions, and mechanical coagulation) – Preliminary account. *Proc. R. Soc. Lond.*, **72**, 156–164.

[66] Pickering, S.U. (1907) CXCVI.—Emulsions. *J. Chem. Soc. Trans.*, **91**, 2001–2021.

[67] Simovic, S., Ghouchi-Eskandar, N. and Prestidge, C.A. (2011) Pickering emulsions for dermal delivery. *J. Drug Deliv. Sci. Technol.*, **21**, 123–133.

[68] Gers-Barlag, H. and Muller, A. (2002) Emulsifier-free finely disperse systems of the oil-in-water and water-in-oil type. US Patent 6,410,035, Beiersdorf AG, Hamburg.

[69] Gers-Barlag, H. and Muller, A. (2003) Emulsifier-free finely disperse systems of the oil-in-water and water-in-oil type. US Patent 6,579,529, Beiersdorf AG, Hamburg.

[70] Gers-Barlag, H. and Muller, A. (2004) Emulsifier-free finely disperse systems of the oil-in-water and water-in-oil type. US Patent 6,692,755, Beiersdorf AG, Hamburg.

[71] Gers-Barlag, H. and Muller, A. (2002) Emulsifier-free finely disperse systems of the oil-in-water and water-in-oil type. US Patent 6,428,796, Beiersdorf AG, Hamburg.

[72] Gers-Barlag, H. and Muller, A. (2003) Emulsifier-free finely disperse systems of the oil-in-water and water-in-oiltype. US Patent 20,030,003,122 Al, Beiersdorf AG, Hamburg.

[73] Gers-Barlag, H. and Muller, A. (2004) Emulsifier-free finely disperse systems of the oil-in-water and water-in-oil type. US Patent 6,703,032, Beiersdorf AG, Hamburg.

[74] Lissant, K.J. (1966) The geometry of high – internal – phase – ration emulsions. *J. Colloid Interface Sci.*, **22**, 462–468.

[75] Sebba, F. (1987) Applications of colloidal gas aphrons, in Foams and Biliquid Foams-Aphrons (ed F. Sebba), John Wiley & Sons Ltd, Chichester, pp. 135–153.

[76] Hoffman, H. and Ebert, G. (1988) Tenside, Micellen und faszinierende Phänomene. *Angew. Chem.*, **100**, 933–944.

[77] Solans, C., Carrera, I., Pons, R. *et al.* (1993) Gel emulsions: Formulating with highly concentrated W/O emulsions. *Cosmet. Toil.*, **108**, 61–64.

[78] Fueller, S., von Thaden, S., Bleckmann, A. *et al.* (30 April 2003) Refreshing cosmetic or dermatological preparation is an O/W gel emulsion containing hydrocolloids together with lipids, emulsifiers and diastereomers of menthol and/or menthol itself. DE10151247 A1, Beiersdorf Ag.

[79] Fry, B.E. (2 November 2005) Organopolysiloxane gels for use in cosmetics. EP 1132430 B1, Wacker-Chemie GmbH.

[80] Alam, M.M. and Aramaki, K. (2014) Liquid crystal-based emulsions: Progress and prospects. *J. Oleo Sci.*, **63**, 97–108.

[81] Massaro, R.C., Zabagli, M.S., Souza, C.R. *et al.* (2003) O/w dispersions development containing liquid crystals. *Boll. Chim. Farm.*, **142**, 264–270.

[82] Suzuki, K. and Sakon, K. (1990) The applications of liposomes to cosmetics. *Cosmet. Toil.*, **105**, 65–78.

[83] Junginger, H.E., Hofland, H.J. and Bouwstra, J.A. (1991) Liposomes and niosomes: Interaction with human skin. *Cosmet. Toil.*, **106**, 45–50.

[84] Nacht, S. (1995) Encapsulation and other topical delivery systems. *Cosmet. Toil.*, **110**, 25–30.

[85] Citernesi, U. and Sciacchitano, M. (1995) Phospholipid/active ingredient complexes. *Cosmet. Toil.*, **110**, 57–68.

[86] Sharma, B. and Sharma, A. (2012) Future prospect of nanotechnology in development of anti-ageing formulations. *Int. J. Pharm. Pharm. Sci.*, **4**, 57–66.

[87] Kaur, I.P., Kapila, M. and Agrawal, R. (2007) Role of novel delivery systems in developing topical antioxidants as therapeutics to combat photoageing. *Ageing Res. Rev.*, **6**, 271–288.

[88] Seiller, M., Puisieux, F. and Grossiord, J.L. (1997) Multiple emulsions in cosmetics, in Surfactants in Cosmetics (eds M.M. Reiger and L.D. Rhein), New York, Marcel Dekker Inc., pp. 139–153.

[89] Yoshida, K., Sekine, T., Matsuzaki, F. *et al.* (1999) Stability of vitamin a in oil-in-water-in-oil-type multiple emulsions. *J. Am. Oil Chem. Soc.*, **76**, 1–6.

[90] Akhtar, N. and Yazan, Y. (2005) Formulation and characterization of a cosmetic multiple emulsion system containing macadamia nut oil and two antiaging agents. *Turk J Pharm Sci*, **2**, 173–185.

[91] Mahmood, T. and Akhtar, N. (2013) Stability of a cosmetic multiple emulsion loaded with green tea extract. *Scientific World Journal*, **153695**.

[92] Jadhav, S.M., Morey, P., Karpe, M. and Kadam, V. (2012) Novel vesicular system: An overview. *J. Appl. Pharm. Sci.*, **2**, 193–202.

[93] Egbaria, K. and Weiner, N. (1990) Liposomes as a topical delivery system. *Adv. Drug Deliv. Rev.*, **5**, 287–300.

[94] Madsen, J.T. and Andersen, K.E. (2010) Microvesicle formulations used in topical drugs and cosmetics affect product efficiency, performance and allergenicity. *Dermatitis*, **21**, 243–247.

[95] El Maghraby, G.M., Barry, B.W. and Williams, A.C. (2008) Liposomes and skin: From drug delivery to model membranes. *Eur. J. Pharm. Sci.*, **34**, 203–222.

[96] Mehl Sr., T.L. and Zaias, N. (23 March 1999) Freeze-dried liposome delivery system for application of skin treatment agents. US Patent 5,885,260 A. http://worldwide.espacenet.com/publicationDetails/originalDocument?CC=US&NR=5885260A&KC=A&FT=D&ND=&date=19990323&DB=&&locale=en_EP (accessed 24 May 2015).

[97] Mehl Sr., T.L. and Zaias, N. (6 February 2001) Method of delivery of skin treatment agents using freeze-dried liposomes. US Patent 6,183,451 B1. http://worldwide.espacenet.com/publicationDetails/originalDocument?CC=US&NR=6183451B1&KC=B1&FT=D&ND=&date=20010206&DB=&&locale=en_EP (accessed 24 May 2015).

[98] Sinico, C., Manconia, M., Peppib, M. *et al.* (2005) Liposomes as carriers for dermal delivery of tretinoin, in vitro evaluation of drug permeation and vesicle–skin interaction. *J. Control. Release*, **103**, 123–136.

[99] Contreras, M.J.F., Soriano, M.M.J. and Dieguez, A.R. (2005) In vitro percutaneous absorption of all-trans retinoic acid applied in free form or encapsulated in stratum corneum lipid liposomes. *Int. J. Pharm.*, **297**, 134–145.

[100] Fang, J.Y., Hwang, T.L., Huanga, Y.L. and Fang, C.L. (2006) Enhancement of the transdermal delivery of catechins by liposomes incorporating anionic surfactants and ethanol. *Int. J. Pharm.*, **310**, 131–138.

[101] Kitagawa, S. and Kamasaki, M. (2006) Enhanced delivery of retinoic acid to skin by cationic liposomes. *Chem. Pharm. Bull.*, **54**, 242–244.

[102] Touitou, E., Dayan, N., Bergelson, L. *et al.* (2000) Ethosomes—novel vesicular carriers for enhanced delivery: Characterization and skin penetration properties. *J. Control. Release*, **65**, 403–418.

[103] Elsayed, M.M.A., Abdallah, O.Y., Naggar, V.F. and Khalafallah, N.M. (2006) Deformable liposomes and ethosomes: Mechanism of enhanced skin delivery. *Int. J. Pharm.*, **322**, 60–66.

[104] Cevc, G. and Vierl, U. (6 January 2009) NSAID formulations, based on highly adaptable aggregates, for improved transport through barriers and topical drug delivery. US Patent 7,473,432 B2, IDEA AG, Munich, Germany.

[105] Cevc, G. and Vierl, U. (18 June 2009) Aggregates with increased deformability, comprising at least three amphipats, for improved transport through semi-permeable barriers and for the non-invasive drug application in vivo, especially through the skin. US Patent 20,090,155,235 A1, IDEA AG, Munich, Germany.

[106] Jain, S., Jain, P., Umamaheshwari, R.B. and Jain, N.K. (2003) Transfersomes – a novel vesicular carrier for enhanced transdermal delivery: Development, characterization and performance evaluation. *Drug Dev. Ind. Pharm.*, **29**, 1013–1026.

[107] Romero, E.L. and Morilla, M.J. (2013) Highly deformable and highly fluid vesicles as potential drug delivery systems: Theoretical and practical considerations. *Int. J. Nanomedicine*, **8**, 3171–3186.

[108] Sachan, R., Parashar, T., Soniya *et al.* (2013) Drug carrier transfersomes: A novel tool for transdermal drug delivery system. *Int. J. Res. Dev. Pharm. Life Sci.*, **2**, 309–316.

[109] Cevc, G. (1996) Transfersomes, liposomes and other lipid suspensions on the skin, permeation enhancement, vesicle penetration, and transdermal drug delivery. *Crit. Rev. Ther. Drug Carrier Syst.*, **13**, 257–388.

[110] Gallarate, M., Chirio, D., Trotta, M. and Carlotti, M.E. (2006) Deformable liposomes as topical formulations containing a-tocopherol. *J. Dispers. Sci. Technol.*, **27**, 703–713.

[111] Saraf, S., Jeswani, G., Kaur, C.D. and Saraf, S. (2011) Development of novel herbal cosmetic cream with *Curcuma longa* extract loaded transfersomes for antiwrinkle effect. *Afr. J. Pharm. Pharmacol.*, **5**, 1054–1062.

[112] Jacob, L. and Anoop, K.R. (2013) A review on surfactants as edge activators in ultradeformable vesicles for enhanced skin delivery. *Int. J. Pharm. Bio Sci.*, **4**, 337–344.

[113] Tyagi, L.K., Kumar, S., Maurya, S.S. and Kori, M.L. (2013) Ethosomes: Novel vesicular carrier for enhanced transdermal drug delivery system. *Bull. Pharm. Res.*, **3**, 6–13.

[114] Nikalje, A.P. and Tiwari, S. (2012) Ethosomes: A novel tool for transdermal drug delivery. *Int. J. Res. Pharm. Sci.*, **2**, 1–20.

[115] Verma, P. and Pathak, K. (2010) Therapeutic and cosmeceutical potential of ethosomes: An overview. *J. Adv. Pharm. Technol. Res.*, **1**, 274–282.

[116] Tadwee, I.K., Gore, S. and Giradkar, P. (2012) Advances in topical drug delivery system: A review. *Int. J. Pharm. Res. Allied Sci.*, **1**, 19–23.

[117] Umashankar, M.S., Sachdeva, R.K. and Gulati, M. (2010) Aquasomes: A promising carrier for peptides and protein delivery. *Nanomed. Nanotechnol. Biol. Med.*, **6**, 419–426.

[118] Vengala, P., Shwetha, D., Sana, A. *et al.* (2012) Aquasomes: A novel drug carrier system. *Int. Res. J. Pharm.*, **3**, 123–127.

[119] Patravale, V.B. and Mandawgade, S.D. (2008) Novel cosmetic delivery systems: An application update. *Int. J. Cosmet. Sci.*, **30**, 19–33.

[120] Lautenschlager, H. (1990) Liposomes in dermatological preparations: Part II. *Cosmet. Toil.*, **105**, 63–72.

[121] Decome, L., De Meo, M., Geffard, A. *et al.* (2005) Evaluation of photolyase (photosome) repair activity in human keratinocytes after a single dose of ultraviolet B irradiation using the comet assay. *J. Photochem. Photobiol. B*, **79**, 101–108.

[122] Alexiades-Armenakas, M. (10 September 2013) Multi-active microtargeted anti-aging skin care cream polymer technology. US Patent 8,529,925 B2, NY Derm LLC. http://worldwide. espacenet.com/publicationDetails/originalDocument?CC=US&NR=8529925B2&KC=B2&FT=D&ND=&date=20130910&DB=&&locale=en_EP (accessed 24 May 2015).

[123] Amit, G., Ashawat, M.S., Shailendra, S. and Swarnlata, S. (2007) Phytosome: A novel approach towards functional cosmetics. *J. Plant Sci.*, **2**, 644–649.

[124] Bombardelli, E. (1991) Phytosome: New cosmetic delivery system. *Boll. Chim. Farm.*, **130**, 431–438.

[125] Bombardelli, E. and Spelta, M. (1991) Phospholipid-polyphenol complexes: A new concept in skin care ingredients. *Cosmet. Toil.*, **106**, 69–76.

[126] Bombardelli, E., Cristoni, A. and Morazzoni, P. (1994) Phytosome in functional cosmetics. *Fitoterapia*, **65**, 147–152.

[127] Ziboh, V.A., Miller, C.C. and Cho, Y. (2000) Metabolism of polyunsaturated fatty acids by skin epidermal enzymes: Generation of anti-inflammatory and antiproliferative metabolites. *Am. J. Clin. Nutr.*, **71**, 361S–366S.

[128] Moussaoui, N., Cansell, M. and Denizot, A. (2002) Marinosomes® marine lipid-based liposomes: Physical characterization and potential applications in cosmetics. *Int. J. Pharm.*, **242**, 361–365.

[129] Cansell, M., Moussaoui, N. and Mancini, M. (2007) Prostaglandin E2 and interleukin-8 production in human epidermal keratinocytes exposed to marine lipid-based liposomes. *Int. J. Pharm.*, **343**, 277–280.

[130] Andersson, S. and Jacob, M. (1995) Structure of cubosomes – A closed lipid bilayer aggregate. *Z. Krist.*, **210**, 315–318.

[131] Bansal, S., Kashyap, C.P., Aggarwal, G. and Harikumar, S.L. (2012) A comparative review on vesicular drug delivery system and stability issues. *Int. J. Res. Pharm. Chem.*, **2**, 704–713.

[132] Boyd, B.J. (2003) Character of drug release from cubosomes using the pressure ultrafiltration method. *Int. J. Pharm.*, **260**, 239–247.

[133] Hewitt-Jones, J.D., Leng, J.F. and Parrot, D.T. (10 November 1994) Antiperspirant compositions, WO 1994024993 A1, Unilever Lv., Unilever Plc. http://worldwide.espacenet.com/publicationDetails/originalDocument?CC=WO&NR=9424993A1&KC=A1&FT=D&ND=&date=19941110&DB=&&locale=en_EP (accessed 24 May 2015).

[134] El-Nokaly, M., Walling, D.W., Vatter, M.L. and Leatherbury, N.C. (31 March 1994) Lipsticks. CA 2144844 A1. http://worldwide.espacenet.com/publicationDetails/originalDocument?CC=CA&NR=2144844A1&KC=A1&FT=D&ND=&date=19940331&DB=&&locale=en_EP (accessed 23 May 2015).

[135] Gilchrest, B.A. and Gordon, P.R. (31 March 1994) Use of diacylglycerols for increasing the melanin content in melanocytes. WO 1994004122 A3, Boston University, Boston, MA. http://worldwide.espacenet.com/publicationDetails/originalDocument?CC=WO&NR=9404122A3&KC=A3&FT=D&ND=&date=19940331&DB=&&locale=en_EP (accessed 24 May 2015).

[136] Tomalia, D.A., Naylor, A.M. and Goddard, W.A. (1990) Starburst dendrimers: Molecular-level control of size, shape, surface chemistry, topology, and flexibility from atoms to macroscopic matter. *Angew. Chem. Int. Ed. Engl.*, **29**, 138–175.

[137] Silpa, R., Shoma, J., Sumod, U.S. and Sabitha, M. (2012) Nanotechnology in cosmetics: Opportunities and challenges. *J. Pharm. Bioallied Sci.*, **4**, 186–193.

[138] Tournilhac, F. and Simon, P. (11 September 2001) Cosmetic or dermatological topical compositions comprising dendritic polyesters. US Patent 6,287,552 B1, L'Oréal, France.

[139] Baecker, S., Clausen, T., Franzke, M. *et al.* (30 May 2000) Cosmetic compositions for hair treatment containing dendrimers or dendrimer conjugates. US Patent 6,068,835 A, Wella Aktiengesellschaft, Germany.

[140] Percec, V., Wilson, D.A., Leowanawat, P. *et al.* (2010) Self-Assembly of Janus dendrimers into uniform dendrimersomes and other complex architectures. *Science*, **328**, 1009–1014.

[141] Percec, V., Hughes, A.D., Leowanawat, P. *et al.* (6 December 2012) Amphiphilic Janus-dendrimers. US Patent 20,120,308,640 A1, The Trustees of The University of Pennsylvania, Pennsylvania.

[142] Caminade, A.-M., Laurent, R., Delavaux-Nicota, B. and Majoral, J.-P. (2012) 'Janus' dendrimers: Syntheses and properties. *New J Chem*, **36**, 217–226.

[143] Zhang, S., Suna, H.-J., Hughes, A.D. *et al.* (2014) Self-assembly of amphiphilic Janus dendrimers into uniform onion-like dendrimersomes with predictable size and number of bilayers. *Proc. Natl. Acad. Sci.*, **111**, 9058–9063.

[144] Newton, J., Stoller, C. and Starch, M. (2004) Silicone technology offers novel methods for delivery active ingredients. *SÖFW J*, **130**, 8–13.

[145] Ostergaard, T., Gomes, A., Quackenbush, K. and Johnson, B. (2004) Silicone quaternary microemulsion: A multifunctional product for hair care. *Cosmet. Toil.*, **119**, 45–52.

[146] Newton, J., Postiaux, S., Stoller, C. *et al.* (2004) Silicone-based vesicle delivery systems. *Cosmet. Toil.*, **119**, 53–60.

[147] Hougeir, F.G. and Kircik, L. (2012) A review of delivery systems in cosmetics. *Dermatol. Ther.*, **25**, 234–237.

[148] Victor, A.R., Gerald, K.S. and Starch, M. (2006) Controlled release of active ingredients from cross-linked silicones. *Cosmet. Toil.*, **120**, 69–72.

[149] Newton, J., Stoller, C., Lin, S. *et al.* (2004) Silicone elastomer delivery systems. *Cosmet. Toil.*, **119**, 24–31.

[150] Smith, S.D., Vatter, M.L. and Doyl, K.L. (11 June 2009) Stable dispersions of solid particles in a hydrophobic solvent for cosmetic uses and methods of preparing the same. WO 2,009,073,384 A1, Procter & Gamble Co., Cincinnati, OH.

[151] Klug, P., Pilz, F.M., Kluth, G. *et al.* (17 December 2009) Cosmetic or pharmaceutical compositions, emulsions comprising modified polysiloxanes with at least one carbamate group. WO 2009149879 A1, Clariant International Ltd, Switzerland.

[152] Araki, H., Iimura, T., Kimura, T. *et al.* (13 August 2009) Cleansing composition. WO 2009099007 A1, Shiseido Company Ltd and Dow Corning Toray Co., Ltd, Japan.

[153] Ichikawa, Y. and Hata, M. (13 August 2009) Oil-in-water type emulsion cosmetics. JP 2009-179588 A, Hoyu Co., Ltd, Japan.

[154] Griat, J., Handjani, R.M., Ribier, A. *et al.* (16 May 1989) Cosmetic and pharmaceutical compositions containing niosomes and water-soluble polyamide, and a process for preparing these compositions. US Patent 4,830,857 A, L'Oréal, France.

[155] Vanlerberghe, G. and Handjani, R.M. (30 June 1975) Procède de fabrication de dispersions aqueuses de spherules lipidiques et nouvelles compositions correspondantes. FR 2315991-A1, L'Oréal, France.

[156] Braunfalco, O., Kortung, H.C. and Maibach, H.I. (1992) Griesbach Conference: Liposomes Dermatitis, vol. **201**, Springer-Verlag, Berlin, pp. 19–23.

[157] Buckton, G. (1995) Interfacial Phenomena in Drug Delivery and Targeting, Harwood Academic Publishers GmbH, Switzerland, pp. 154–155.

[158] Hunt, C.A. and Tsang, S. (1981) α-Tocopherol retards autoxidation and prolongs the shelf-life of liposomes. *Int. J. Pharm.*, **8**, 101–110.

[159] Birman, M. and Lawrence, N. (2002) Liposome stability via multi-walled delivery systems. *Cosmet. Toil.*, **117**, 51–58.

[160] Hadgraft, J., Peck, J., Willians, D.G. *et al.* (1996) Mechanisms of action of skin penetration enhancers/retarders: Azone and analogues. *Int. J. Pharm.*, **141**, 17–25.

[161] Rubio, L., Alonso, C., Rodríguez, G. *et al.* (2013) Bicellar systems as new delivery strategy for topical application of flufenamic acid. *Int. J. Pharm.*, **444**, 60–69.

[162] Lee, S.J., Jo, B.K., Lee, Y.J. and Lee, C.M. (21 June 2005) Cosmetic material containing triple-encapsulated retinol. US Patent 6,908,625 B2, Coreana Cosmetics Co., Ltd, South Korea.

[163] Pathak, Y. and Thassu, D. (2009) Drug delivery nanoparticles formulation and characterization, in Drugs and the Pharmaceutical Sciences, vol. **191** (ed J. Swarbeck), Informa Healthcare, New York, pp. 239–251.

[164] Lee, R.W., Shenoy, D.B. and Sheel, R. (2009) Micellar nanoparticles: Applications for topical and passive transdermal drug delivery, in Handbook of Non-Invasive Drug Delivery Systems (ed V.S. Kulkarni), New York, William Andrew Publishing, pp. 37–58.

[165] Vert, M., Doi, Y., Hellwich, K.-H. *et al.* (2012) Terminology for biorelated polymers and applications (IUPAC Recommendations 2012). *Pure Appl. Chem.*, **84** (377–410).

[166] Torrado, S., Torrado, J.J. and Cadorniga, R. (1992) Topical application of albumin microspheres containing vitamin A: Drug release and availability. *Int. J. Pharm.*, **86**, 147–149.

[167] Mora-Huertas, C.E., Fessi, H. and Elaissari, A. (2010) Polymer-based nanocapsules for drug delivery. *Int. J. Pharm.*, **385**, 113–142.

[168] Parison, V. (1993) Active delivery from nylon particles. *Cosmet. Toil.*, **108**, 97–100.

[169] Rogers, K. (1999) Controlled release technology and delivery systems. *Cosmet. Toil.*, **114**, 53–60.

[170] Iannuccelli, V., Coppi, G., Sergi, S. and Cameroni, R. (2001) Preparation and *in vitro* characterization of lipospheres as a carrier for the cosmetic application of glycolic acid. *J. Appl. Cosmetol.*, **19**, 113–119.

[171] Scalia, S. and Mezzena, M. (2009) Incorporation of quercetin in lipid microparticles: Effect on photo- and chemical-stability. *J. Pharm. Biomed. Anal.*, **49**, 90–94.

[172] Albertini, B., Mezzena, M., Passerini, N. *et al.* (2009) Evaluation of spray congealing as technique for the preparation of highly loaded solid lipid microparticles containing the sunscreen agent, avobenzone. *J. Pharm. Sci.*, **98**, 2759–2769.

[173] Parikh, B.N., Gothi, G.D., Patel, T.D. *et al.* (2010) Microsponge as novel topical drug delivery system. *J. Global Pharma Technol.*, **2**, 17–29.

[174] Nacht, S. and Katz, M. (1990) The microsponge: A novel topical programmable delivery system, in Topical Drug Delivery Formulations (eds D.W. Osborne and A.H. Amann), New York, Marcel Dekker Inc., pp. 322–323.

[175] Blume, G. and Jung, K. (2010) New nanocapsules with high loading of UV-filters. *SÖFW J.*, **136**, 50–55.

[176] Shefer, A., Ng, C. and Shefer, S. (2004) Nanotechnology enhances bioadhesion and release after rinse-off. *Cosmet. Toil.*, **119**, 57–60.

[177] Imai, T. and Fukuda, I. (20 August 2009) Method for producing solid powdery cosmetic. JP 2009-184944A, Kao Corp., Japan.

[178] Kwon, S.S., Nam, Y.S., Lee, J.S. *et al.* (2002) Preparation and characterization of coenzyme Q10-loaded PMMA nanoparticles by a new emulsification process based on microfluidization. *Colloids Surf. A Physicochem. Eng. Asp.*, **210**, 95–104.

[179] Guterres, S.S., Alves, M.P. and Pohlmann, A.R. (2007) Polymeric nanoparticles, nanospheres and nanocapsules, for cutaneous applications. *Drug Target Insights*, **2**, 147–157.

[180] Keck, C.M., Kovačević, A., Müller, R.H. *et al.* (2014) Formulation of solid lipid nanoparticles (SLN): The value of different alkyl polyglucoside surfactants. *Int. J. Pharm.*, **474**, 33–41.

[181] Battaglia, L. and Gallarate, M. (2012) Lipid nanoparticles: State of the art new preparation methods and challenges in drug delivery. *Expert Opin. Drug Deliv.*, **9**, 497–508.

[182] Müller, R.H., Shegokar, R. and Keck, C.M. (2011) 20 years of lipid nanoparticles (SLN and NLC): Present state of development and industrial applications. *Curr. Drug Discov. Technol.*, **8**, 207–227.

[183] Kalepu, S., Manthina, M. and Padavala, V. (2013) Oral lipid-based drug delivery systems – an overview. *Acta Pharm. Sin. B*, **3**, 361–372.

[184] Ekambaram, P., Sathali, A.A.H. and Priyanka, K. (2012) Solid lipid nanoparticles: A review. *Sci. Rev. Chem. Commun.*, **2**, 80–102.

[185] Souto, E.B. and Müller, R.H. (2008) Cosmetic features and applications of lipid nanoparticles (SLN®, NLC®). *Int. J. Cosmet. Sci.*, **30**, 157–165.

[186] Lippacher, A., Muller, R.H. and Mader, K. (2004) Liquid and semisolid SLN™ dispersions for topical application, rheological characterization. *Eur. J. Pharm. Biopharm.*, **58**, 561–567.

[187] Dingler, A., Blum, R.P., Niehus, H. *et al.* (1999) Solid lipid nanoparticles (SLN™/Lipopearls™) a pharmaceutical and cosmetic carrier for the application of vitamin E in dermal products. *J. Microencapsul.*, **16**, 751–767.

[188] Jenning, V. and Gohla, S.H. (2001) Encapsulation of retinoids in solid lipid nanoparticles (SLN). *J. Microencapsul.*, **18**, 149–158.

[189] Wissing, S.A., Lippacher, A. and Müller, R.H. (2001) Investigations on the occlusive properties of solid lipid nanoparticles (SLN™). *J. Cosmet. Sci.*, **52**, 313–323.

[190] Uner, M. (2006) Preparation, characterization and physico-chemical properties of solid lipid nanoparticles (SLN) and nanostructured lipid carriers (NLC), their benefits as colloidal drug carrier systems. *Pharmazie*, **61**, 375–386.

[191] Jenning, V., Gysler, A., Schafer-Korting, M. and Gohla, S.H. (2000) Vitamin-A loaded solid lipid nanoparticles for topical use, occlusive properties and drug targeting to the upper skin. *Eur. J. Pharm. Biopharm.*, **49**, 211–218.

[192] Jenning, V., Schafer-Korting, M. and Gohla, S. (2000) Vitamin A-loaded solid lipid nanoparticles for topical use, drug release properties. *J. Control. Release*, **66**, 115–126.

[193] Liu, J., Hub, W., Chena, H. *et al.* (2007) Isotretinoin-loaded solid lipid nanoparticles with skin targeting for topical delivery. *Int. J. Pharm.*, **328**, 191–195.

[194] Wissing, S.A. and Muller, R.H. (2001) A novel sunscreen system based on tocopherol acetate incorporated into solid lipid nanoparticles. *Int. J. Cosmet. Sci.*, **23**, 233–243.

[195] Wissing, S.A. and Müller, R.H. (2001) Solid lipid nanoparticles (SLN™) – A novel carrier for UV blockers. *Pharmazie*, **56**, 783–786.

[196] Souto, E.B., Muller, R.H. and Gohla, S. (2005) A novel approach based on lipid nanoparticles (SLN) for topical delivery of a-lipoic acid. *J. Microencapsul.*, **22**, 581–592.

[197] zur Mühlen, A., Schwarz, C. and Mehnert, W. (1998) Solid lipid nanoparticles (SLN) for controlled drug delivery – drug release and release mechanism. *Eur. J. Pharm. Biopharm.*, **45**, 149–155.

[198] Muller, R.H., Mader, K. and Gohla, S. (2000) Solid lipid nanoparticles (SLN) for controlled drug delivery – A review of the state of the art. *Eur. J. Pharm. Biopharm.*, **50**, 161–177.

[199] Dahms, G., Jung, A. and Seidel, H. (6 May 2014) Compositions for the targeted release of fragrances and aromas. US Patent 8,716,214 B2, Otc GmbH, Germany.

[200] Müller, R.H., Petersen, R.D., Hommoss, A. and Pardeike, J. (2007) Nanostructured lipid carriers (NLC) in cosmetic dermal products. *Adv. Drug Deliv. Rev.*, **59**, 522–530.

[201] Ruktanonchai, U., Bejrapha, P., Sakulkhu, U. *et al.* (2009) Physicochemical characteristics, cytotoxicity, and antioxidant activity of three lipid nanoparticulate formulations of alpha-lipoic acid. *AAPS PharmSciTech*, **10**, 227–234.

[202] De Vringer, T. (16 September 1997) Topical preparation containing a suspension of solid lipid particles. US Patent 5,667,800, Yamanouchi Europe B.V., Zoetermeer, the Netherlands.

[203] Keck, C. and Muller, R.H. (2006) Drug nanocrystals of poorly soluble drugs produced by high pressure homogenization. *Eur. J. Pharm. Biopharm.*, **62**, 3–16.

[204] Petersen, R. (16 September 2009) Nanocrystals for use in topical cosmetic formulations and method of production thereof, EP 2099420 A1, Abbott GmbH & Co. KG, USA.

[205] Murugan, M.A., Gangadharan, B. and Mathur, P.P. (2002) Antioxidative effect of fullerenol on goat epididymal spermatozoa. *Asian J. Androl.*, **4**, 149–152.

[206] Ali, S.S., Hardt, J.I., Quick, K.L. *et al.* (2004) A biologically effective fullerene (C60) derivative with superoxide dismutase mimetic properties. *Free Radic. Biol. Med.*, **37**, 1191–1202.

[207] Chen, Y.W., Hwang, K.C., Yen, C.C. and Lai, Y.L. (2004) Fullerene derivatives protect against oxidative stress in RAW 264.7 cells and ischemia reperfused lungs. *Am. J. Physiol. Regul. Integr. Comp. Physiol.*, **287**, R21–R26.

[208] Takada, H., Mimura, H., Xiao, L. *et al.* (2006) Innovative anti-oxidant, fullerene (INCI #, 7587) is as "radical sponge" on the skin. Its high level of safety, stability and potential as premier anti-aging and whitening cosmetic ingredient. *Fullerenes Nanotubes Carbon Nanostruct.*, **14**, 335–341.

[209] Amann, M. and Dressnandt, G. (1993) Solving problems with cyclodextrins in cosmetics. *Cosmet. Toil.*, **108**, 90–95.

[210] Duchene, D., Wouessidjewe, D. and Poelman, M.C. (1999) Cyclodextrins in cosmetics, in Novel Cosmetic Delivery Systems (eds S. Magdassi and E. Touitou), Marcel Dekker Inc., New York, pp. 275–278.

]211] Simeoni, S., Scalia, S., Tursilli, R. and Benson, H. (2006) Influence of cyclodextrin complexation on the in vitro human skin penetration and retention of the sunscreen agent, oxybenzone. *J. Incl. Phenom. Macrocyclic Chem.*, **54**, 275–282.

[212] Cal, K. and Centkowska, K. (2008) Use of cyclodextrins in topical formulations: Practical aspects. *Eur. J. Pharm. Biopharm.*, **68**, 467–478.

[213] Basavaraj, K.H., Johnsy, G., Navya, M.A. *et al.* (2010) Biopolymers as transdermal drug delivery systems in dermatology therapy. *Crit. Rev. Ther. Drug Carrier Syst.*, **27**, 155–185.

[214] Amin, S., Rajabnezhad, S. and Kohli, K. (2009) Hydrogels as potential drug delivery systems. *Sci. Res. Essays*, **3**, 1175–1183.

[215] Siemann, U. (2005) Solvent cast technology – A versatile tool for thin film production. *Progr. Colloid Polym. Sci.*, **130**, 1–14.

[216] Konis, Y. and Kalay, A. (27 September 2007) Cosmetic and pharmaceutical foam carrier. US Patent 20,070,224,143 A1, Kamedis Ltd., Israel.

[217] Zhao, Y., Jones, S.A. and Brown, M.B. (2010) Dynamic foams in topical drug delivery. *J. Pharm. Pharmacol.*, **62**, 678–684.

[218] Fotinos, S.A. (2001) Cometics patches, in Handbook of Cosmetic Science and Technology (eds A.O. Barel, M. Paye and H.I. Maibach), New York, Marcel Dekker Inc., pp. 233–243.

[219] Singh, T.K., Tiwari, P., Singh, C.S. and Prasad, R.K. (2013) Cosmeceuticals: Enhance the health and beauty of the skin. *World J. Pharm. Res.*, **2**, 1475–1485.

[220] Brandt, F.S., Cazzaniga, A. and Hann, M. (2011) Cosmeceuticals: Current trends and market analysis. *Semin. Cutan. Med. Surg.*, **30**, 141–143.

[221] Malerich, S. and Berson, D. (2014) Next generation cosmeceuticals: The latest in peptides, growth factors, cytokines, and stem cells. *Dermatol. Clin.*, **32**, 13–21.

[222] Kimball, A.B., Grant, R.A., Wang, F. *et al.* (2012) Beyond the blot: Cutting edge tools for genomics, proteomics and metabolomics analyses and previous successes. *Br. J. Dermatol.*, **166**, 1–8.

[223] Luque de Castro, M.D. (2011) Cosmetobolomics as an incipient '-omics' with high analytical involvement. *Trends Anal. Chem.*, **30**, 1365–1371.

[224] Giacomoni, P.U. (2005) Ageing, science and the cosmetics industry. *EMBO Rep.*, **6** (Suppl), S45–S48.

[225] Nasto, B. (2007) Biotech at the beauty counter. *Nat. Biotechnol.*, **25**, 617–619.

[226] Padamwar, M.N. and Pokharkar, V.B. (2006) Development of vitamin loaded liposomal formulations using factorial design approach: Drug deposition and stability. *Int. J. Pharm.*, **320**, 37–44.

[227] Mu, L. and Sprando, R.L. (2010) Application of Nanotechnology in Cosmetics. *Pharm. Res.*, **27**, 1746–1749.

[228] Singh, R., Tiwari, S. and Tawaniya, J. (2013) Review on nanotechnology with several aspects. *Int. J. Res. Comput. Eng. Electron.*, **2**, 1–8.

[229] Gao, X.-H., Zhang, L., Wei, H. and Chen, H.-D. (2008) Efficacy and safety of innovative cosmeceuticals. *Clin. Dermatol.*, **26**, 367–374.

[230] Elmarzugi, N.A., Keleb, E.I., Mohamed, A.T. *et al.* (2013) The relation between sunscreen and skin pathochanges: Mini review. *Int. J. Pharm. Sci. Invent.*, **2**, 43–52.

[231] Nohynek, G.J., Dufour, E.K. and Roberts, M.S. (2008) Nanotechnology, cosmetics and the skin: Is there a health risk? *Skin Pharmacol. Physiol.*, **21**, 136–149.

[232] Locascio, L.E., Reipa, V., Zook, J.M. and Pleus, R.C. (2011) Nanomaterial toxicity: Emerging standards and efforts to support standards development, in Nanotechnology Standards. Nanostructure Science and Technology (eds V. Murashov and J. Howard), Springer, New York, pp. 179–208.

[233] Saint-Leger, D. (2012) 'Cosmeceuticals'. Of men, science and laws…. *Int. J. Cosmet. Sci.*, **34**, 396–401.

[234] Kumar, S. (2005) Exploratory analysis of global cosmetic industry: Major players, technology and market trends. *Technovation*, **25**, 1263–1272.

11

Commercial and Regulatory Considerations in Transdermal and Dermal Medicines Development

Marc. B. Brown[1,2], Jon Lenn[3], Charles Evans[1,2] and Sian Lim[1]

[1]*MedPharm Ltd, R&D Centre, Surrey Research Park, Guildford, UK*
[2]*School of Pharmacy, University of Hertfordshire, Hatfield, UK*
[3]*Stiefel, A GSK Company, Research Triangle Park, NC, USA*

11.1   Introduction

Over the last 20 years, pharmaceutical research and development (R&D) productivity has experienced what at best can be described as a static period for the rate of approval of new chemical entities (NCE) in medicines by the Food and Drug Administration (FDA) [1]. Although this may be on the mend (between 2010 and 2012, an increase in the number of new molecular entities both applied for and approved by the FDA has been observed (23–41 and 21–39, respectively [2])), attrition rates, development times and expenditure have all increased significantly [3, 4]. For example, over the past 40 years, the US pharmaceutical industry's inflation-adjusted R&D spend has increased from \$2.5 to \$27 billion, which equates to a current average cost of \$843 million per NCE. In addition, only 2 out of 10 marketed products end up providing a return on investment. Many reasons have been put forwards for this lack of productivity, including a more stringent and tightly controlled regulatory environment, increased competition and patent issues. It is also important to note that the effects of the recent multinational mergers and the resultant so-called pipeline consolidation are yet to be seen. However,

Novel Delivery Systems for Transdermal and Intradermal Drug Delivery, First Edition.
Ryan F. Donnelly and Thakur Raghu Raj Singh.
© 2015 John Wiley & Sons, Ltd. Published 2015 by John Wiley & Sons, Ltd.

a striking observation is that this decline in productivity has coincided with multinational pharmaceutical companies beginning to search for, and spending enormous amounts of money on, the identification of increasingly complex 'druggable' disease targets. This often relies on genomics and the resultant development and validation of techniques such as high-throughput screening, robotics, combinatorial chemistry and bioinformatics. In addition, although such techniques are producing an increasing number of potential drug candidates and the attrition rates of such molecules as a result of problematic physicochemical properties, safety and efficacy are also escalating. For example, at present, it is estimated that for every 10 000 molecules screened during the discovery process, only one will gain regulatory approval as a medicine. The pharmaceutical development industry often forgets that it is not a drug that is given to a patient but a medicine and the art of formulation development in producing a medicinal product that meets the relevant regulatory authority's criteria of acceptable quality, safety and efficacy is being lost.

However, this lack of success in NCE discovery means that pharmaceutical companies are looking at improving the efficacy or reducing the toxicity of existing drugs or biopharmaceuticals to satisfy their shareholders. As such, the global market for drug delivery systems in 2010 was \$131.6 billion. The market is expected to rise at a compound annual growth rate (CAGR) of 5% and reach nearly \$175.6 billion by 2016.

As has been detailed in this book, transdermal delivery is one way of improving the efficacy or reducing the toxicity of drugs. At the time of writing, it is just over 30 years since the first transdermal patch for systemic delivery was launched and subsequently approximately 20 other products have been approved [5]. Thus, in 2010, the transdermal delivery market was worth \$21 billion (patch market value of \$12.5 billion) and is estimated to be worth \$32 billion by 2015 [6]. This compares well with the global dermatology market which was valued at \$17 billion in 2010 with sales dominated by anti-infectives (23% of the market) and treatments for psoriasis (18% of market) [7].

Despite these sales and growth, up until relatively recently, the successes for passive delivery to and across the skin have been based on the '500 Da rule' [8]. This rule is based on the evolution of the barrier to protect the human body from the external environment and the patch test series from the International Contact Dermatitis Research Group (ICDRG) used for the diagnosis of contact allergy [9]. Very few, if any, compounds with a molecular weight close to and above 1000 Da have shown clinical efficacy from topical delivery without physical disruption of the barrier (Figure 11.1). Overcoming this size exclusion phenomenon could present a major disruptive breakthrough in dermatology and topical drug delivery. Tacrolimus and pimecrolimus (MW = 804 and MW = 810, respectively) are the two most well-known compounds that seem to go against this rule. However, the development of novel protein therapeutics or biologics has gained significant momentum in the biopharmaceutical sector in recent years with many of these actives subject to poor oral bioavailability due to their large molecular size and first-pass metabolism. Hence, despite the issue of molecular weight, there has been significant interest in the skin as an alternate route of delivery with the most common strategies to achieve the required efficacy being 'active' delivery or the implementation of a form of barrier disruption via platform technologies such as microneedles, jet injections, dermabrasion, thermal ablation, iontophoresis, electroporation or sonophoresis as exemplified in this book and other reviews.

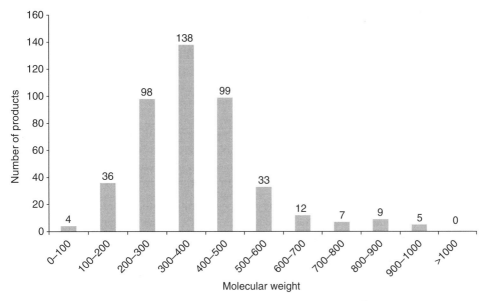

Figure 11.1 *Marketed products organised by molecular weight obtained from Citeline's Pipeline Database. Reproduced with permission from Citeline [10]. Pipeline Database. 2014; Available from http://www.citeline.com/products/pipeline/.*

11.2 Dermal and Transdermal Product/Device Development

Prior to development of a new product, it is important that the critical attributes of the product are clearly defined. In the pharmaceutical industry, these product attributes are referred to as the quality target product profile (QTPP). Ideally, the QTPP describes how the product will be utilised by the end user and will help identify project goals and potential risks and can serve as the basis to link regulatory and commercial requirements before and during development. As discussed earlier, since the development of pharmaceutical products is a lengthy and expensive exercise, it is critical to ensure that the right product is being developed. As such, the QTPP is a key tool to help ensure the eventual product meets all expectations and is often used as a tool for discussion with the regulatory authorities (e.g. FDA and is considered a 'critical path' tool by most agencies).

Inevitably, the first thing considered by a pharmaceutical company is whether any medicine or device has the potential to be profitable. Often, this can be decided as a paper exercise; less frequently, a decision will be made once development has started and problems are encountered which make the end product less financially viable. Such considerations are always a priority for the pharmaceutical industry and organisations such as NICE. Although it is not the focus of this chapter to discuss the pharmacoeconomical justification, the starting point of any commercial consideration is to have a good understanding of any commercial opportunity in the pharmaceutical and medical industry, market competition and patient needs. Nevertheless, when pharmacoeconomical justification is found, other pharmaceutical-related commercial aspects which are equally important should not be forgotten. Such aspects include the API/dosage form selection up to development and scale-up/manufacture, costs, patient compliance/safety and regulatory requirements.

11.2.1   Drug Candidate Selection

Once the clinical target has been chosen, one or more drug candidates may be available. To streamline the selection process in order to identify a lead candidate for clinical development, a number of early processes to identify molecules which possess suitable characteristics to make acceptable medicines have been developed. These include mathematical models to predict skin permeation through to assay development, pre-formulation, high-throughput screening and so on.

In addition to the physicochemical properties and pharmaceutical considerations, several other issues should be considered if they have not already been addressed in the candidate selection process before a decision is made to take a drug into transdermal or dermal product development, and this will include one or more of the following:

- Has the pharmacological activity of the drug been demonstrated or predicted and what is the IC50 or minimum concentration required to exert a therapeutic effect?
- Have the pharmacological models used in assessing and/or predicting the pharmacological activity of the drug been appropriate? Is the target site definitely known? Is the drug metabolised in the skin?
- If relevant for certain transdermal systems, is the drug compatible with or stable to the device and/or method achieving enhanced delivery?
- Commercial aspects such as scalability of drug manufacturing process for GMP and subsequently commercial supplies.

For example, when targeting a drug to the skin or systemic circulation, a highly potent drug with a low permeability coefficient (k_p) may not necessarily be the most efficacious when compared to a less potent drug with a higher k_p as it is a combination of a drug's potency and ability to permeate the skin which is important. Thus, such parameters should be well defined and understood during drug selection and during any pre-formulation work since this information will allow proper evaluation of dosage form/device type, dose/drug concentration and the selection of the final device/excipients. For example, in patches, such calculations can mean that when the dose required per square cm is calculated, the patch will end up the size of dinner plate which would not be a viable option.

11.2.2   Dosage/Device Form

Obviously, there are many types of dermal and transdermal formulations and devices that could be developed all having their own distinct intricacies. Transdermal systems have been generally divided into three generations [11, 12]. First-generation transdermal systems are responsible for the majority of such systems on the market to date. In this group, delivery is often limited by the barrier properties of the *stratum corneum* although they still rely on passive diffusion such as patches, metered sprays or gels. As stated previously, there have been 20 or so drugs/drug combinations developed in this form over the last 30 years with the majority being patches. Initially, these were in the form of drug in adhesive and/or reservoir systems with more recently multilayer matrix system. Such developments in the technologies involved along with improved patient acceptability resulted in a surge in approvals in the mid-2000s. However, this has slowed down in the present day as the number of drugs with the required physicochemical properties, the required poor oral bioavailability and the need for less frequent dosing and steady-state delivery associated with the more conventional oral delivery diminishes.

Second-generation transdermal systems rely on permeability enhancement in some form, the most common approach being absorption enhancers. Some of the classes of excipient enhancers examined include azones, pyrrolidones, fatty acids, alcohols, glycols, surfactants and phospholipids, to name a few. Among these, fatty acids, alcohols, glycols and surfactants usually have a dual role as a solubiliser in addition to their potential enhancing capabilities. The choice and selection of such enhancers depend upon the formulation type and nature of the drug. However, consideration should also be given to the enhancers' potential pharmacological activity, toxicity, duration of action, enhancing mechanism (and reversibility), stability and cosmetic acceptability. Alternative methods of enhancement include active systems such as iontophoresis which along with increasing skin permeability by disrupting the *stratum corneum* lipids should also provide an added driving force for drug absorption while avoiding deeper tissue injury. However, this has proved to be somewhat problematic with lack of success despite the investment spent on developing such systems.

Third-generation transdermal delivery systems are the most exciting development, and the techniques involved have their effect on the *stratum corneum* alone. Such approaches produce greater disruption of the *stratum corneum* and thus more effective transdermal delivery allowing the potential delivery of drugs without the normal limitations in physicochemical properties including biologics, for example, proteins, peptides, siRNA, antisense and vaccines. The active techniques involved include iontophoresis (again), electroporation, ultrasound, microneedles, thermal ablation, microdermabrasion and thermophoresis and combinations thereof. Nevertheless, the issues of miniaturisation, patient compliance and acceptability, safety and cost of goods, as with all the other generations, remain.

Whatever the transdermal system selected, the process of development has many similarities (drug candidate selection, pre-formulation, formulation development and optimisation, scale-up, GMP production, etc.) but inevitably becomes more difficult when the delivery system that is to be designed becomes more complex. However, ultimately, when the product development scientist starts the long and often painful process of developing a topical or transdermal formulation or device, there are a plethora of issues that need to be considered. The pharmaceutical history is littered with examples of drugs that could have been the next blockbuster if only a formulation or device had been developed that delivered the drug safely and efficaciously to the pathological site in a cost-effective manner. Such problems are exacerbated even further in dermal drug delivery in that patients often care what they apply to or use on the skin and may even have a choice. For example, the dermal products developed many years ago that were greasy, malodorous and stained clothes are no longer acceptable. Cosmetics and aesthetics of the final product are, in some cases, almost as important as the product's efficacy. Thus, the intricacies of dermal formulation development are considerably different to those for a tablet, for example. It is even more the case for transdermal devices and systems. Not only does a formulation have to be optimised, characterised and developed but so do the device itself and the combination thereof ensuring these are sufficient, safe and efficacious while being of the quality expected by the regulatory authorities.

11.2.3   Pre-formulation and Formulation/Device Development

Pre-formulation is an R&D stage where the drug's physicochemical properties and desired dosage and device form along with the drug's mechanism of action and target disease are considered. Pre-formulation studies typically involve solubility, stability and compatibility studies. Such studies are conducted to identify any parameters which may affect the development of the final product and may include poor drug solubility, inherent drug instability and potential excipient or device/drug incompatibility among others. In addition, another aim of pre-formulation studies is to develop and explore methodologies to improve these defined issues such that the QTPP can be achieved. Apart from pharmaceutical considerations, the selection of excipient types based on regulatory suitability for the final intended use should also be considered. The acceptability of excipients and levels of use may vary from one market territory to another, and therefore, such factors should be taken into consideration at this stage. A transdermal patch, as with most transdermal delivery systems, is classified by the FDA as a combination product, consisting of a medical device combined with a drug or biological product that the device is designed to deliver. As such, the selection of acceptable liners and backing layers should also be taken into consideration in addition to excipients and adhesive itself in the case of a drug in adhesive patch. From a commercial perspective, the source and availability of any excipients or components used to develop the final product should also be chosen with care since certain components may only be available from a limited supplier(s) which may in turn be exclusive to certain territories. Lastly, the cost of goods for the final product must satisfy the demands of its particular market.

Obviously, there are many types of dermal and transdermal formulations/systems available. A detailed review for dermal product development is provided in Ref. [13], and thus, as they are the most commonly available transdermal system on the market, first-generation patches will be the focus herein.

In 1997, three scientific organisations, the American Association of Pharmaceutical Scientists (AAPS), the FDA and the United States Pharmacopeia (USP), organised a workshop to explore Scale-Up and Post-approval Change (SUPAC) principles for adhesive transdermal drug delivery systems (TDS). The findings of this workshop included commentary on the impact of formulation or compositional changes, process variable changes, process scale changes and process site changes on the finished quality parameters of such transdermal products. Although the findings were published [14], unlike the outcome of other meetings, no guidance document was produced; however, the findings were updated in 2012 [15]. As such, various International Conference on Harmonisation (ICH) guidelines have been published to help globally harmonise the manufacture and use of APIs and drug products. Of special interest for transdermal systems are the following guidelines/documents:

- ICH Q8(R2) Pharmaceutical Development [16] – this guideline encourages the establishment of a detailed understanding of the manufacturing process using quality-by-design (QbD) principles (during the design and development phase) in order to define appropriate design space through assessment of process parameters. Based on this understanding, manufacturers should be able to assess how any variation of critical material attributes and these processes within their ranges (design space) influences product safety, quality and efficacy.

- ICH Q9 Quality Risk Management [17] – this defines the principles intended to describe the development and use of systematic processes for the assessment, control communication and review of quality risks throughout a product's lifecycle.
- ICH Q10 Pharmaceutical Quality System [18] – the principles described encourage manufacturers to identify, establish and maintain a state of control for process performance and product quality to help facilitate and control continual improvements and changes to a product attribute or process parameter.

Utilising the principles behind these guidelines should provide opportunities for flexible regulatory approaches in TDS development and product changes. As such, the QTPP is critical because along with all the other areas detailed previously, this will also detail the required quality characteristics of the drug product that should be achieved to ensure the desired quality, safety and efficacy of the product which may include the following:

- Clinical use
  For example, route of administration, dosage form, posology, residual drug, patient instructions and safety advice (any safety precautions related to the use of the product) and container closure system
- Quality attributes of the drug product
  For example, physical attributes, identity, strength, assay, uniformity, crystalline form, particle size, purity, impurity profile, stability and microbiological tests
- API release or delivery and attributes affecting pharmacokinetic characteristics. For example dissolution, release and permeation.

In the QTPP, a critical quality attribute (CQA) is selected based on its potential impact on safety and efficacy and is a physical, chemical, biological or microbiological property or characteristic that should be within limit, range or distribution to ensure the desired product quality. For a TDS, some examples include residual drug, adhesion, cohesion, cold flow, TDS component compatibility, rate controlling membrane, API release and patch size/shape (some of which are covered in the following).

The primary role of the adhesive is to affix the TDS to the skin, while it can also act as a carrier or a component of the formulation matrix for the API. Typically, the adhesive is laminated as a continuous adhesive layer on the TDS although it can also only be replaced around the edges. As such, the adhesive in the TDS is critical to the safety, efficacy and quality of the product since it is in intimate contact with the patient and the API and other excipients that may alter the adhesive properties (resulting in improper dosing) and/or may influence the API release rate (by API/excipient interactions, filler composition and porosity, tortuosity and thickness). As such, adhesive performance (adhesion, cohesion and tack), functionality, monomer content and impurities, drug and/or excipient solubility/stability, backing and/or release liner compatibility are all properties that need to be assessed and monitored throughout the product shelf life and if any post-approval changes are made. CQAs include adhesion, tack or shear and some of the most common test methods are detailed below:

- Adhesion force to remove adhesive from a defined substrate
  - Peel adhesion (PSTC 101, ASTM D3330/D3330 M0-4)
  - Release force (PSTC 4)

- Tack capacity of an adhesive to form a bond with another surface upon brief contact
  - Rolling ball tack (PSTC 6)
  - Probe tack (ASTM D2979-01)
  - Loop tack (PSTC 16)
  - Texture analyser 'tack'
- Shear strength measure of the internal or cohesive strength of an adhesive film
  - Static shear (PSTC 107, ASTM D3654/D3654 M-06)
  - Dynamic shear
- Rheology – measure of the viscoelastic properties (deformation and flow) of a material
- Viscosity – measure of shear thinning and/or resistance to flow
- Creep resistance – measure of cold flow property

Identity, appearance, non-volatile and volatile components and residual solvents are also other properties of the adhesive to be assessed and monitored. In addition, the effect of the adhesive and any subsequent changes on skin irritation and sensitisation always need to be considered.

Obviously, the formulation development of a TDS is a complex and often empirical process, and although the specifics of a QTPP for a TDS will vary depending on its ultimate purpose, there are key aspects of most such profiles that are the same for most formulation types. The most basic of these is the use of approved excipients and packaging materials, where the type/concentration, grade, etc. used should be acceptable from a regulatory perspective (as discussed earlier). When various prototypes that meet the QTPP are developed along with measuring/screening their CQAs, comparison of their performance and potential further optimisation needs to be assessed. The extent and rate of *in vitro* release of drug from a TDS or formulation should be well understood, and *in vitro* release rates are a useful assessment of this parameter and can also serve as a valuable QC release tool in monitoring formulation changes on storage. Clearly, in addition, it should be demonstrated that the formulation/TDS should deliver the drug into/across the skin at the required concentration and to the required site of action. Ultimately, a TDS optimised for drug release and permeation is often more efficacious and may require a lower concentration of drug which may reduce the cost of the final product and drug irritation potential and also maximise clinical efficacy. The physical and chemical stability of the drug/TDS must yield adequate shelf life. It is also important to ensure that the developed formulation can be manufactured at commercial scales. Lastly, the cost of goods for the product must satisfy the demands of its particular market. Ultimately, it is always important to remember a general rule that the simpler a formulation or medicine is, the less things there are to go wrong.

Once a series of prototype TDS are produced from the excipients evaluated at the pre-formulation stage, they are placed on short-term accelerated and real-time stability studies performed to identify a narrower range of lead formulation candidates for further performance testing.

11.2.4   Performance Testing

Some currently marketed TDS products retain up to 95% of the initial API loading after their intended use period. Although this is often necessary to achieve the delivery required, the high levels of drug left in a patch can have a potential impact on the product's safety, quality and efficacy. This is especially true for TDS containing fentanyl or methylphenidate, for example, where the residual drug has been extracted and illicitly injected. As such, it is necessary to ensure that a scientific approach is used to design and develop these products to ensure minimum residual drug. The choice and design of the formulation and packaging components may provide potential to optimise delivery and minimise residual. Possible approaches include the type of patch system, use of penetration enhancers, self-depleting solvent systems and adhesive choice and thickness. An example of such an improvement is when the Duragesic reservoir patch was replaced with a new Duragesic matrix patch that was designed to increase safety and adherence and decrease side effects and misuse of the reservoir patch [19, 20].

The performance of TDS products can be assessed with different models ranging from compendial methods to Franz cell-type models.

11.2.4.1   Compendial Methods for Drug Release

There are a number of different compendial methods detailed in both the EP and the USP which include the following:

Paddle-over-disk/disk assembly method (USP apparatus 5/Ph. Eur 2.9.4.1)
Cylinder/rotating cylinder method (USP apparatus 6/Ph. Eur 2.9.4.3)
Reciprocating holder (USP apparatus 7)
Cell method (with extraction cell, Ph. Eur 2.9.4.2)

The paddle-over-disk method is the most widely used method because it is simple and easy to reproduce. Along with the paddle and vessel of the USP 2, this apparatus includes a stainless steel disk assembly which holds the transdermal system at the bottom of the vessel. The disk assembly is positioned in such a way that the release surface is parallel with the bottom of the paddle blade and exposed to the medium and ensures the patch is prevented from floating during the testing period. The test used needs to be reproducible and reliable and must be capable of detecting drug release from the finished product. The EP also describes [21] the use of a membrane (inert porous or silicones), which does not affect the rate of release of the active substance from the drug product, as an alternative to the dissolution test. Such a test is described in more detail in the following and has the advantage of the membrane ensuring the product and the receptor medium are kept separate and distinct unlike the compendial dissolution test where the patch is entirely submerged in the receptor medium. The concept described in the EP is also similar to the Franz cell-type models.

11.2.4.2   In Vitro *Drug Release/Dissolution Studies*

Absorption of drugs into or through the skin depends upon a number of factors including composition of the formulation, type and condition of the skin and factors such as temperature, humidity and occlusion. However, one factor that has a major influence on the rate or extent of percutaneous absorption is the thermodynamic activity of the drug in the formulation on the skin surface. This is obviously strongly influenced by the physicochemical properties of the drug and ultimately by its solubility in the solvents/excipients in which it is formulated.

Synthetic membranes have been investigated as a readily available and easy-to-use tool to study the *in vitro* release profiles of drugs from topical/transdermal formulations in order to

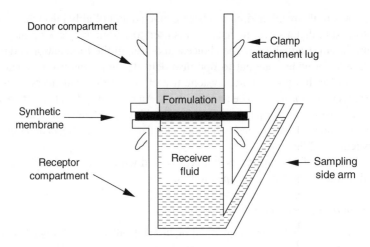

Figure 11.2   *Schematic representation of a Franz cell.*

ascertain batch-to-batch uniformity. Examples of artificial membranes used include silicone, polycarbonate, cellulose and combinations of these. Such membranes can provide additional information contributing to the understanding of mechanistic aspects of skin permeability. For example, the thermodynamic effects of drug solubility, partition coefficient, pH, drug–excipient interactions, and so on can sometimes be better understood by using synthetic membranes. When a formulation/TDS is applied to the skin, the drug must first partition into and diffuse across the formulation, and the thermodynamic activity must be sufficient for the drug to be released and partition into the *stratum corneum*. As such, this method is routinely used in order to optimise formulations for drug release. The assessment of drug release should be performed using previously validated methodology based on the principles of the FDA's SUPAC-SS guidelines [22]. A typical *in vitro* release method for topical or transdermal dosage forms is based on an open chamber diffusion cell system such as a Franz cell system, fitted usually with a synthetic membrane (Figure 11.2). The test product is placed on the upper side of the membrane in the open donor chamber of the diffusion cell, and a sampling fluid is placed on the other side of the membrane in a receptor cell. Diffusion of drug from the product across the membrane is monitored by assay of sequentially collected samples from the receptor fluid. Aliquots removed from the receptor phase can be analysed for drug content by high-pressure liquid chromatography (HPLC) or other analytical methodology as appropriate. A plot of the amount of drug released per unit area ($\mu g/cm^2$) against the square root of time yields a straight line, the slope of which is representative of the drug release rate.

An alternative to the methods described earlier is based on classic dissolution testing; however, whatever the method used, it must be validated and capable of distinguishing between different processing and formulation variables.

11.2.4.3   Ex Vivo *Skin Drug Permeation and Penetration Testing*

Extensive guidelines on conducting permeation and penetration experiments have been made available by the OECD [23], European Food Safety Authority [24] and (currently under review) European Medicines Agency [25]. A validated, reliable and reproducible test procedure defining the use of human or animal skin is essential for meaningful interpretation of drug permeability. As with *in vitro* drug release experiments, *ex vivo* skin permeation experiments involve the use of a diffusion cell designed to mimic the physiological and anatomical conditions of skin *in situ*. Full-thickness skin, epidermal membrane or dermatomed skin instead of a synthetic membrane is positioned between the two halves with the *stratum corneum* facing the donor compartment. A finite dose (1–10 mg/cm²) of formulation is applied to the surface of the skin membrane. The receptor compartment of the Franz cells is filled with a suitable receiver fluid to maintain sink conditions and the cells fixed in a water bath maintained at 37°C. The receptor chamber content is continuously agitated by small magnetic followers, and at regular time intervals, samples of receiver fluid are taken from the receptor compartment, replaced with fresh receiver medium and assayed by a suitable assay method such as HPLC.

Ultimately, the objective of performance testing is to select the formulations that closely meet the criteria QTPP and to mitigate the risk of failure during clinical investigation. However, once all such assessments and formulation optimisations have been completed, one lead and preferably at least one backup formulation/TDS should be selected for further development and clinical evaluation.

11.3   Product Scale-Up and Process Optimisation, Validation and Stability Testing

11.3.1   Product Scale-Up, Process Optimisation and Specification Development

As discussed previously, the product development scientist develops, refines and optimises a product through CQAs. The methods used will be refined (especially with regard to robustness) as the product progresses through preclinical and clinical evaluation. The ICH guidelines [16] outline how the pharmaceutical development of a drug product should evolve from initial stages of development right through the manufacture and scale-up process by building in QbD and not by later testing. At this time, process scale-up commences in order to develop a product that will yield a product that meets the CQAs previously identified. This, as with any necessary formulation optimisation is a continuous process, and process analytical technologies (PAT) are recommended as these have the advantages of identification and control of critical processing parameters (CPPs) and the operating ranges for the process parameters that ensure the CQAs identified are met. Designed experiments can be used to correlate CPPs to CQAs and can take the form of factorial design, single factor or systematic approaches. Factorial designs are probably the most commonly used during product development, particularly when different variables are thought to interact significantly. As the most significant CPPs are identified, it can also be beneficial to look at ways of identifying PAT approaches to analyse and control these CPPs and to demonstrate their link to the CQA of the finished product. Process validation requirements for the TDS generally focus on the product quality.

11.3.2   Analytical Method Validation

Analytical method validation is performed to ensure the drug (and other essential components such as preservative/antioxidant) content and level of impurities in the drug product can be accurately and reproducibly determined at drug product release and over the duration of the drug product shelf life. Typically, validation is performed based on ICH guidelines [26], and the attributes of the analytical method that are validated are linearity (and range), accuracy, precision (repeatability and intermediate precision of drug substance and drug product preparations), specificity and robustness. In addition to the analytical method, the drug product extraction method is also validated (typically at three different drug levels, 80, 100 and 120% of the drug target concentration) where accuracy, precision (repeatability and intermediate precision of drug substance and drug product preparations) and specificity are determined. Forced degradation and stability studies are also performed on the drug product to characterise degradation of the drug product and verify that the method is stability indicating and suited to its intended purpose in the analysis of drug product.

11.3.3   ICH Stability Testing

The purpose of stability testing is to provide evidence on how the quality of a drug product varies with time under the influence of a variety of environmental factors such as temperature, humidity and light. The choice of test conditions is largely based on the analysis of the effect of the climatic conditions where the product is to be marketed and the world can be divided into four climatic zones, I–IV. Once the lead prototype TDS are identified, these are placed on long-term accelerated and real-time stability studies performed under ICH conditions. In general, a drug product should be evaluated under storage conditions that test its thermal stability as well as its final packaging for sensitivity to moisture or potential for solvent loss in the case of a TDS. The long-term testing should cover a minimum of 24 months' duration at the desired drug product storage temperature. Typical real-time and accelerated storage conditions comprise of 25° C and 60% relative humidity (RH), 30°C and 65% RH and 40°C and 75% RH [27] with measurement of all CQAs being performed at regular intervals (e.g. 0, 1, 3, 6, 9, 12, 18 and 24 months). The ICH guidelines (ICH Q6A) describe the following general tests for a drug product:

Description
Identification
Drug assay
Drug impurities
Uniformity of dosage units
Microbial limits
Antioxidant/preservative content
Leak test

In addition, for the purpose of a transdermal drug product, an adhesive test (as described in the adhesives section: peel, tack and shear tests) should be performed.

11.4   The Commercial Future of Transdermal Devices

The previous sections have described some of the preclinical hurdles and considerations to overcome when developing what is a relatively simple transdermal drug delivery device. The commercial, scale-up, manufacturing/fabrication, quality, safety and efficacy issues that arise when clinical investigations and then marketing authorisation/NDA preparation and submission commence cannot also be underestimated. As such, it is interesting to note that at this moment in time, although there have been approximately 20 different drugs approved in passive transdermal drug delivery systems as reviewed by Watkinson [5], the same cannot be said for the active/device forms of delivery. Though clinical trials are still ongoing to entirely evaluate the safety profile and the efficacy of several active strategies [28, 29], few products have reached and maintained commercial value on the market. One example is an iontophoretic delivery system containing lidocaine and epinephrine (from Iomed) approved by the FDA in 1995 which has been discontinued. Another example is LidoSite™ (Vyteris), a lidocaine topical iontophoretic patch; this product is believed to provide dermal anaesthesia six-fold faster compared to the leading topical anaesthetic cream EMLA (2.5% lidocaine and 2.5% prilocaine) [30]. Despite such therapeutic benefit, the elevated cost of LidoSite limited its commercial success, thus leading the manufacturer to withdraw it from the market after only 2 years [31]. Similarly, a fentanyl iontophoretic patch (Ionsys) has been withdrawn from the European market after an issue was identified with corrosion of the system that could cause a lethal overdose [29, 31]. In mid-2011, the FDA issued a Complete Response Letter (CRL) to NuPathe for their sumatriptan iontophoretic patch (Zelrix) for migraine. The FDA accepted the efficacy of the system but stated that some CMC and safety questions still needed answering. The Zelrix system, now known as ZECUITY, was finally approved in early 2013 but as of writing has not been launched. Another interesting area is the use of heat in combination with a patch which is sometimes known as the thermophoretic system. One such licensed system is known as Synera in the United States and Rapydan in Europe. The patch consists of a eutectic drug reservoir of lidocaine (70 mg) and tetracaine (70 mg) and uses the Controlled Heat-Assisted Drug Delivery (CHADD®) technology manufactured by Zars. Despite the faster onset that was clinically shown, the uptake of Synera in the US market remains poor due to its higher cost compared to the other conventional topical products [31] although it does appear to remain on the market. Lidocaine seems to be a popular drug to develop in active forms of transdermal drug delivery. For example, in 2004, SonoPrep, a device using ultrasound to enhance lidocaine delivery, was approved by the FDA but again due to poor market uptake was withdrawn from the market in 2007. Around a similar time, Zingo, a needleless powder injection system for lidocaine, was approved by the FDA, but again, its uptake by the market was poor and it was withdrawn from the market in 2008 although in 2013 there are reports to suggest it will again be marketed in the United States.

It is obvious that the preceding text suggests that the success of active transdermal delivery devices has been somewhat limited and is disappointing when the scale of dollar investment in these technologies is considered. Nevertheless, perhaps now is the time that these devices will come to the fore with the recent advances in miniaturisation and electronics and the recovery from the recent financial crisis. However, as stated throughout this chapter, whatever the technology, the cost, design, manufacturing/fabrication, indication, market need, safety, quality, efficacy and end user all need to fit together to make the end product commercially viable.

References

[1] http://www.fda.gov/downloads/AboutFDA/Transparency/Basics/UCM247465 (accessed 10 March 2015).
[2] U.S. Food and Drug Administration Center for Drug Evaluation and Research (January 2014) Novel New Drugs 2013 Summary.

[3] Pammoli, F., Magazzini, L. and Riccaboni, M. (2011) The productivity crisis in pharmaceutical R and D. *Nat. Rev. Drug Discov.*, **10**, 428–438.

[4] Brown, M.B. (2005) It is not a molecule you give to a patient but a medicine, the lost of science of formulation. *Drug Discov. Today*, **10**, 1405–1407.

[5] Watkinson, A.C. (2012) Transdermal and topical Delivery today, in Topical and Transdermal Drug Delivery Principles and Practice (eds H. Benson and A.C. Watkinson), *John Wiley & Sons, Inc, Hoboken*, NJ, pp. 357–366.

[6] Paudel, K.S., Milewski, M., Swadley, C.L., Brogden, N.K., Ghosh, P. and Stinchcomb, A.L. (2010) Challenges and opportunities in dermal/transdermal delivery. *Ther. Deliv.*, **1** (1), 109–131.

[7] SCRIP Business Insights (2011) The dermatology market outlook to 2016 – Competitive landscape, pipeline analysis, and growth opportunities, document no. 840436279.

[8] Kasting, G.B. (2013) Lipid solubility and molecular weight: Whose idea was that. *Skin Pharmacol. Physiol.*, **26** (4–6), 295–301.

[9] Bos, J.D. and Meinardi, M.M. (2000) The 500 Dalton rule for the skin penetration of chemical compounds and drugs. *Exp. Dermatol.*, **9** (3), 165–169.

[10] Citeline (2014) *Pipeline Database*, http://www.citeline.com/products/pipeline/ (accessed 10 March 2015).

[11] Prausnitz, M.R. and Langer, R. (2008) Trandermal drug delivery *Nat. Biotechnol.*, **26**, 1261–1268.

[12] Hughes, P.J., Freeman, M.K. and Wensel, T.M. (2013) Appropriate use of transdermal drug delivery systems. *J. Nurse Educ. Pract.*, **3**, 129–138.

[13] Brown, M.B. and Lim, S.T. (2012) Topical product development, in Transdermal and Topical Drug Delivery: Principles and Practice (eds H. Benson and A.C. Watkinson), *John Wiley & Sons, Inc, Hoboken*, NJ, pp. 255–286.

[14] Van Buskirk, G.A., González, M.A., Shah, V.P. *et al.* (1997) Scale up of adhesive transdermal drug delivery systems. *Pharm. Res.*, **14** (7), 848–852.

[15] Van Buskirk, G.A., Arsulowicz, D., Basu, P. *et al.* (2012) Passive transdermal Systems white paper incorporating current chemistry, manufacturing and controls (CMC) development principles. *AAPS PharmSciTech*, **13**, 218–230.

[16] ICH Expert Working Group (2009) ICH International Conference on Harmonization Guideline ICH Q8(R2), Pharmaceutical Development. European Union, Japan.

[17] ICH Expert Working Group (2005) ICH International Conference on Harmonization Guideline ICH Q9, Quality Risk Management. European Union, Japan.

[18] ICH Expert Working Group (2008) ICH International Conference on Harmonization Guideline ICH Q10 Pharmaceutical Quality Systems. European Union, Japan.

[19] Van Nimmen, N.F. and Veulemans, H.A. (2007) Validated GC–MS analysis for the determination of residual fentanyl in applied Durogesic reservoir and Durogesic reservoir® and Durogesic d-Trans® matrix transdermal fentanyl patches. *J. Chromatogr. B*, **846**, 264–272.

[20] Firestone, M., Goldman, B. and Fischer, B. (2009) Fentanyl use among street drug users in Toronto, Canada: Behavioural dynamics and public health implications. *Int. J. Drug Policy*, **20**, 90–92.

[21] European Pharmacopoeia 8th edition, 1011, Patches, *Transdermal*, 01/2008:1011.

[22] FDA (CDER) (1997) Guidance for Industry – SUPAC-SS Non-sterile Semisolid Dosage Form, Scale-up and post approval changes: Chemistry, manufacturing and controls; *in vitro* release testing and *in vivo* bioequivalence documentation. U.S. Department of Health and Human Services, Rockville, MD.

[23] OECD (2004) 428, European Commission Guidance document on dermal absorption, rev 7.

[24] European Food Safety Authority (2012) Guidance on Dermal Absorption. *EFSA J*, **10** (4), 2665.

[25] European Medicines Agency (EMA) (2014) EME/CHMP/QWP/608924/2014.

[26] ICH Expert Working Group (1994) ICH International Conference on Harmonization guideline ICH Q2 (R1), Validation of Analytical procedures, Text and Methodology. European Union, Japan.

[27] ICH (2003) Q1A(R2): Stability Testing of New Drug Substances and Products. Proceedings of the International Conference on Harmonisation. US FDA Federal Register.

[28] Paudel, K.S., Milewski, M., Swadley, C.L. *et al.* (2010) Challenges and opportunities in dermal/transdermal delivery. *Ther. Deliv.*, **1**, 109–131.

[29] Perumal, O., Murthy, S. and Kalia, Y. (2013) Turning theory into practice: The development of modern transdermal drug delivery systems and future trends. *Skin Pharmacol. Physiol.*, **26**, 331–342.

[30] Cada, D.J., Arnold, B., Levien, T.L. and Baker, D.E. (2006) Formulary drug reviews – Lidocaine/Tetracaine patch. *Hosp. Pharm.*, **41**, 265–273.

[31] Yeoh, T. (2012) Current landscape and trends in transdermal drug delivery systems. *Ther. Deliv.*, **3**, 295–297.

Index

Novel Delivery Systems for Transdermal and Intradermal Drug Delivery, First Edition.
Ryan F. Donnelly and Thakur Raghu Raj Singh.
© 2015 John Wiley & Sons, Ltd. Published 2015 by John Wiley & Sons, Ltd.